Bootstrap Methods

Bootstrap Methods

A Practitioner's Guide

MICHAEL R. CHERNICK
Novo Nordisk
Diamond Bar, California

A Wiley-Interscience Publication
JOHN WILEY & SONS, INC.
New York · Chichester · Weinheim · Brisbane · Singapore · Toronto

For ordering and customer service, call 1-800-CALL-WILEY

Library of Congress Cataloging in Publication Data:
Chernick, Michael R.
    Bootstrap methods : a practitioner's guide / by Michael R. Chernick.
        p.   cm. -- (Wiley series in probability and statistics)
    "A Wiley-Interscience publication."
    Includes bibliographical references and indexes.
    ISBN 0-471-34912-7 (cloth : alk. paper)
    1. Bootstrap (Statistics)   I. Title.   II. Series.
QA276.8.C48   1999
519.5′44--dc21   99-21924

Printed in the United States of America

10 9 8 7 6 5

In memory of my parents
Jack and Norma Chernick
who encouraged me in my education
of mathematics and statistics

Contents

Preface

The bootstrap is a resampling procedure. It is named that because it involves resampling from the original data set. Some resampling procedures similar to the bootstrap go back a long way. The use of computers to do simulation goes back to the early days of computing in the late 1940s. However, it was Efron (1979a) that unified ideas and connected the simple nonparametric bootstrap, which "resamples the data with replacement," with earlier accepted statistical tools for estimating standard errors, such as the jackknife and delta method.

The purpose of this book is (1) to provide an introduction to the bootstrap for readers who do not have an advanced mathematical background, (2) to update some of the material in the recent Efron and Tibshirani (1993) book by presenting results on improved confidence set estimation, estimation of error rates in discriminant analysis and applications to a wide variety of hypothesis testing and estimation problems, (3) to exhibit counterexamples to the consistency of bootstrap estimates so that the reader will be aware of the limitations of the methods, (4) to connect it with some older and more traditional resampling methods including the permutation tests described by Good (1994), and (5) to provide a bibliography that is extensive on the bootstrap and related methods up through 1992 with key additional references from 1993 through 1998, including many new applications.

The objectives of this book are very similar to those of Davison and Hinkley (1997) especially (1) and (2). However, I differ in that this book does not contain exercises for students but it does include a much more extensive bibliography.

This book is not a classroom text. It is intended to be a reference source for statisticians and other practitioners of statistical methods. It could be used as a supplement in an undergraduate or graduate course on resampling methods for an instructor, who wants to incorporate some real world applications and supply additional motivation for the students.

The book is aimed at an audience similar to the one addressed by Efron and Tibshirani (1993) and does not develop the theory and mathematics to the extent of Davison and Hinkley (1997). Mooney and Duval (1993) and Good

(1998) are elementary accounts but they do not provide enough development to help the practitioner gain a great deal of insight into the methods.

The spectacular success of the bootstrap in error rate estimation for discriminant functions with small training sample sizes along with my detailed knowledge of the subject justifies the extensive coverage given to this topic in Chapter 2. A text that provides a detailed treatment of the classification problem and is the only other text to include a comparison of bootstrap error rate estimates with other traditional methods is McLachlan (1992).

Mine is the first text to provide extensive coverage of real world applications for practitioners in many diverse fields. I also provide the most detailed guide yet available to the bootstrap literature. This I hope will motivate research statisticians to make theoretical and applied advances in bootstrapping.

Several books (at least thirty) deal in part with the bootstrap in specific contexts but none of these are totally dedicated to the subject [Sprent (1998) devotes Chapter 2 to the bootstrap and provides discussion of bootstrap methods throughout his book]. Schervish (1995) provides an introductory discussion on the bootstrap in Section 5.3 and cites Young (1994) as an article that provides a good overview of the subject. Babu and Feigelson (1996) address applications of statistics in astronomy. They refer to it as astrostatistics. Chapter 5, (pp. 93–103) of the Babu–Feigelson text covers resampling methods emphasizing the bootstrap. At this point there are about a half dozen other books devoted to the bootstrap but of these only four others (Davison and Hinkley, 1997; Manly, 1997; Hjorth, 1994; Efron and Tibshirani, 1993) are not highly theoretical.

Davison and Hinkley (1997) give a good account of the wide variety of applications and provide a coherent account of the theoretical literature. They do not go into the mathematical details to the extent of Shao and Tu (1995) or Hall (1992a). Hjorth (1994) is unique in that it provides detailed coverage of model selection applications.

Although many authors are now including the bootstrap as one of the tools in a statistician's arsenal (or for that matter in the tool kit of any practitioner of statistical methods), they deal with very specific applications and do not provide a guide to the practitioner on the variety of uses and limitations of the technique. This book is intended to present the practitioner with a guide to the use of the bootstrap while at the same time providing him or her with an awareness of its known limitations. As an additional bonus I provide an extensive guide to the research literature on the bootstrap.

This book is aimed at two audiences. The first consists of applied statisticians, engineers, scientists, and clinical researchers who need to use statistics in their work. For them, I have tried to maintain a low mathematical level. Consequently, I do not go into the details of stochastic convergence or the Edgeworth and Cornish–Fisher expansions that are important to the rate of convergence for various estimators.

However, I do not avoid discussion of these topics. Readers should bear with me. There is a need to understand the role of these techniques and the corresponding bootstrap theory in order to get an appreciation and under-standing of how, why, and when the bootstrap works. This audience should have background in statistics (at least one elementary course) but need not have had calculus or advanced mathematics, advanced probability, or math-ematical statistics.

The second primary audience is the mathematical statistician who has done research in other areas but wants to learn what the bootstrap is all about. For him or her, my historical notes and extensive references to applications and theoretical papers will be helpful. This second audience may also appreciate the way I try to tie things together with a somewhat objective view.

To a lesser extent, the serious bootstrap researcher may find value in this book and the bibliography in particular. I do attempt to be technically accurate and the bibliography is extensive with many applied papers that could motivate further research. It is more extensive than one obtained simply by using a key word search for "bootstrap" and "resampling" in the *Current Index to Statistics* CD ROM. However, I would not try to claim that such a search could not uncover articles that I have missed.

I invite readers to notify me of any errors or omissions in the book, particularly regarding references. There are many more papers listed in the bibliography than are referenced in the text. In order to make it clear which references are cited in the text I put an asterisk next to the cited references along with a numbering according to alphabetical order.

MICHAEL R. CHERNICK

Diamond Bar, California
January 1999

Acknowledgments

Peter Hall was kind enough to send an advance copy of his book, *The Bootstrap and Edgeworth Expansion*, which was published by Springer-Verlag (Hall, 1992a). Peter is one of the key contributors to bootstrap research and his book has been helpful in providing insight into the accuracy of bootstrap confidence intervals through Edgeworth expansions. His *Annals of Statistics* paper in 1988 on this subject was included in Johnson and Kotz's *Breakthroughs in Statistics Volume III*. The Edgeworth expansions and the related Cornish–Fisher expansions unify the theory for several bootstrap applications (particularly related to the variations on confidence interval estimates).

Brad Efron taught me mathematical statistics when I was a graduate student at Stanford. He has helped me over the years as a teacher, mentor, and colleague. His contributions to the bootstrap are enormous and obvious throughout this book.

This book was originally planned to be half of a two-volume series on resampling methods that Phillip Good and I started. Phil completed the volume on permutation tests that was published by Springer-Verlag (Good, 1994). In the early drafts of this book, I prepared the bootstrap volume with Phil doing some proofreading and continually reminding me to bring out the main points first. Eventually, since Phil was busy with other efforts, I decided to complete the book on my own.

The book was delayed due to my illness in 1993 and inability to find time to revise it based on a reviewer's comments. Subsequent publication of books on the bootstrap and many new research articles from 1993 to 1998 made it a difficult task to bring the book up to date.

Recently, I used the services of NERAC, a company that helps with literature searches. They found numerous articles that I had missed, particularly those in applied journals over the period 1994 to 1998. My thanks to Gerri Beth Potash of NERAC for her help with the bibliography. Also, Professor Robert Newcomb at the School of Social Sciences at the University of California at Irvine helped me identify references on bootstrap through the *Current Index to Statistics*. He and his staff at the UCI Statistical Consulting

Center (especially Mira Hornbacher) were helpful with a few other particular requests.

I am indebted to many typists who have helped over the years in producing the manuscript for this book. They include Sally Murray from Nichols Research Corporation, Cheryl Larsson from the University of California at Irvine, and Jennifer Del Villar from Pacesetter.

I also owe a great deal to my wife, Ann, for allowing me time away from home to write the book during a very busy and important period in the growth of our family. She played a major role in helping me through my illness and encouraging me to go back and complete the book. I also owe a debt of gratitude to several anonymous reviewers who suggested that I add some applied examples and demonstrate the ideas through the examples.

MICHAEL R. CHERNICK

CHAPTER 1

What Is Bootstrapping?

1.1. BACKGROUND

The bootstrap is a form of a larger class of methods that resample from the original data set and thus are called resampling procedures. Some resampling procedures similar to the bootstrap go back a long way [e.g., the jackknife goes back to Quenouille (1949) and permutation methods to Fisher and Pitman in the 1930s]. Use of computers to do simulation also goes back to the early days of computing in the late 1940s.

However, it was Efron (1979a) who unified ideas and connected the simple nonparametric bootstrap, which "resamples the data with replacement," with earlier accepted statistical tools for estimating standard errors such as the jackknife and delta method. It was only after the later papers by Efron and Gong (1983), Efron and Tibshirani (1986), and Diaconis and Efron (1983) and the monograph by Efron (1982a) that the statistical and scientific community began to take notice of many of these ideas and recognize their importance.

Since the publication of *The Jackknife, the Bootstrap, and Other Resampling Plans* by Bradley Efron (1982a), research activity on the bootstrap has grown exponentially. There have been many theoretical developments on the asymptotic consistency of bootstrap estimates and some counterexamples (cases where the bootstrap is inconsistent).

Many real world applications are beginning to appear along with numerous Monte Carlo studies on the performance of the bootstrap and its competitors for various problems. It is also becoming clearer that the bootstrap has significant practical value but it also has some limitations.

A special conference of the Institute of Mathematical Statistics was held in Ann Arbor, Michigan, in May 1990, where many of the prominent bootstrap researchers presented papers exploring the applications and limitations of the bootstrap. The publication *Exploring the Limits of Bootstrap* edited by LePage and Billard is a compilation of papers presented at that conference. It was published by John Wiley & Sons, Inc., in 1992.

A second similar conference was held in Tier, Germany, also in 1990. The European conference dealt with Monte Carlo methods, bootstrap confidence bands and prediction intervals, hypothesis tests, time series methods, linear models, special topics, and applications. Limitations of the methods were not addressed in the European conference. The proceedings were edited by Jockel, Rothe, and Sendler and were published by Springer-Verlag also in 1992.

Although Efron introduced his version of the bootstrap in a 1977 Stanford University Statistics Department technical report [later published in a well-known paper in the *Annals of Statistics*, (Efron, 1979a)], the procedure has been slow to catch on. Many of the applications have just begun to appear and be mentioned in textbooks over the past eight years.

Initially, there was a great deal of distrust regarding the bootstrap methodology. As mentioned in Davison and Hinkley (1997, p. 3): "In the simplest nonparametric problems we do literally sample from the data, and a common initial reaction is that this is a fraud. In fact it is not." This feeling by practitioners continued even after the publication of the *Scientific American* article (Diaconis and Efron, 1983), which tried to explain the bootstrap and its wide range of applications in a layperson's terms but, unfortunately, succeeded in increasing the skepticism of scientists and engineers.

Other efforts to popularize the techniques that were partially successful include the SIAM monograph by Efron (1982a), Efron and Gong (1981), the *American Statistician* article by Efron and Gong (1983), Efron's *SIAM Review* article (Efron, 1979b), and the *Statistical Science* article by Efron and Tibshirani (1986). These publications all helped to some extent. The one by Efron and Tibshirani (1986) was perhaps the most effective. Still, this mainly aided the understanding of the statistical community. It was only the *Scientific American* article that got significant exposure among scientists and engineers.

My experience with the scientists and aerospace engineers who were my colleagues at the Aerospace Corporation was that the *Scientific American* article generally led them to misconceptions about bootstrapping, due to the way the concepts were oversimplified. Some supported it because they saw it as a way to use simulation in place of sampling (a clear misunderstanding). Others, who thought that it claimed to allow inference based on no assumptions, using simulations in place of data, viewed it as phony science (which it would be if that was really what it was about).

I found it necessary to explain the bootstrap to the engineering community at the Aerospace Corporation. That got me involved in thinking about and doing research in bootstrap methods. It also led me to write this book. As a consequence of this experience, it is my belief that any effort to popularize bootstrap methods for scientists and engineers, which ignores all the important theory and mathematics, is doomed to failure.

The monograph by Mooney and Duval (1993) is one such attempt. At least it partially tries to describe some theory. The book by Efron and Tibshirani (1993) was the first text that tried to reach social scientists and other practitioners. It did a very good job of describing the subject for this audience. Yet, I have heard that Bradley Efron felt that the book was not simple enough to reach all of his intended audience.

There is a fine line to draw here. On the one hand, we want to reach those who do not have the mathematical background necessary for a deep understanding of the bootstrap theory. Yet, on the other hand, if we eliminate all discussion of stochastic convergence and the tools of probability theory that provide insight into the technique, we run the risk of perpetuating many of the common misconceptions about the bootstrap.

Even for those of us who understood the idea behind the bootstrap during the early research, the major impact it would have on the field of statistics was not at all appreciated. Back in 1977, I was a graduate student at Stanford University in the Statistics Department and was looking for a dissertation topic. Because of my interest in stochastic processes and since two faculty members (Mittal and Resnick) and one visitor (Leadbetter) were doing interesting research in extreme value theory, I chose that area.

Why didn't I give the bootstrap more consideration? At that time, the bootstrap seemed so simple and straightforward. Many of us did not see it as part of a revolution in statistical methods that many of its advocates now see it as. Also, key results on the asymptotic properties of the bootstrap appeared to be difficult. The key papers by Bickel and Freedman (1981) and Singh (1981) had not yet appeared and Edgeworth and Cornish–Fisher expansions were not yet recognized as useful tools. Many of my fellow Stanford graduate students felt the same way.

Gail Gong was the first Stanford graduate student to do a dissertation on the bootstrap a few years after my graduation in 1978. The situation has changed considerably since that time. Many students have now done dissertations on bootstrap. One well-known statistician from the Stanford graduates is Rob Tibshirani. Stanford dissertations in statistics by Therneau (1983) and Hesterberg (1988) deal with variance reduction methods applied to the Monte Carlo approximation of the bootstrap estimates. At Stanford and Berkeley there was a great deal of research activity in bootstrap in the 1980s and it continues today.

My interest in the bootstrap began seriously in 1983, after reading Efron's paper on bias adjustment in error rate estimation for classification problems. It was then that I recognized the value it had relative to the classification problems that I had been working on at the Aerospace Corporation. Phillip Good became interested in it in the late 1980s when he was looking for a permutation test for the test of equality of variances and realized that the bootstrap could answer the question. He also recognized that the bootstrap had many of the same characteristics as permutation methods. It was our recognition of the value of the bootstrap and the need to explain it to the practitioner that led to the development of this book. Both bootstrap and permutation methods could be computer-intensive methods depending on the application. They both are forms of resampling. Both avoid unverified parametric assumptions, by relying solely on the original sample. Both seem to work based on general assumptions of exchangeability of observations (a slightly weaker assumption than that of independent and identically distributed observations). Formally, the term exchangeable sequence means that the joint distribution of any k consecutive observations does not change when the order of the observations is changed through a permutation.

The importance of the bootstrap has now been noted in the additional material in the supplemental volume of the *Encyclopedia of Statistical Sciences* (1989 Bootstrapping—II by David Banks, pp. 17–22). The bootstrap has also been recognized for its importance by the inclusion of Efron's 1979 *Annals of Statistics* paper in *Breakthroughs in Statistics Volume II Methodology and Distribution*, S. Kotz and N. L. Johnson, Editors (1992, pp. 565–595 with introduction by R. Beran).

In Volume III of *Breakthroughs in Statistics*, edited by S. Kotz and N. L. Johnson (1997), Peter Hall's 1988 *Annals of Statistics* paper, "Theoretical Comparison of Bootstrap Confidence Intervals," is honored and is introduced by Enno Mammen. This is another indication of the impact that the bootstrap has had on statistical methodology. The bootstrap is also referenced prominently in the new *Encyclopedia of Biostatistics* with two entries in Volume 1: (1) "Bootstrap Methods," by DeAngelis and Young (1998) and (2) "Bootstrapping in Survival Analysis," by Sauerbrei (1998).

The bibliography in this text contains 1650 references, including many applied papers that appeared in the 1990s. Of these, 619 are referenced directly in the text. They are noted with an asterisk and citation number after them in the bibliography.

The idea of sampling with replacement from the original data did not begin with Efron. Also, even earlier, there were a few related techniques that are now often referred to as resampling techniques. Some of these we have already mentioned. These other resampling techniques predate Efron's bootstrap. Among these are the jackknife, cross-validation, random subsampling, and the permutation test procedures as described in Good (1994), Edgington (1980, 1987, 1995), and Manly (1991, 1997).

The idea of resampling from the empirical distribution to form a Monte Carlo approximation to the bootstrap estimate may have been thought of and used by others prior to Efron. Simon (1969) takes credit for using this idea as a teaching tool in elementary statistics classes. Bruce and Simon have been instrumental in popularizing the bootstrap approach in recent years with their Resampling Stats software. They also use the Monte Carlo approximation to the bootstrap as a tool for introducing statistical concepts in a first course in statistics (see Simon and Bruce, 1991, 1995).

It is clear, however, that widespread use of the method (particularly by professional statisticians) and many significant theoretical developments did not occur until after Efron's 1979 work. That paper (Efron, 1979a) connected this simple bootstrap idea to established methods for estimating the standard error of an estimator, namely, the jackknife, cross-validation, and the delta method, thus providing the theoretical underpinnings that were later further developed by many researchers.

Among the other procedures that have also been referred to as bootstrap, I shall mention two in Section 1.4. Whenever I refer to the bootstrap in this book, I will mean Efron's version. But even Efron's bootstrap has a number of modified versions. Among these are the smoothed bootstrap, the double bootstrap, the Bayesian bootstrap (applied in a missing data problem in Section 8.7), and the parametric bootstrap (discussed in Chapter 6). Some variants, used in the estimation of error rates in discriminant analysis, are discussed in Section 2.1.2 (especially the .632 estimator and a convex bootstrap that I introduced).

In May of 1998 a conference was held at Rutgers University in New Jersey, which had been set up by Professor Kesar Singh, a prominent bootstrap researcher on the Rutgers University Statistics Department faculty. The purpose of the conference was to provide a collection of papers on recent bootstrap developments by key researchers and a celebration for the past 20 years of bootstrap research. It was considered to be the 20th anniversary of Efron's initial work on the bootstrap (though Efron's paper appeared in the *Annals of Statistics* in 1979, some earlier technical reports and presentations go back at least to 1977). Abstracts of the papers presented were available on the Rutgers University web site for the Statistics Department.

Although no proceedings were produced from the meeting, many of the papers were made available to me by request (by mail or e-mail to the authors). Presenters at the meeting included Michael Sherman, Brad Efron, Gutti Babu, C. R. Rao, Kesar Singh, Alastair Young, Dimitris Politis, J.-J. Ren, and Peter Hall. Papers that I received are included in the bibliography. They are Babu, Pathak, and Rao (1998), Sherman and Carlstein (1997), Efron and Tibshirani (1998), and Babu (1998).

This book is organized as follows. Chapter 1 introduces the key ideas and describes briefly the wide range of applications. Chapter 2 deals with estimation and particularly application to the estimation of bias of an estimator. It includes a detailed account of the discriminant analysis problem and related simulation studies comparing error rate estimates. It shows how the bootstrap performs relative to other traditional methods, when the training sample sizes are small. This is followed with discussion of estimation of medians, standard errors, and quantiles. Chapter 3 deals with confidence intervals and hypothesis testing. An application to the determination of the efficacy of the Tendril DX steroid lead is given.

Chapter 4 covers regression problems, both linear and nonlinear. Chapter 5 addresses forecasting and time series. Chapter 6 provides a comparison with other resampling methods and recommends a favored approach in cases where there is evidence in the literature showing an advantage to a particular approach. This is a unique feature of this book.

Chapter 7 deals with simulation methods, emphasizing the variety of available variance reduction techniques and showing the applications for which they can effectively be applied. Chapter 8 gives an account of a variety of miscellaneous topics. These include kriging (a form of analysis of spatial data), survey sampling, subset selection in stepwise procedures including both re-gression and discriminant analysis, analysis of censored data, p-value adjust-ment for multiplicity, process capability indices (measures of manufacturing process performance used in quality assurance work), and application of the Bayesian bootstrap in problems involving missing data. Chapter 9 describes the examples in the literature where ordinary bootstrapping fails. In some cases remedies exist and are mentioned.

Bootstrap diagnostics (in particular the jackknife-after-bootstrap approach), which are techniques that help us to discover when trouble exists (much like regression diagnostics identify problems with statistical assumptions in re-gression analysis), are also covered in Chapter 9. Chapter 9 differs from the other chapters in that it is somewhat brief but it does go deeper into the mathematics than do the other chapters. This chapter can be skipped by the practitioners who do not have the necessary background in probability theory. They need only be aware of the types of problems where naive use of the bootstrap may cause trouble.

Following in the footsteps of Hall (1992a), Efron and Tibshirani (1993), and Davison and Hinkley (1997), with the exception of Chapter 6, I provide a section of historical notes on research developments at the end of each chapter. Although related references are cited throughout the text, the histori-cal notes are intended to provide a perspective regarding when certain approaches were originally proposed and how important developments fol-lowed chronologically.

1.2. INTRODUCTION

Two of the most important problems in applied statistics are the determination of an estimator for a particular parameter of interest and the evaluation of the accuracy of that estimator through estimates of the standard error of the estimator and the determination of confidence intervals. Efron, when introduc-ing his version of the "bootstrap" (Efron, 1979a), was particularly motivated by these two problems. Most important was the estimation of the standard error of the parameter estimator, particularly when the estimator was complex and standard approximations such as the delta method were either not appropriate or too inaccurate.

Because of the bootstrap's generality, it has been applied to a much wider class of problems than just the estimation of standard errors and confidence intervals. Applications include error rate estimation in discriminant analysis, subset selection in regression, logistic regression and classification problems, density estimation, quantile estimation, cluster analysis, kriging (i.e., a form of spatial modeling), nonlinear regression, time series analysis, complex surveys and other finite population problems, p-value adjustment in multiple testing problems, estimating process capability indices, handling missing data prob-lems, and survival and reliability analysis.

Although this is a long list, I am sure there are other applications that I have not thought to include here. It has been applied in various disciplines including psychology, physics, geology, ecology, ornithology, econometrics, biology, meteorology, genetics, signal and image processing, medicine, engineering, reliability, chemistry, and accounting.

It is my purpose to describe some of these applications in detail for the practitioner in order to exemplify the bootstrap's usefulness and illustrate its limitations. In some cases the bootstrap will offer a solution that may not be very good but may still be used for lack of an alternative approach.

Before providing a formal definition of the bootstrap, here is an informal description of how it works. In its most general form, we have a sample of size n and we want to estimate a parameter or determine the standard error or a confidence interval for the parameter or even test a hypothesis about the parameter. If we do not make any parametric assumptions, we may find this difficult to do. The bootstrap provides a way to do this.

We look at the sample and consider the empirical distribution. The empirical distribution is the probability distribution that has probability $1/n$ assigned to each sample value. The bootstrap idea is simply to replace the unknown population distribution with the known empirical distribution.

Properties of the estimator such as its standard error are then determined based on the empirical distribution. Sometimes these properties can be determined analytically, but more often they are approximated by Monte Carlo methods (i.e., we actually sample with replacement from the empirical distribution).

Now here is a more formal definition. Efron's bootstrap (or the nonparametric bootstrap) is defined as follows: Given a sample of n independent identically distributed random vectors $\mathbf{X}_1, \mathbf{X}_2, \ldots, \mathbf{X}_n$ and a real-valued estimator $\theta(\mathbf{X}_1, \mathbf{X}_2, \ldots, \mathbf{X}_n)$ (denoted by $\hat{\theta}$) of the distribution parameter θ, a procedure (the bootstrap) to assess the accuracy of $\hat{\theta}$ is defined in terms of the empirical distribution function F_n. This empirical distribution function assigns probability mass $1/n$ to each observed value of the random vectors $\mathbf{X}_i$ for $i = 1, 2, \ldots, n$.

The empirical distribution function is the maximum likelihood estimator of the distribution for the observations when no parametric assumptions are made. The bootstrap distribution for $\hat{\theta} - \theta$ is the distribution obtained by generating $\hat{\theta}$ values by sampling independently with replacement from the empirical distribution F_n. The bootstrap estimate of the standard error of $\hat{\theta}$ is then the standard deviation of the bootstrap distribution for $\hat{\theta} - \theta$.

It should be noted here that almost any parameter of the bootstrap distribution may serve as a "bootstrap" estimate of the corresponding population parameter. We could consider the skewness, the kurtosis, the median, or the 95th percentile of the bootstrap distribution for $\hat{\theta}$.

Practical application of the technique usually requires the generation of bootstrap samples or resamples (i.e., samples obtained by independently sampling with replacement from the empirical distribution). From the bootstrap sampling a Monte Carlo approximation of the bootstrap estimate is obtained. The procedure is straightforward.

1. Generate a sample of size n (where n is the original sample size) with replacement from the empirical distribution (a bootstrap sample).
2. Compute θ^*, the value of $\hat{\theta}$ obtained by using the bootstrap sample in place of the original sample.
3. Repeat steps 1 and 2 k times.

For standard error estimation, k is recommended to be at least 100. This recommendation can be attributed to Efron (1987). It has recently been challenged in a paper by Booth and Sarkar (1998). Further discussion can be found in Chapter 7.

By replicating steps 1 and 2 k times, we obtain a Monte Carlo approximation to the distribution of θ^*. The standard deviation of this Monte Carlo distribution of θ^* is the Monte Carlo approximation to the bootstrap estimate of the standard error for $\hat{\theta}$. Often this estimate is simply referred to as the bootstrap estimate, and for k very large (e.g., 500) there is very little difference between the bootstrap estimator and this Monte Carlo approximation.

We really would like to know the distribution of $\hat{\theta} - \theta$. What we have is a Monte Carlo approximation to the distribution of $\theta^* - \hat{\theta}$. The key idea of the bootstrap is that for n sufficiently large we expect the two distributions to be nearly the same. The key idea that the distribution of $\theta^* - \hat{\theta}$ behaves almost like the distribution of $\hat{\theta} - \theta$ is sometimes referred to as the bootstrap principle.

In a few cases, we are able to compute the bootstrap estimator directly without the Monte Carlo approximation. For example, in the case of the estimator being the mean of the distribution of a real-valued random variable, Efron states (1982a, p. 2) that the bootstrap estimate of the standard error of θ is $\hat{\sigma}_{\text{boot}} = [(n - 1)/n]^{1/2} \hat{\sigma}$, where $\hat{\sigma}$ is defined as

$$\hat{\sigma} = \left[\frac{1}{n(n - 1)} \sum_{i=1}^{n} (x_i - \bar{x})^2 \right]^{1/2}$$

where x_i is the value of the ith observation and $\bar{x}$ is the mean of the sample.

In the case of testing the hypothesis of equality of distributions for censored matched pairs (i.e., observations whose values may be truncated), the bootstrap test applied to paired differences is equivalent to the sign test and the distribution under the null hypothesis is binomial with $p = \frac{1}{2}$. So no bootstrap sampling is required to determine the critical region for the test.

The bootstrap is often referred to as a *computer-intensive* method. It gets that label because in most practical problems where it is deemed to be useful the estimation is complex and bootstrap samples are required. In the case of confidence interval estimation and hypothesis testing problems, this may mean at least 1000 bootstrap replications (i.e., $k = 1000$). In Section 7.1, we address the important practical issue of what value to use for k.

Methods for reducing the computer time by more efficient Monte Carlo sampling are discussed in Section 7.2. The examples above (i.e., standard error of sample mean and censored matched pairs test for equality of distributions) illustrate that there are cases for which the bootstrap is not *computer intensive* at all!

Another point worth emphasizing here is that a typical bootstrap sample differs from the original sample because some of the observations will be repeated once, twice, or more in a bootstrap sample. There will also be some observations that will not appear at all in a particular bootstrap sample. Consequently, the values for σ^* will vary from one bootstrap sample to the next.

The actual probability that a particular X_i will appear j times in a bootstrap sample for $j = 0, 1, 2,\ldots, n$ can be determined using the multinomial distribution, or alternatively it can be derived based on classical occupancy theory. For the latter approach see Chernick and Murthy (1985). Efron (1983) calls these probabilities the repetition rates and discusses them in motivating the use of the .632 estimator (a particular bootstrap-type estimator) for classification error rate estimation. A general account of the classical occupancy theory can be found in Johnson and Kotz (1977).

The basic idea behind the bootstrap is that the variability of θ^* (based on F_n) around $\hat{\theta}$ will be similar to (or mimic) the variability of $\hat{\theta}$ (based on the true population distribution F) around σ (i.e., the bootstrap principle). There is good reason to believe this will be true for large sample sizes, since as n gets larger and larger F_n comes closer and closer to F and so sampling with replacement from F_n is almost like random sampling from F.

The strong law of large numbers for independent identically distributed random variables implies that with probability one, F_n converges to F pointwise [for the mathematically curious reader see Chung (1974, pp. 131–132) for details]. Strong laws pertaining to the bootstrap can be found in Athreya (1983). A stronger result, the Glivenko–Cantelli theorem (see Chung, 1974, p. 133), asserts that the empirical distribution converges uniformly with probability one to the distribution F when the observations are independent and identically distributed.

Although not stated explicitly in the bootstrap literature, this fundamental theoretical result lends credence to the bootstrap approach. The theorem was extended in Tucker (1959) to the case of a random sequence from a strictly stationary stochastic process (a sequence of observations that may not be independent but have probability distributions that do not change over time).

In addition to the Glivenko–Cantelli theorem, the validity of the bootstrap requires that the estimator (a functional of the empirical distribution function) converge to the "true parameter value" (i.e., the functional for the "true" population distribution). A functional is simply a mapping that assigns a real value to a function. Most commonly used parameters of distribution functions can be expressed as functionals of the distribution, including the mean, the variance, the skewness, and the kurtosis of the distribution.

Interestingly, sample estimates such as the sample mean can be expressed as the same functional applied to the empirical distribution. For more discussion of this see Chernick (1982), who deals with a form of a functional derivative called an influence function. The influence function idea was first introduced by Hampel (1974) as a way of comparing robust estimators of location.

Influence functions have had uses in robust statistical methods and in the detection of outlying observations in data sets. Formal treatment of statistical functional derivatives such as the influence function can be found in Fernholtz (1983). There also are connections with the jackknife and bootstrap [see Efron (1982a) for some details on this].

Convergence (or consistency of bootstrap estimates) requires some sort of smoothness conditions on the functional corresponding to the estimator. In particular, conditions given in Hall (1992a) employ asymptotic normality for the functional and further allow for the existence of an Edgeworth expansion for its distribution function. So there is more needed. For independent and identically distributed observations we require (1) the convergence of F_n to F (satisfied by the Glivenko–Cantelli theorem), (2) an estimate that is the corresponding functional of the empirical distribution (satisfied for sample means, standard deviations, variances, medians, and other sample quantiles), and (3) a smoothness condition on the functional. Often in cases where the bootstrap principle fails (i.e., bootstrap estimates are inconsistent) conditions (1) and (2) are satisfied but (3) fails to hold (e.g., extreme order statistics such as the minimum and maximum values).

These Edgeworth expansions along with the Cornish–Fisher expansions not only assure the consistency of the bootstrap but also provide asymptotic rates of convergence. Examples where the bootstrap fails asymptotically due to a lack of smoothness of the functional are given in Chapter 9.

Also, the original bootstrap idea applies to independent identically distributed observations and is guaranteed to work only in large samples. Using the Monte Carlo approximation, bootstrapping can be applied to many practical problems such as parameter estimation in time series, regression and analysis of variance problems, and even to problems involving small samples.

For some of these problems, we may be on shaky ground, particularly when small sample sizes are involved. Nevertheless, it has been discovered through the extensive research that has taken place in the 1980s and 1990s that the bootstrap sometimes works better than conventional approaches even in small samples (e.g., the case of error rate estimation for linear discriminant functions to be discussed in Section 2.1.2).

There is also a strong temptation to apply the bootstrap to a number of complex statistical problems where there is no resort to classical theory. At least for some of these problems, the practitioner should try the bootstrap. Only for cases where there is theoretical evidence that the bootstrap leads one astray would I advise against its use.

The determination of variability in subset selection for regression, logistic regression, or discriminant analysis problems provides an example of such a complex problem. Another is the determination of the variability of spatial contours based on the method of kriging. The bootstrap and alternatives in spatial data problems are treated in Cressie (1991). Other books dealing with spatial data are Mardia, Kent, and Bibby (1979) and Hall (1988c). Tibshirani (1992) provides some examples of the usefulness of the bootstrap in complex problems.

Diaconis and Efron (1983) demonstrate, with just five bootstrap sample contour maps, the value of the bootstrap approach in uncovering the variability in the contours based on kriging. These problems, which can be addressed by the bootstrap approach, are discussed in more detail in Chapter 8.

1.3. WIDE RANGE OF APPLICATIONS

As mentioned at the end of the last section, there is a great deal of temptation to apply the bootstrap in a wide number of settings. In the regression case, for example, we may treat the vector including the dependent variable and the explanatory variable as an independent random vector, or alternatively we may compute residuals and bootstrap them. These are two distinct approaches to bootstrapping in regression problems that will be discussed in detail in Chapter 5.

In the case of estimating the error rate for a linear discriminant function, Efron (1982a, pp. 49–58; 1983) showed that the bootstrap could be used to (1) estimate the bias of the "apparent error rate" estimate (also referred to as resubstitution) and (2) produce an improved error rate estimate by adjusting for the bias. The resulting bootstrap estimate is superior to the popular cross-validation or leave-one-out estimator.

The most attractive feature of the bootstrap described here, in Efron and Tibshirani (1993), and in Davison and Hinkley (1997) and the permutation tests described in Good (1994), in Manly (1991, 1997), and in Edgington (1980, 1987, 1995) is the freedom they provide from restrictive parametric assumptions and simplified models. There is no need to force Gaussian or other parametric distributional assumptions on our data.

In many problems, the data may be skewed or have a heavy-tailed distribution or may even have a multimodal distribution. The model does not need to be simplified to some "linear" approximation and the estimator itself can be complicated.

We do not require an analytic expression for the estimator. The bootstrap Monte Carlo approximation can be applied as long as there is a computational method for deriving the estimator. This means that we can even numerically integrate using iterative schemes to calculate the estimator. The bootstrap doesn't care. The only price we pay for such complications is in the time and cost for the computer usage (which is becoming faster and cheaper).

Another feature that makes the bootstrap approach attractive is its simplicity. We can formulate bootstrap simulations for almost any conceivable problem. Once we program the computer to carry out the bootstrap replications we let the computer do all the work. A danger to this approach is that a practitioner might bootstrap at will, without consulting a statistician (or considering the statistical implications) and without giving careful thought to the problem. This book will aid practitioners in the proper use of the bootstrap by acquainting them with its advantages and limitations, lending theoretical support where available and Monte Carlo results where the theory is not yet available. Theoretical counterexamples to the consistency of bootstrap estimates also provide guidelines to its limitations and warn the practitioner when not to apply the bootstrap. Some simulation studies also provide such negative results.

Nevertheless, I believe that most simulation studies indicate that the bootstrap can safely be applied (directly or in a modified form) to a large number of problems where strong theoretical justification does not yet exist. For many problems where realistic assumptions make other statistical approaches impossible or at least intractable, the bootstrap at least provides a solution even if it is not a very good one. For some, in certain situations, even a poor solution is better than no solution.

Another problem that creates difficulties for the scientist, engineer, or clinician is that of missing data. In designing an experiment or a survey, we may strive for balance in the design and choose specific sample sizes in order to make the planned inferences from the data. The correct inference can be made only if we observe the complete data set.

Unfortunately in the real world, the cost of experimentation, faulty measurement, or lack of response from those selected for the survey may lead to incomplete and possibly unbalanced designs. Milliken and Johnson (1984) refer to such problem data as messy data.

Milliken and Johnson (1984, 1989) provide ways to analyze messy data. When data are missing or censored, bootstrapping provides another approach for dealing with the messy data (see Section 8.4 for more details on censored data and Section 8.7 for an application to missing data).

The bootstrap alerts the practitioner to variability in the data of which he or she may not be aware. In regression, logistic regression, or discriminant analysis, stepwise subset selection is a commonly used method available in most statistical computer packages. The computer does not tell the user how arbitrary the final selection actually is. When a large number of variables or features are included, and many are correlated or redundant, there can be a great deal of variability to the selection. The bootstrap samples enable the user to see how the chosen variables or features change from bootstrap sample to bootstrap sample and provide some insight as to which variables or features are really important and which ones are correlated and easily substituted for by others. This is particularly well illustrated by the logistic regression problem studied in Gong (1986). This problem is discussed in detail in Section 8.2.

In the case of kriging, spatial contours of features such as pollution concentration are generated based on data at monitoring stations. The method is a form of interpolation between the stations based on certain statistical spatial modeling assumptions. However, the contour maps themselves do not provide the practitioner with an understanding of the variability of these estimates. Kriging plots for different bootstrap samples provide the practitioner with a graphical display of this variability and at least warn of variability in the data and analytic results. Diaconis and Efron (1983) make this point convincingly and I will demonstrate this application in Section 8.1. The practical value of this cannot be underestimated!

Babu and Feigelson (1996) discuss applications in astronomy. They devote a whole chapter (Chap. 5, pp. 93–103) to resampling methods emphasizing the importance of the bootstrap.

In clinical trials, sample sizes are determined based on achieving a certain power for a statistical hypothesis of efficacy of the treatment. In Section 3.3, I show an example of a clinical trial for a pacemaker lead (Pacesetter's Tendril DX model). In this trial, the sample sizes for both the treatment leads and the control leads were chosen to provide an 80% chance of detecting a clinically significant improvement (decrease of 0.5 volt) in the average capture threshold at the three-month follow-up for the experimental Tendril DX lead (model 1388T) compared to the respective control lead (Tendril model 1188T) when applying a one-sided significance test at the 5% significance level. This was based on the standard normal distribution theory. In the study, nonparametric methods were also considered. Bootstrap confidence intervals (based on Efron's percentile method) are applied to form a hypothesis test without requiring parametric modeling assumptions. The Wilcoxon rank sum test (another nonparametric procedure) was also considered.

A similar study for a passive fixation lead, the Passive Plus DX lead, was conducted to get FDA approval of the steroid eluting version of that type of lead. In addition to comparing the investigational lead (the steroid eluting lead) with the nonsteroid control lead, through the percentile method boot-strap and the Wilcoxon test, I also considered a possibly more accurate bootstrap confidence interval (the bootstrap percentile t method). Results were very similar and conclusive for all three methods. The percentile t (or bootstrap t) method for confidence interval estimation is described in Section 3.1.5.

However, the statistical conclusion for such a trial is based on a single test at the three-month follow-up after all 99 experimental and 33 control leads have been implanted and the patients had threshold tests at the three-month follow-up. In practice the physicians do not want to wait for all the patients to reach their three-month follow-up before doing the analysis. Consequently, it is quite common to do interim analyses at some point or points in the trial (it could be one in the middle of the trial or two at the one-third and two-thirds points in the trial). Also, separate analyses are sometimes done on subsets of the population. Multiple testing implies a need to adjust the p-value for the test. Probability bounds, such as Bonferroni, can be used to give conservative estimates of the *p*-value or simultaneous inference methods can be used [see Miller (1981b) for a thorough treatment of this subject].

An alternative approach would be to estimate the *p*-value adjustment by bootstrapping. This idea has been exploited by Westfall and Young and is described in detail in Westfall and Young (1993). The application of bootstrap *p*-value adjustment to the Passive Plus DX clinical trial data is briefly discussed in Section 8.5, although it was not necessary from the practical viewpoint. In Section 8.5 I also provide an example of where there is a benefit to using the bootstrap adjustment over the highly conservative Bonferroni bound.

In concluding this section, I must emphasize that the bootstrap is not a panacea. There are certainly practical problems where classical parametric methods are reasonable and provide either more efficient estimates or more powerful hypothesis tests. Even for some parametric problems, the parametric bootstrap, as discussed by Davison and Hinkley (1997, p. 3) and illustrated by them on pages 148 and 149, can be useful.

What the bootstrap does do is free the scientist from restrictive modeling and distributional assumptions or complicated algebra by using the power of the computer to replace difficult analysis. In an age when computers are becoming more and more powerful, inexpensive, fast, and easy to use, the future looks bright for additional use of these so-called computer-intensive statistical methods.

1.4. HISTORICAL NOTES

It should be pointed out that bootstrap research began in the late 1970s although many key related developments can be traced back to earlier times. Most of the important theoretical development took place in the 1980s after Efron (1979a).

Regarding this seminal paper (by Efron), Davison and Hinkley (1997) write: "The publication in 1979 of Bradley Efron's first article on bootstrap methods was a major event in Statistics, at once synthesizing some of the earlier resampling ideas and establishing a new framework for simulation-based statistical analysis. The idea of replacing complicated and often inaccurate approximations to biases, variances, and other measures of uncertainty by computer simulations caught the imagination of both theoretical researchers and users of statistical methods."

As mentioned earlier in the chapter a number of related techniques are often referred to as resampling techniques. These other resampling techniques predate Efron's bootstrap. Among these are the jackknife, cross-validation, random subsampling, and the permutation test procedures described in Good (1994), Edgington (1980, 1987, 1995), and Manly (1991, 1997).

Makinodan, Albright, Peter, Good, and Heidrick (1976) apply permutation test methods to study the effects of age in mice on the mediation of immune response. Due to the fact that an entire factor combination was missing, the model and the permutation test provided a clever way to deal with the imbalance in the data. A detailed description is given in Good (1994, pp. 58–59).

Efron himself points to some of the early work of R. A. Fisher (in the 1920s) on maximum likelihood estimation as the inspiration for many of the basic ideas. The jackknife was introduced by Quenouille (1949) and popularized by Tukey (1958). Miller (1974) provides an excellent review of jackknife methods. Extensive coverage of the jackknife can be found in the book by Gray and Schucany (1972).

Bickel and Freedman (1981) and Singh (1981) presented the first results demonstrating the consistency of the bootstrap under certain mathematical conditions. Bickel and Freedman (1981) also provide a counterexample for consistency of the nonparametric bootstrap and this is also illustrated by Schervish (1995, p. 330, Example 5.80). Gine and Zinn (1989) provide necessary conditions for the consistency of the bootstrap for the mean. Athreya (1987), Knight (1989), and Angus (1993) all provide examples where the bootstrap failed to be consistent due to its inability to meet certain necessary mathematical conditions. Hall, Hardle, and Simar (1993) show that estimators of the bootstrap distribution can be inconsistent.

The general subject of empirical processes is related to the bootstrap and can be used as a tool to demonstrate consistency (see Csorgo, 1983; Shorack and Wellner, 1986; van der Vaart and Wellner, 1996). Fernholtz provides the mathematical theory of statistical functionals and functional derivatives (such as influence functions), which relate to bootstrap theory. Quantile estimation via bootstrapping appears in Helmers, Janssen, and Veraverbeke (1992) and Falk and Kaufman (1991). Csorgo and Mason (1989) bootstrap empirical functions and Tu (1992) uses jackknife pseudovalues to approximate the distribution of a general standardized functional statistic.

Subsampling methods began with Hartigan (1969, 1971, 1975) and McCarthy (1969). These papers are discussed briefly in the development of bootstrap confidence intervals in Chapter 3. A more recent account is given by Babu (1992).

Young and Daniels (1990) discuss the bias that is introduced in Efron's nonparametric bootstrap by the use of the empirical distribution as a substitute for the true unknown distribution.

Diaconis and Holmes (1994) show how to avoid the Monte Carlo approximation to the bootstrap by cleverly enumerating all possible bootstrap samples using Gray codes.

The term bootstrap has been used in other similar contexts that predate Efron's work, but these methods are not the same and some confusion occurs. When I gave a presentation on the bootstrap at the Aerospace Corporation in 1983 a colleague, Dr. Ira Weiss, mentioned that he used the bootstrap in 1970 long before Efron coined the term. After looking at Ira's paper, I realized that it was a different procedure with a similar idea.

Apparently control theorists came up with a procedure for applying Kalman filtering with an unknown noise covariance, which they too named the bootstrap. Like Efron, they were probably thinking of the old adage "picking yourself up by your own bootstraps." This idea was attributed to the fictional Baron von Munchausen, who supposedly performed this trick to climb out from the bottom of a lake. Now the control theorists probably applied the term to an estimation procedure that avoids a priori assumptions and uses only the data at hand. This would be similar to Efron's notion. A survey and comparison of procedures for dealing with the problem of unknown noise covariance including this other bootstrap technique is given in Weiss (1970).

The term "bootstrap" has also been used, in a totally different context, by computer scientists. A careful review of literature in various disciplines might reveal other procedures, different from Efron's bootstrap, which were named bootstrap before Efron coined the term in the statistics literature.

An entry on bootstrapping in the *Encyclopedia of Statistical Science* (1982 Vol. 1, p. 301) is provided by the editors and is very brief. In 1982 when that volume was published, the true value of bootstrapping was not fully appreciated. The editors subsequently remedied this with an article in the Supplemental Volume.

The point, however, is that the original entry cited only three references. The first, Efron's *SIAM Review* article (Efron, 1979b), was one of the initial published works describing Efron's bootstrap. The second article from *Technometrics* by Fuchs (1978) does not appear to deal with the bootstrap at all! The third article by LaMotte (1978) and also in *Technometrics* does refer to a bootstrap but does not mention any of Efron's ideas and appears to be discussing a different bootstrap.

Because of these other bootstraps, I have tried to refer to the bootstrap as "Efron's bootstrap" and a few others have done the same, but it has not caught on. In the statistical literature, reference to the bootstrap will almost always mean Efron's bootstrap or some derivative of it. In the engineering literature, an ambiguity may exist and we really need to look at the description of the procedure in detail to determine precisely what the author means.

The term "bootstrap" also commonly appears in the computer science literature and I understand that mathematicians use the term to describe certain types of numerical solutions to partial differential equations. Still it is my experience that if I search for articles in mathematical or statistical indices in the present (1999) using just the keyword "bootstrap," I would find the vast majority of the articles referred to Efron's bootstrap or a variant of it. These other bootstraps are now much less common in the mathematics literature.

The same may not be true in computer science or engineering. Efron's bootstrap now has a presence there also. In the case of computer science this is because of its nature as a computer-intensive method and because it is a common topic in the annual meetings on the interface between computer science and statistics. In engineering, the rapid growth of applications to real engineering problems is the reason for its presence.

Efron (1983) compared several variations to the bootstrap estimate. He considered simulation of Gaussian distributions for the two-class problem (with equal covariances for the classes) and small sample sizes (e.g., a total of 14 to 20 training samples split equally among the two populations). For linear discriminant functions, he showed that the bootstrap and, in particular, the .632 estimator are superior to the commonly used leave-one-out estimate (also called cross-validation by Efron). Subsequent simulation studies will be summarized in Section 2.1.2 along with guidelines for the use of some of the bootstrap estimates.

There have since been a number of interesting simulation studies that show the value of certain bootstrap variants when the training sample size is small (particularly the estimator referred to as the .632 estimate). In a series of simulation studies, Chernick, Murthy, and Nealy (1985, 1986, 1988a, b) confirmed the results in Efron (1983). They also showed that the .632 estimator was superior when the populations were not Gaussian but had finite first moments. In the case of Cauchy distributions, and other heavy-tailed distributions from the Pearson VII family of distributions, which do not have finite first moments, they showed that other bootstrap approaches were better than the .632 estimator.

Other related simulation studies include Chatterjee and Chatterjee (1983), McLachlan (1980), Snapinn and Knoke (1984, 1985a, b, 1988), Jain, Dubes, and Chen (1987), and Efron and Tibshirani (1997a). We summarize the results of these studies and provide guidelines to the use of the bootstrap procedures for linear and quadratic discriminant functions in Section 2.1.2. McLachlan (1992) also gives a good summary treatment to some of this literature. Additional theoretical results can be found in Davison and Hall (1992). Hand (1986) is another good survey article on error rate estimation. The 632+ estimator proposed by Efron and Tibshirani (1997a) was applied to an ecological problem in Furlanello, Merler, Chemini, and Rizzoli (1998). Ueda and Nakano (1995) apply the bootstrap and cross-validation to error rate estimation for neural network type classifiers. Hand (1981, p. 189; 1982, pp. 178–179) discusses the bootstrap approach to estimating the error rates in discriminant analysis.

In recent years, there have been a number of books that cover some aspect of bootstrapping at least partially. A book by Noreen (1989) deals with the bootstrap in very elementary ways for hypothesis testing only.

There are now several survey articles on bootstrapping in general, including Babu and Rao (1993), Young (1994), Stine (1992), Efron (1982b), Efron and LePage (1992), Efron and Tibshirani (1985, 1986, 1996a, 1997b), Hall (1994), Manly (1993), Gonzalez-Mantiega, Prada-Sanchez, and Romo (1994), Politis (1998), and Hinkley (1984, 1988). Overviews on the bootstrap or special aspects of bootstrapping include Beran (1984b), Leger, Politis, and Romano (1992), Pollack, Simon, Bruce, Borenstein, and Lieberman (1994), and Fiellin and Feinstein (1998) on the bootstrap in general; Babu and Bose (1989), DiCiccio and Efron (1996), and DiCiccio and Romano (1988, 1990) on confidence intervals; Efron (1988b) on regression; Falk (1992a) on quantile estimation; and DeAngelis and Young (1992) on smoothing. Lanyon (1987) reviews the jackknife and bootstrap for applications to ornithology. Efron (1988c) gives a general discussion of the value of bootstrap confidence intervals aimed at an audience of psychologists.

The latest edition of *Kendall's Advanced Theory of Statistics, Volume I*, deals with the bootstrap as a tool for estimating standard errors in Chapter 10 (see Stuart and Ord, 1994, pp. 365–368).

The use of the bootstrap to compute standard errors for estimates and to obtain confidence intervals for multilevel linear models is given in Goldstein (1995, pp. 60–63). Waclawiw and Liang (1994) give an example of parametric bootstrapping using generalized estimating equations. Other works involving the bootstrap and jackknife in estimating equation models include Lele (1991a, b).

Lehmann and Casella (1998) mention the bootstrap as a tool in reducing the bias of an estimator (p. 144) and in the attainment of higher order efficiency (p. 519). Lehmann (1999, Sec. 6.5, pp. 420–435) presents some details on the asymptotic properties of the bootstrap.

In the context of generalized least squares estimation of regression parameters, Carroll and Ruppert (1988, pp. 26–28) describe the use of the bootstrap to get confidence intervals. Briefly, Nelson (1990) mentions the bootstrap as a potential tool in regression models applicable to right censored data for applications in accelerated testing. Srivastava and Singh (1989) deal with multiplicative models. Bickel and Ren (1996) apply an m out of n bootstrap for goodness of fit tests with doubly censored data.

McLachlan and Basford (1988) discuss the bootstrap in a number of places in their book as an approach for determining the number of distributions or modes to be included in a mixture model. Another good text dealing with mixture models is Titterington, Smith, and Makov (1985). Efron and Tibshirani (1996b) take a novel approach to bootstrapping that can be applied to the determination of the number of modes in a density function and the number of variables in a model. In addition to determining the number of modes, the bootstrap has been used to estimate the location of a mode (Romano, 1988c).

Linhart and Zucchini (1986, pp. 22–23) describe how the bootstrap can be used for model selection. Thompson (1989, pp. 42–43) mentions the use of bootstrap techniques for estimating parameters in growth models (i.e., a nonlinear regression problem). McDonald (1982) shows how smoothed or ordinary bootstrap samples can be drawn to obtain regression estimates.

Rubin (1987, pp. 44–46) discusses his "Bayesian" bootstrap for problems of imputation. The original paper on the Bayesian bootstrap is Rubin (1981). Banks (1988) provides a modification to the Bayesian bootstrap. Other papers on the Bayesian bootstrap include Lo (1987, 1988, 1993a) and Weng (1989). Geisser (1993) discusses the bootstrap with respect to predictive distributions (another Bayesian concept). Ghosh and Meeden (1997, pp. 140–149) discuss application of the Bayesian bootstrap to finite population sampling. The Bayesian bootstrap is often applied to imputation problems. Rubin (1996) is a survey article detailing the history of multiple imputation that was over 18 years old at the time of the article.

Rey (1983) devotes Chapter 5 to the bootstrap. He is using it in the context of robust estimation. His discussion is particularly interesting because he mentions both the pros and the cons and is critical of some of the early claims made for the bootstrap [particularly in Diaconis and Efron (1983)].

Staudte and Sheather (1990) deal with the bootstrap as an approach to estimating standard errors of estimates. They are particularly interested in the standard errors of robust estimators. Although they do deal with hypothesis testing, they do not use the bootstrap for any hypothesis testing problems. Their book includes a computer disk that has Minitab macros for bootstrapping in it. Minitab code for these macros is presented in Appendix D of their book.

Barnett and Lewis (1995) discuss the bootstrap as it relates to checking modeling assumptions in the face of outliers. Agresti (1990) discusses the bootstrap as it applies to categorical data.

McLachlan and Krishnan (1997) discuss the bootstrap as a tool in robust estimation of a covariance matrix. Beran and Srivastava (1985) provide bootstrap tests for functions of a covariance matrix. Other papers presenting theoretical results related to robust estimators are Babu and Singh (1984b) and Arcones and Gine (1992). Lahiri (1992a) does bootstrapping of M-estimators (a type of robust location estimator).

The book by van der Vaart and Wellner (1996) is devoted to weak convergence and empirical processes. Empirical process theory can be applied to obtain important results in bootstrapping and van der Vaart and Wellner illustrate this in Section 3.6 of their book (14 pages devoted to the subject of bootstrapping, pp. 345–359).

Hall (1992a) considers functionals that admit Edgeworth expansions. Edgeworth expansions and the related Cornish–Fisher expansions provide insight into the accuracy of bootstrap confidence intervals, the value of bootstrap hypothesis tests, and use of the bootstrap in nonparametric regression. These expansions also provide guidance to the practitioner regarding the variants of the bootstrap and the Monte Carlo approximations. Some articles relating Edgeworth expansions to uses of the bootstrap include Abramovitch and Singh (1985), Bhattacharya and Qumsiyeh (1989), Babu and Singh (1989), Bai and Rao (1991), and Bai and Rao (1992).

Chambers and Hastie (1992) discuss applications of statistical models through the use of the S language. They discuss the bootstrap in various places.

The bootstrap has also been applied to problems in multivariate analysis. See Gifi (1990) for some applications. Diaconis and Efron (1983) apply the bootstrap to principal component analysis. Greenacre (1984) deals with correspondence analysis, a branch of multivariate analysis.

A classic text on multivariate analysis is Anderson (1984). The bootstrap approach is mentioned in this revised edition. Flury (1997) provides a very recent account of multivariate methods and Flury (1988) is a text devoted to the technique of principal components as is Jolliffe (1986). Seber (1984), Gnanadesikan (1977, 1997), Hawkins (1982), and Mardia, Kent, and Bibby (1979) deal with mutivariate data.

Scott (1992, pp. 257–260) discusses the bootstrap for estimating standard errors and confidence intervals in the context of multivariate density estimation. Other articles on the use of bootstrap in density estimation include Faraway and Jhun (1990), Falk (1992b), Marron (1992), and Taylor and Thompson (1992).

Applications in survival analysis include Burr (1994), Hsieh (1992), Leblanc and Crowley (1993), and Gross and Lai (1996a). An application of the double bootstrap is McCullough and Vinod (1998). Application to the estimation of correlation coefficients can be found in Lunneborg (1985) and Young (1988a).

General discussion of bootstrapping related to nonparametric procedures include Romano (1988a), Romano (1989b), and Simonoff (1986) (regarding goodness of fit in sparse multinomial data problems). Tu, Burdick, and Mitchell (1992) do nonparametric rank estimation using bootstrap resampling.

Hahn and Meeker (1991) briefly discuss bootstrap confidence intervals. Frangos and Schucany (1990) discuss the technical aspects of estimating the Efron acceleration constant for the BC_a confidence interval method. Bickel and Krieger (1989) use the bootstrap to obtain confidence bands for a distribution function and Wang and Wahba (1995) get bootstrap confidence bands for smoothing splines and compare them to Bayesian intervals.

Bailey (1992) provides a form of bootstrapping for order statistics and other random variables whose distribution can be represented as convolutions of distributions. By substituting the empirical distributions for the unknown distributions in the convolution, a "bootstrap" distribution for the random variable is derived.

Beran (1982) compares the bootstrap with various competitive methods in estimating sampling distributions. Bau (1984) does bootstrapping for statistics with linear combinations. Parr (1983) is an early reference comparing the bootstrap, jackknife, and delta method in the context of bias and variance estimation. Hall (1988d) deals with the rate of convergence for bootstrap approximations.

Applications to directional data include Fisher and Hall (1989) and Ducharme, Jhun, Romano, and Troung (1985). Applications to finite population sampling include Chao and Lo (1985), Booth, Butler, and Hall (1994), Kuk (1987, 1989), and Sitter (1992b).

Applications have recently appeared in a variety of disciplines. These include Choi, Nam, and Park (1996) for process capability indices; Jones, Wortberg, Kreissig, Hammock, and Rockel (1996) in engineering; Bajgier (1992), Seppala, Moskowitz, Plante, and Tang (1995) and Liu and Tang (1996) for process control; Chao and Huwang (1987) in reliability; Coakley (1996) for image processing; Bar-Ness and Punt (1996) for communications; and Zoubir and Iskander (1996) and Zoubir and Boashash (1998) in signal processing. Ames and Muralidhar (1991) and Biddle, Bruton, and Siegel (1990) provide applications for auditing. Robeson (1995) applies the bootstrap in meteorology, Tambour and Zethraecus (1998) in economics, and Tran (1996) in sports medicine. Roy (1994) and Schafer (1992) provide applications in chemistry, Rothery (1985) and Lanyon (1987) in ornithology. Das Peddala and Chang (1992) give an application in physics. Mooney (1996) covers the use of the bootstrap in political science. Adams, Gurevitch, and Rosenberg (1997) and Shipley (1996) apply the bootstrap to problems in ecology; Andrieu, Caraux, and Gascuel (1997) in evolution; and Aastveit (1990), Felsenstein (1985), Sanderson (1989, 1995), Sitnikova (1996), Sitnikova, Rzhetsky, and Nei (1995), Leal and Ott (1993), Tivang, Nienhuis, and Smith (1994), Schork (1992), Zharkikh and Li (1992), and Zharkikh and Li (1995) in genetics. Lunneberg (1987) gives applications in the behavioral sciences. Abel and Berger (1990) and Brey (1990) give applications in biology. Aegerter, Muller, Nakache, and Boue (1994), Baker and Chu (1990), Barlow and Sun (1989), Mapleson (1986), Rosen and Cohen (1995), Shen and Iglewicz (1994), Smith and Sielken (1988), Tsodikov, Hasenclever, and Loeffler (1998), and Wahrendorf and Brown (1980) all apply the bootstrap to a variety of medical problems.

The first monograph on the bootstrap was Efron (1982a). In the 1990s there have been a number of books introduced that are dedicated to bootstrapping and/or related resampling methods. These include Beran and Ducharme (1991), Davison and Hinkley (1997), Efron and Tibshirani (1993), Hall (1992a), Helmers (1991b), Hjorth (1994), Janas (1993), Mammen (1992b), Manly (1997), Good (1994, 1998), Mooney and Duval (1993), Shao and Tu (1995), and Westfall and Young (1993). Schervish (1995) devotes a section and Sprent (1998) a full chapter to the bootstrap. The bootstrap is discussed throughout Sprent (1998) as it is one of a few data-driven statistical method that is the theme for his book.

Efron has demonstrated the value of the bootstrap in a number of applied and theoretical contexts. In Efron (1988a) he provides three examples of the value of computer-intensive inference. In Efron (1992b) he shows how the bootstrap has impacted theoretical statistics by raising six basic theoretical questions.

Davison and Hinkley (1997) provide a computer diskette with a library of useful SPLUS functions that can be used to implement bootstrapping in a variety of problems. These routines can be used with the commercial Version 3.3 of SPLUS and they are described in Chapter 11 of the book. Barbe and Bertail (1995) deal with weighted bootstraps.

1.5. SUMMARY

In this chapter I have provided the reader with a basic explanation of Efron's nonparametric bootstrap. I have followed this up with explanations as to why the procedure can be expected to work in a wide variety of applications and have provided some historical perspective. I also have pointed the reader to later chapters and other references that provide more detail.

The discussion has been in friendly terms with each concept given a simple definition. I needed to mention advanced concepts such as statistical functionals, influence functions, convergence in probability, Edgeworth and Cornish–Fisher expansions, and stationary stochastic processes.

Since these topics involve advanced mathematics and probability, I deliberately avoid a mathematical development of the concepts and theorems that have implications in bootstrapping. The mathematically curious reader can consult the references I have given for such development. In addition, Serfling (1980) is a good reference for asymptotic statistical theory.

Other readers of this text should not be bothered by this type of discussion. They should, however, be aware of the existence of these results. They do not need to worry about the details.

The approach that I take is really no different than mentioning the concept of the central limit theorem to justify the use of the normal distribution to approximate the sampling distribution for averages in moderately large samples. In elementary courses, this is done without a detailed description of weak convergence (i.e., convergence in distribution), triangular arrays, and the Lindeberg–Feller conditions.

CHAPTER 2

Estimation

In this chapter, we deal with problems involving point estimates. Section 2.1 covers the estimation of the bias of an estimator by the bootstrap technique. After showing you how to use the bootstrap to estimate bias in general, we will focus on the important application to the estimation of error rates in the classification problem.

This will require that you first be given an introduction to the classification problem and the difficulties with the classical estimation procedures when the size of the training set is small. Another application to classification problems — the determination of a subset of features to be included in the classification rule — will be discussed in Section 8.2.

Section 2.2 explains how to bootstrap to obtain point estimates of location and dispersion parameters. When the distributions have finite second moments the mean and the standard deviation are the common measures. However, we sometimes have to deal with distributions that do not even have first moments (the Cauchy distribution is one such example).

Such distributions come up in practice when taking ratios or reciprocals of random variables where the random variable in the denominator can take on the value zero or values close to zero. The commonly used location parameter is the median and the interquartile range R is a common measure of dispersion where $R = L_{75} - L_{25}$ for L_{75} the 75th percentile of the distribution and L_{25} the 25th percentile of the distribution.

2.1. ESTIMATING BIAS

2.1.1. How to Do It by Bootstrapping

Let $E(X)$ denote the expected (or mean) value of a random variable X. For an estimator $\hat{\theta}$ of a parameter θ, we consider the random variable $\hat{\theta} - \theta$ for our X. The bias of an estimator $\hat{\theta}$ for θ is defined to be $b = E(\hat{\theta} - \theta)$. As an example, the sample variance,

$$S^2 = \sum_{i=1}^{n} \frac{(x_i - \bar{x})^2}{n-1}$$

based on a sample of n independent and identically distributed random variables, $X_1, X_2, \ldots, X_n$, from a population distribution with a finite variance, is an unbiased estimator for σ^2, the population variance, where

$$\bar{x} = \sum_{i=1}^{n} \frac{x_i}{n}$$

On the other hand, for Gaussian populations the maximum likelihood estimator for σ^2 is equal to

$$(n-1)S^2/n$$

It is a biased estimator with the bias equal to

$$-\sigma^2/n \quad \text{since} \quad E((n-1)S^2/n) = (n-1)\sigma^2/n$$

The bootstrap estimator B^* of b is then $E(\theta^* - \hat{\theta})$, where θ^* is an estimate of θ based on a bootstrap sample. A Monte Carlo approximation to B^* is obtained by doing k bootstrap replications as described in Section 1.1.

For the ith bootstrap replication, we denote the estimate of θ by θ_i^*. The Monte Carlo approximation to B^* is the average of the differences between the bootstrap sample estimate θ_i^* of θ and the original sample estimate $\hat{\theta}$,

$$B_{\text{Monte}} = \sum_{i=1}^{k} (\theta_i^* - \hat{\theta})/k$$

Generally, the purpose of estimating bias is to improve a biased estimator by subtracting an estimate of its bias from it. In Section 2.1.2, we shall see that Efron's definition of the bias is essentially the negative of the definition given here [i.e., $B^* = E(\hat{\theta} - \theta^*)$] and consequently we will add the bias to the estimator rather than subtract it.

Bias correction was the original idea in a related resampling method, the jackknife [dating back to Quenouille (1949) and Tukey (1958)]. In the next section, we find an example of an estimator that generally has a large bias but not a very large variance in small samples. For this problem, error rate estimation in linear discriminant analysis, the bootstrap bias correction provides a spectacular success!

2.1.2. Error Rate Estimation in Discrimination

First you'll be given a brief description of the two-class discrimination problem. Then, some of the traditional procedures for estimating the expected conditional error rate (i.e., the expected error rate given a training set) will be described. Next, a description is provided of some of the various bootstrap-type estimators that have been applied. Finally, results are summarized for some of the simulation studies that compared the bootstrap estimators with the resubstitution and leave-one-out (or cross-validation) estimators.

I again emphasize that this particular example is one of the big success stories for the bootstrap. It is a case where there is strong empirical evidence for the superiority of bootstrap estimates over traditional methods, particularly when the sample sizes are *small!*

In the two-class discrimination problem you are given two classes of objects. A common example is the case of a target and some decoys that are made to look like the target. The data consist of a set of values for variables that are usually referred to as features.

We hope that the values of the features for the decoys will be different from the values for the targets. We shall also assume that we have a training set (i.e., a sample of features for decoys and a separate sample of features for targets where we know which values correspond to targets and which correspond to decoys). We need the training set in order to learn something about the unknown feature distributions for the target and the decoy.

We shall briefly mention some of the theory for the two-class problem. The interested reader may want to consult Duda and Hart (1973), Srivastava and Carter (1983, pp. 231–253), Fukunaga (1990), or McLachlan (1992) for more details.

Before considering the use of training data, for simplicity, let us suppose that we know exactly the probability density of the feature vector for the decoys and also for the targets. These densities shall be referred to as the class-conditional densities.

Now suppose that someone discovers a new object and does not know whether it is a target or a decoy but does have measured or derived values for that object's features. Based on the features, we want to decide whether it is a target or a decoy.

This is a classical multivariate hypothesis testing problem. There are two possible decisions: (1) to classify the object as a decoy and (2) to classify the object as a target. Associated with each possible decision is a possible error: (a) we can decide (1) when the object is a target or (b) we can decide (2) when the object is a decoy.

Generally, there are costs associated with making the wrong decisions. These costs need not be equal. If the costs are equal, Bayes's theorem provides us with the decision rule that minimizes the cost.

For the reader who is not familiar with Bayes's theorem, it will be presented in the context of this problem, after we define all the necessary terms. Even with unequal costs, we can use Bayes's theorem to construct the decision rule that minimizes the expected cost. This rule is called the Bayes rule and it follows our intuition.

For equal costs, we classify the object as a decoy if the *a posteriori* probability of a decoy, given that we observe the feature vector **x**, is higher for the decoy than the *a posteriori* probability of a target given that we observe feature **x**. We classify it as a target otherwise.

Bayes's theorem gives us a way to compute these *a posteriori* probabilities. If our *a priori* probabilities are equal (i.e., before collecting the data we assume that the object is as likely to be a target as it is to be a decoy), the Bayes rule is equivalent to the likelihood ratio test.

The likelihood ratio test classifies the object as the type that has the greater likelihood for **x** (i.e., the larger class-conditional density). For more discussion see Duda and Hart (1973, p. 16.)

Many real problems have unequal *a priori* probabilities; sometimes we can determine these probabilities. In the target versus decoy example, we may have intelligence information that the enemy will put out nine decoys for every real target. In that case, the *a priori* probability for a target is .1, whereas the *a priori* probability for a decoy is .9.

Let $P_D(\mathbf{x})$ be the class-conditional density for decoys and $P_T(\mathbf{x})$ be the class-conditional density for targets. Let C_1 be the cost of classifying a decoy as a target, C_2 the cost of classifying a target as a decoy, P_1 the *a priori* probability for a target, and P_2 the *a priori* probability for a decoy.

Let $P(D|\mathbf{x})$ and $P(T|\mathbf{x})$ denote, respectively, the probability that an object with feature vector **x** is a decoy and the probability that an object with feature vector **x** is a target. For the two-class problem, it is obvious that $P(T|\mathbf{x}) = 1 - P(D|\mathbf{x})$, since the object must be one of these two types. By the same argument $P_1 = 1 - P_2$ for the two-class problem. Bayes's theorem states that

$$P(D|\mathbf{x}) = P_D(\mathbf{x})P_2/[P_D(\mathbf{x})P_2 + P_T(\mathbf{x})P_1]$$
$$= P_D(\mathbf{x})P_2/[P_D(\mathbf{x})P_2 + P_T(\mathbf{x})(1 - P_2)]$$

The Bayes rule minimizes expected cost and is defined as follows:

Classify the object as a decoy if $P_D(\mathbf{x})/P_T(\mathbf{x}) > K$

Classify the object as a target if $P_D(\mathbf{x})/P_T(\mathbf{x}) \leqslant K$

where $K = C_2P_1/C_1P_2$. See Duda and Hart (1973, pp. 10–15) for a derivation of this result.

Notice that we have made no assumptions about the form of the class-conditional densities. The Bayes rule works for any probability densities. Of course, the form of the decision boundary and the associated error rates depend on these *known* densities. If we make the further assumption that the densities are both multivariate Gaussian with different covariance matrices, then the Bayes rule has a quadratic decision boundary (i.e., the boundary is a quadratic function of **x**).

If the densities are Gaussian and the covariance matrices are equal, the Bayes rule has a linear boundary (i.e., the boundary is a linear function of **x**). Both of these results are derived in Duda and Hart (1973, pp. 22–31). The possible decision boundaries for Gaussian distributions with unequal covariances and two-dimensional feature vectors are illustrated in Fig. 2.1, which was taken from Duda and Hart (1973, p. 31).

The circles and ellipses in the figure represent, say, the one sigma equal probability contours, corresponding to the respective covariances. These covariances are taken to be diagonal without any loss of generality. The shaded region R_2 is the region in which class 2 is accepted.

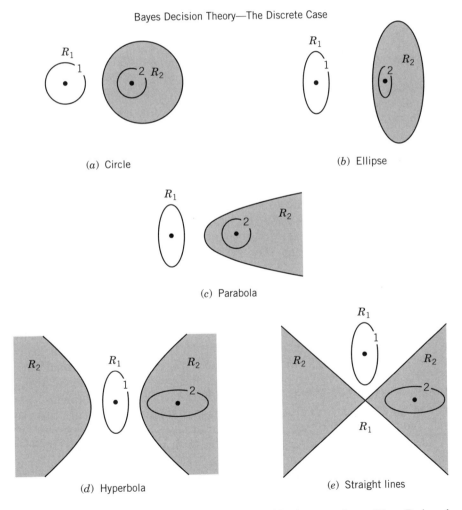

Figure 2.1 Forms for decision boundaries for the general bivariate normal case. [From Duda and Hart (1973, p. 31), with permission from John Wiley & Sons, Inc.]

In many practical problems the class-conditional densities are *not known*. If we assume the densities to be Gaussian, the training samples can be used to estimate the mean vectors and covariance matrices (i.e., the parameters required to determine the densities). If we have no knowledge of the form of the underlying densities, we may use available data whose classes are known (such data are referred to as training data) to obtain density estimates.

One common approach is to use the kernel density estimation procedure. The rule used in practice replaces the Bayes rule (which is not known) with an approximation to it based on the replacement of the class-conditional densities in the Bayes rule with the estimated densities.

Although the resulting rule does not have the optimal properties of the Bayes rule, we argue that it is approximately the optimal rule since as the training set gets larger and larger for both classes the estimated densities come closer and closer to the true densities and the rule comes closer and closer to the Bayes rule. For small sample sizes, it at least appears to be a reasonable approach. We shall call this procedure the estimated decision rule. To learn more about kernel discrimination consult Hand (1981, 1982).

For known class-conditional densities, the Bayes rule can be applied and the error rates calculated by integrating these densities in the region in which a misclassification would occur. In parametric problems, "plug-in" methods compute these integrals using the estimated densities, gotten by plugging in the parameter estimates for their unknown values. These plug-in estimates of the error rates are known to be optimistically biased [i.e., they tend to underestimate the actual expected error rates; see Hills (1966)].

When we are unable to make any parametric assumptions, a naive approach is to take the estimated decision rule, apply it to the training data, and then count how many errors of each type would be made. We divide the number of misclassified objects in each class by their respective number of training samples to get our estimates of the error rates. This procedure is referred to as resubstitution, since we are substituting training samples for possible future cases and these training samples were already used to construct the decision rule.

In small to moderate sample sizes, the resubstitution estimator is generally a poor estimator because it too tends to have a large optimistic bias (actually the magnitude of bias is a function of the true error rate and will be small when the true error rate is very small even in the case of small sample sizes). Intuitively, the optimistic bias of the plug-in and resubstitution estimators are due to the fact that in both cases the training data are used to construct the rule and then reused to estimate the error rates.

Ideally, it would be better to estimate the error rates based on an independent set of data with known classes. This, however, creates a dilemma. It is wasteful to throw away the information in the independent set, since the independent set could be used to enlarge the training set and hence provide better estimates of the class-conditional densities. On the other hand, the holdout estimator, obtained by the separation of this independent data set for error rate estimation from the training set, eliminates the optimistic bias of the resubstitution estimate.

Lachenbruch (1967) (see also Lachenbruch and Mickey, 1968) provided the leave-one-out estimate to overcome this dilemma. Each training vector is used in the construction of the rule. To estimate the error rate, the rule is reconstructed n times, where n is the total number of training vectors. In the ith reconstruction, the ith training vector is the only training vector left out of the construction.

We then count the ith vector as misclassified if the reconstructed rule would misclassify it. We take the total number of misclassified vectors in each class and divide it by the number of cases in the respective class to obtain the error rates.

This procedure is referred to as leave-one-out or cross-validation and the estimators are called the leave-one-out or U-estimates. Because the observations are left out one at a time, some have referred to it as the jackknife estimator, but Efron (1982a, pp. 53–58) defines another bias correction estimator to be the jackknife estimator (see also Efron, 1983).

Now you'll be shown how to apply the bootstrap in this application. Essentially, we will apply the bootstrap bias-correction procedure that we learned about in Section 2.1.1 to the resubstitution estimator.

The resubstitution estimator, although generally poor in small samples, has a large bias that can be estimated by bootstrapping (e.g., see Chernick, Murthy, and Nealy, 1985, 1986). Cross-validation (i.e., the leave-one-out estimator) suffers from a large variance for small training sample sizes. Despite this large variance, cross-validation has been traditionally the method of choice.

Glick (1978) was one of the first authors to recognize this problem and he proposed certain "smooth" estimators as an alternative. Glick's approach has since been followed up by Snapinn and Knoke (1984, 1985a).

Efron (1982a, 1983) showed that the bootstrap bias correction can produce an estimator that is nearly unbiased (the bias is small though not quite as small as for the leave-one-out estimator) and has a far smaller variance than the leave-one-out estimator. Consequently, the bootstrap is superior in terms of mean square error (a common measure of statistical accuracy).

As a guideline to the practitioner, I believe that the simulation studies to date indicate that for most applications (cases where the distributions for the feature populations do not have very heavy tails), the .632 estimator is to be preferred. For very heavy-tailed cases I recommend the ordinary bootstrap estimator. What follows is a description of the research studies to date that provide the evidence to support this general guideline. I shall now describe the various bootstrap estimators that were studied in Efron (1983) and Chernick, Murthy, and Nealy (1985, 1986, 1988a,b).

It is important to clarify here what error rate we are estimating. It was pointed out by Sorum (1972) that when training data are involved there are at least three possible error rates to consider [see also Page (1985) for a more recent account].

In the simulation studies that we review here, only one error rate is considered. It is the expected error rate conditioned on the training set of size n. This averages the two error rates (weighing each equally). It is the natural estimator to consider since in the classification problem the training set is fixed and we need to predict the class for new objects based solely on our prior knowledge and the particular training set at hand.

A slightly different and less appropriate error rate would be the one obtained by averaging these conditional error rates over the distribution of possible training sets of size n. Without carefully defining the error rate to be estimated, confusion can arise and some comparisons may be inappropriate.

The resubstitution and cross-validation estimators have already been defined. The standard bootstrap (obtained using 100 to 200 bootstrap samples in the simulations of Efron and Chernick, Murthy, and Nealy) uses the bootstrap sample analog to Eq. 2.10 of Efron (1983, p. 317) to correct the bias. Define the estimated bias as

$$\omega_h = E_* \left\{ \sum_i (n^{-1} - P_i^*) Q[y_i, \eta(t_i, X^*)] \right\}$$

where E_* denotes the expectation under the bootstrap random sampling mechanism (i.e., sampling with replacement from the empirical distribution), $Q[y_i, \eta(t_i, X^*)]$ is the indicator function defined to be equal to zero if $y_i = \eta(t_i, X^*)$ and one if $y_i \neq \eta(t_i, X^*)$, y_i is the ith observation of the response, t_i is the vector of predictor variables, and η is the prediction rule.

X^* is the vector of a bootstrap sample (of length n) and P_i^* is the ith repetition frequency (i.e., the proportion of cases in a particular bootstrap sample X^*, where the ith sample value occurs). The bootstrap estimate is then $e_{boot} = \text{err}_{app} + \omega_h$, where err_{app} is the resubstitution estimate of the error rate (also referred to as the apparent error rate) and ω_h is the bootstrap estimate of bias as defined above.

This is technically slightly different from the simple bias correction procedure described in Section 2.1.1, but it is essentially the same. Using the convention given in Efron (1983), this bias estimate is added to the apparent error rate to produce the bootstrap estimate as shown in the equation above.

To be more explicit, let $X_1, X_2, \ldots, X_n$ denote the n training vectors where, say, for convenience, $n = 2m$ for m an integer; $X_1, X_2, \ldots, X_m$ come from class 1 and $X_{m+1}, X_{m+2}, \ldots, X_n$ come from class 2. A bootstrap sample is generated by sampling with replacement from the empirical distribution for the pooled data $X_1, X_2, \ldots, X_n$.

Although different, this is almost the same as taking m samples with replacement from $X_1, X_2, \ldots, X_m$ and another $n - m = m$ samples with replacement from $X_{m+1}, X_{m+2}, \ldots, X_n$. In the latter case, each bootstrap sample contains m vectors from each class, whereas in the former case the number in each class varies according to a binomial distribution, where N_1, the number from class 1, is binomial with parameters n and p (with $p = \frac{1}{2}$) and N_2, the number from class 2, equals $n - N_1$. $E(N_1) = E(N_2) = n/2 = m$.

The approach used in Efron (1983) and Chernick, Murthy, and Nealy (1985, 1986, 1988a, b) is essentially the former approach except that the original training set itself is also selected in the same way as the bootstrap samples. So when, for example, $n = 14$, it is possible to have 7 training vectors from class 1 and 7 from class 2, but we may also have 6 from class 1 and 8 from class 2, and so forth.

Once a bootstrap sample has been selected, we treat the bootstrap sample as though it were the training set. We construct the discriminant rule (linear for the simulations under discussion, but the procedure can apply to other forms such as the quadratic rule) based on the bootstrap sample and subtracting the fraction of the observations in the bootstrap sample that would be misclassified by the same rule (where each observation is counted as many times as it occurs in the bootstrap sample).

The first term is a bootstrap sample estimate of the "true" error rate while the second term is a bootstrap sample estimate of the apparent error. The difference is a bootstrap sample estimate of the optimistic bias in the apparent error rate. Averaging these estimates over the k Monte Carlo replications provides a Monte Carlo approximation to the bootstrap estimator.

An explicit formula for the bootstrap estimator and its Monte Carlo approximation, given in Efron (1983, p. 317), is just as we have described it above. Although the formulas are explicit, the notation is somewhat complicated, as we have seen.

The e_0 estimator was introduced as a variant to the bootstrap in Chatterjee and Chatterjee (1983), although the name came later in Efron (1983). For the e_0 estimator, we simply count the total number of training vectors misclassified in each bootstrap sample. The estimate is then obtained by summing over all bootstrap samples and dividing by the total number of training vectors not included in the bootstrap samples.

The .632 estimator is obtained by the following formula:

$$\text{err}_{632} = 0.368\text{err}_{\text{app}} + 0.632e_0$$

where err_{app} denotes the apparent error rate and e_0 is as defined in the previous paragraph. With only the exception of the very heavy-tailed distributions, the .632 estimator is the clear-cut winner over the other variants.

Some heuristic justification for this is given in Efron (1983) (see also Chernick and Murthy, 1985). Basically, the .632 estimator appropriately balances the optimistic bias of the apparent error rate with the pessimistic bias of e_0. The reason for this weighting is that 0.368 is a decimal approximation to $1/e$, the asymptotic expected percentage of training vectors not included in a bootstrap sample.

Chernick, Murthy, and Nealy (1985) devised a variant called the MC estimator. This estimator is obtained just as the standard bootstrap. The difference is that a controlled bootstrap sample is generated in place of the ordinary bootstrap sample. In this procedure, the sample is restricted to include observations with replication frequencies as close as possible to the asymptotic expected replication frequency.

Another variant, due to Chernick, Murthy, and Nealy (1985), is the convex bootstrap. In the convex bootstrap, the bootstrap sample contains linear combinations of the observation vectors. This smoothes the sampling by allowing a continuum of possible observations instead of just the original discrete set.

The theoretical difficulty with this method is that the bootstrap distribution does not converge to the true distribution since the observations are weighted according to λ, which is chosen uniformly on [0,1]. This means that the "resamples" will not behave in large samples exactly like the original samples from the class-conditional densities. We can therefore not expect the estimated error rates to be correct for the given classification rule.

To avoid the inconsistency problem, Chernick, Murthy, and Nealy (1988b) introduced a modified convex bootstrap that concentrates the weight closer and closer to one of the samples, as the training sample size, n, increases. They also introduced a modification to the 632 estimator that they called the adaptive 632.

It was hoped that the modification of adapting the weights would improve the .632 estimator and increase its applicability, but results were disappointing. Efron and Tibshirani (1997a) introduced 632+ that also tries to modify the .632 estimator so that it works well for an even wider set of classification problems and a variety of class-conditional densities.

In Efron (1983) other variants—the double bootstrap, the randomized bootstrap, and the randomized double bootstrap—are also considered. The reader is referred to Efron (1983) for the formal definitions of these estimators.

Of these, only the randomized bootstrap showed significant improvement over the ordinary bootstrap and so these other variants were not considered. Follow-up studies did not include the randomized bootstrap.

The randomized bootstrap applies only to the two-class problem. The idea behind the randomized bootstrap is the modification of the empirical distributions for each class by allowing for the possibility that the observed training vectors for class 1 come from class 2 and vice versa.

Efron allowed a probability of occurrence of 0.1 to the opposite class in the simple version. After modifying the empirical distributions, bootstrap sampling is applied to the modified distributions rather than the empirical distributions and the bias is estimated and then corrected for, just as in the standard bootstrap. In a way, the randomized bootstrap smoothes the empirical distributions, an idea similar in spirit to the convex bootstrap.

Implementation of the randomized bootstrap by Monte Carlo is straightforward. We sample at random from the pooled training set (i.e., training data from both classes are mixed together) and then choose a uniform random number U.

If $U \leqslant 0.9$, we assign the observation vector to its correct class. If not, we assign it to the opposite class. To learn more about the randomized bootstrap and other variations see Efron (1983, p. 320).

For Gaussian populations and small training sample sizes (14 to 29) the .632 estimator was clearly superior in all the studies in which it was considered, namely, Efron (1983), Chernick, Murthy, and Nealy (1985, 1986), and Jain, Dubes, and Chen (1987).

A recent paper by Efron and Tibshirani (1997a) looks at the .632 estimator and a variant called 632+ that is similar in concept to our adaptive .632 estimator. They treat more general classification problems than just the linear (equal covariance) case that we focus on here.

Chernick, Murthy, and Nealy (1988a, b) consider multivariate (two-dimensional, three-dimensional, and five-dimensional) distributions. Uniform, exponential, and Cauchy distributions are considered. The uniform distribution provides shorter-than-Gaussian tails to the distribution, the bivariate exponential provides an example of skewness, and the autoregressive family of Cauchy distributions provides for heavier-than-Gaussian tails to the distribution.

Chernick, Murthy, and Nealy (1988a, b) found that for the uniform and exponential distributions the .632 estimator is again superior. As long as the tails are not heavy, the .632 estimator provides an appropriate weighing to balance the opposite biases of e_0 and the apparent error rate.

However, for the Cauchy distribution, e_0 no longer has a pessimistic bias and both the e_0 and the convex bootstrap outperform the .632 estimator. They conjectured that the result would generalize to any distribution with heavy tails. They also believed that the skewness and other properties of the distributions that depart from the Gaussian would have little effect on the relative performance of the estimators.

In Chernick, Murthy, and Nealy (1988b), the Pearson VII family of distributions was simulated for a variety of values of the parameter m. The probability density function is defined as

$$f(x) = \frac{\Gamma(m) |\Sigma|^{1/2}}{\Gamma(m - (p/2))\pi^{p/2}[1 + (x - \mu)'\Sigma^{-1}(x - \mu)]^m}$$

where μ is a location vector, Σ is a scaling matrix, m is a parameter that affects the dependence and controls the tail behavior, p is the dimension, and Γ is the gamma function. The symbol $| \; |$ denotes the determinant of the matrix.

The Pearson VII distributions are all elliptically contoured (i.e., contours of constant probability density are ellipses). An elliptically contoured density is a property that the Pearson VII family shares with the Gaussian family of distributions.

Only $p = 2$ was considered in Chernick, Murthy, and Nealy (1988b). The parameter m was varied from 1.3 to 3.0. For $p = 2$, second moments exist only for m greater than 2.5 and first moments exist only for m greater than 1.5.

Chernick, Murthy, and Nealy (1988b) found that when $m \leqslant 1.6$ the pattern observed for the Cauchy distributions in Chernick, Murthy, and Nealy (1988a) pertained, namely, the e_0 and the convex bootstrap were the best. As m

decreases from 2.0 to 1.5, the bias of the e_0 estimator decreases and eventually it changes sign (i.e., goes from a pessimistic to an optimistic bias). For m greater than 2.0 the results are similar to the Gaussian and the light-tailed distributions where the .632 estimator is the clear winner.

Table 2.1 is taken from Chernick, Murthy, and Nealy (1988b). It summarizes for various values of m the relative performance of the estimators. The totals represent the number of cases for which the estimators ranked first, second, and third among the seven considered. The cases vary over the range of the "true" error rates that varied from about 0.05 to 0.50.

Table 2.2 is a similar summary taken from Chernick, Murthy, and Nealy (1986), which summarizes the results for the various Gaussian cases considered.

Again, we point out that for most applications, the .632 estimator is preferred. It is not yet clear whether or not the smoothed estimators are as good as the best bootstrap estimates.

Snapinn and Knoke (1985b) claim that their estimator is a better estimator than the .632 estimator. In their study they simulated both Gaussian distributions and a few non-Gaussian distributions.

They also show that the bias correction applied to the smoothed estimators by resampling procedures may be as good as their own smoothed estimators. This has not yet been confirmed in the published literature. Some results comparing their estimator with the .632 and other estimates in two-class cases can be found in Hirst (1996).

For very heavy-tailed distributions, use the ordinary bootstrap or the convex bootstrap. It may sometimes be possible, based on knowledge as to how the data are generated, for the practitioner to determine something about the nature of the tails of the distribution. In many cases it may not be possible.

I now give an example to show how it may be possible for the practitioner to know that he or she has a heavy-tailed distribution. If a feature is the ratio of two random variables and the denominator is known to be approximately Gaussian with zero mean, we will know that the feature has a distribution with tails like the Cauchy.

This is based on the known result that the ratio of two independent Gaussian random variables with zero mean and the same variance has the standard Cauchy distribution. The heavy tails come about because the Gaussian random variable in the denominator has mean zero. Therefore, the denominator has a good chance to be very close to zero and hence the ratio can be large in absolute value.

As the sample size becomes much larger, it makes little difference which estimator is used as the various bootstrap estimates and cross-validation are asymptotically equivalent (with the exception of the convex bootstrap). Even the apparent error rate may work well in very large samples where its bias is much reduced, although never zero. Exactly how large is large is difficult to say since the known studies have not yet adequately varied the size of the training sample.

Table 2.1 Summary Comparison of Estimators Using Root Mean Square Error (Number of Simulations on Which Estimator Attained Top Three Ranks)

	.632	MC	e_0	Boot	Conv	U	App	Total
M = 1.3								
First	0	0	2	0	10	0	0	12
Second	3	0	0	9	0	0	0	12
Third	0	9	0	1	2	0	0	12
Total	3	9	2	10	12	0	0	36
M = 1.5								
First	6	1	8	5	12	0	1	33
Second	8	4	0	14	7	0	0	33
Third	3	15	2	4	8	0	1	33
Total	17	20	10	23	27	0	2	99
M = 1.6								
First	1	1	2	1	5	0	2	12
Second	4	3	0	5	0	0	0	12
Third	0	4	0	4	4	0	0	12
Total	5	8	2	10	9	0	2	36
M = 1.7								
First	2	1	2	1	2	1	3	12
Second	3	3	1	4	1	0	0	12
Third	4	2	0	3	2	0	1	12
Total	9	6	3	8	5	1	4	36
M = 2.0								
First	18	1	3	0	1	0	7	30
Second	10	4	4	2	5	2	3	30
Third	1	9	3	8	5	0	3	30
Total	29	14	10	10	11	2	13	90
M = 2.5								
First	21	0	8	1	0	0	3	33
Second	10	3	4	5	4	2	5	33
Third	1	13	1	6	10	0	2	33
Total	32	16	13	12	14	2	10	99
M = 3.0								
First	21	0	6	0	0	0	3	30
Second	9	3	5	3	2	2	6	30
Third	0	8	1	8	11	1	1	30
Total	30	11	12	11	13	3	10	90

Source: Chernick, Murthy, and Nealy (1988b).

Table 2.2 Summary Comparison

Rank	.632	MC	e_0	Boot	Conv	U	App	Total
First	72	1	29	6	0	0	1	109
Second	21	13	27	23	11	1	13	109
Third	7	20	8	25	37	7	5	109
Total	100	34	64	54	48	8	19	

Source: Chernick, Murthy, and Nealy (1986).

2.1.3. Error Rate Estimation: An Illustrative Problem

In this problem, we have five bivariate normal training vectors from class 1 and five from class 2. For class 1, the mean vector is $\begin{pmatrix} 0 \\ 0 \end{pmatrix}$ and the covariance matrix is

$$\begin{bmatrix} 1 & 0 \\ 0 & 1 \end{bmatrix}$$

For class 2, the mean vector is $\begin{pmatrix} 1 \\ 1 \end{pmatrix}$ and the covariance matrix is also

$$\begin{bmatrix} 1 & 0 \\ 0 & 1 \end{bmatrix}$$

The training vectors generated by random sampling from the above distributions are as follows:

For Class 1

$$\begin{pmatrix} 2.052 \\ 0.339 \end{pmatrix}, \quad \begin{pmatrix} 1.083 \\ -1.320 \end{pmatrix}, \quad \begin{pmatrix} 0.083 \\ -1.542 \end{pmatrix}, \quad \begin{pmatrix} 1.278 \\ -0.459 \end{pmatrix}, \quad \begin{pmatrix} -1.226 \\ -0.606 \end{pmatrix}$$

For Class 2

$$\begin{pmatrix} 1.307 \\ 2.268 \end{pmatrix}, \quad \begin{pmatrix} -0.548 \\ 1.741 \end{pmatrix}, \quad \begin{pmatrix} 2.498 \\ 0.813 \end{pmatrix}, \quad \begin{pmatrix} 0.832 \\ 1.409 \end{pmatrix}, \quad \begin{pmatrix} 1.498 \\ 2.063 \end{pmatrix}$$

We generate four bootstrap samples of size 10 and calculate the standard bootstrap estimate of the error rate. We also calculate e_0 and the apparent error rate in order to compute the .632 estimator. We denote by the indices 1, 2, 3, 4, and 5, the respective five bivariate vectors from class 1 and by the indices 6, 7, 8, 9, and 10, the respective five bivariate vectors from class 2.

A bootstrap sample can be represented by a random set of 10 indices sampled with replacement from the integers 1 to 10. In this instance our four bootstrap samples are [9, 3, 10, 8, 1, 9, 3, 5, 2, 6], [1, 5, 7, 9, 9, 9, 2, 3, 3, 9], [6, 4, 3, 9, 2, 8, 7, 6, 7, 5], and [5, 5, 2, 7, 4, 3, 6, 9, 10, 1].

Bootstrap sample numbers 1 and 2 have five observations from class 1 and five from class 2; bootstrap sample number 3 has four observations from class 1 and six from class 2; and bootstrap sample number 4 has six observations from class 1 and four from class 2. We also observe that in bootstrap sample number 1 indices 3 and 9 repeat once and indices 4 and 7 do not occur. In bootstrap sample number 2 index 9 occurs three times and index 3 twice while indices 4, 6, and 10 do not appear at all. In bootstrap sample number 3 indices 6 and 7 are repeated once while 1 and 10 do not appear.

Finally, in bootstrap sample number 4 only index 5 is repeated and index 8 is the only one not to appear. These samples are fairly typical of the behavior of bootstrap samples (i.e., sampling with replacement from a given sample) and they indicate how the bootstrap samples can mimic the variability due to sampling (i.e., the sample-to-sample variability).

Table 2.3 shows how the observations in the bootstrap sample were classified by the classification rule obtained using the bootstrap sample. We see that only in bootstrap samples 1 and 2 were any of the bootstrap observations misclassified. So for bootstrap samples 3 and 4 the bootstrap sample estimate of the apparent error is zero.

In both bootstrap sample 1 and sample 2 only observation number 1 was misclassified and in each sample observation number 1 appeared one time. So for these two bootstrap samples, the estimate of apparent error is 0.1.

Table 2.4 shows the resubstitution counts for the original sample. Since none of the observations were misclassified, the apparent error rate or resubstitution estimate is also zero.

Table 2.3 Truth Table for the Four Bootstrap Samples

True Class	Sample #1 Classified As		Sample #2 Classified As	
	Class 1	Class 2	Class 1	Class 2
Class 1	4	1	4	1
Class 2	0	5	0	5

True Class	Sample #3 Classified As		Sample #4 Classified As	
	Class 1	Class 2	Class 1	Class 2
Class 1	4	0	6	0
Class 2	0	6	0	4

Table 2.4   Resubstitution Truth Table for Original Data

	Sample #1 Classified As	
True Class	Class 1	Class 2
Class 1	5	0
Class 2	0	5

In the first bootstrap sample, observation number 1 was the one misclassified. Observation numbers 4 and 7 did not appear. They both would have been classified correctly since their discriminant function values were 0.030724 for class 1 and -1.101133 for class 2 for observation 4 and -5.765286 for class 1 and 0.842643 for class 2 for observation 7. Observation 4 is correctly classified as coming from class 1 since its class 1 discriminant function value is larger than its class 2 discriminant function value. Similarly, observation 7 is correctly classified as coming from class 2.

In the second bootstrap sample, observation number 1 was misclassified and observation numbers 4, 6, and 10 were missing. Observation 3 was correctly classified as coming from class 1 and observations 60 and 10 were correctly classified as coming from class 2. Table 2.5 provides the coefficients of the linear discriminant functions for each of the four bootstrap samples.

It is an exercise for the reader to calculate the discriminant function values for observation numbers 4, 6, and 10 to see that the correct classifications would be made with bootstrap sample number 2.

In the third bootstrap sample, none of the bootstrap sample observations were misclassified but observation numbers 1 and 10 were missing. Using Table 2.5 we see that for class 1 observation number 1 has a discriminant function value of -3.8587, whereas for class 2 it has a discriminant function value of 2.6268.

Consequently, observation 1 would have been misclassified by the discrimination rule based on bootstrap sample number 3. The reader may easily check this and also may check that observation 10 would be classified correctly as coming from class 2 since its discriminant function value for class 1 is -9.6767 and 13.1749 for class 2.

In the fourth bootstrap sample, none of the bootstrap sample observations are misclassified and only observation number 8 is missing from the bootstrap sample. We see, however, by again computing the discriminant functions, that observation 8 would be misclassified as coming from class 1 since its class 1 discriminant function value is -2.1756 while its class 2 discriminant function value is -2.4171.

Another interesting point to notice from Table 2.5 is the variability of the coefficients of the linear discriminants. This variability in the estimated coeffi-

Table 2.5 Linear Discriminant Function Coefficient for Bootstrap Samples

True Class	Constant Term	Variable No. 1	Variable No. 2
	Bootstrap Sample No. 1		
Class 1	−1.793	0.685	−2.066
Class 2	−3.781	1.027	2.979
	Bootstrap Sample No. 2		
Class 1	−1.919	0.367	−2.481
Class 2	−3.353	0.584	3.540
	Bootstrap Sample No. 3		
Class 1	−2.343	0.172	−3.430
Class 2	−6.823	1.340	6.549
	Bootstrap Sample No. 4		
Class 1	−1.707	0.656	−2.592
Class 2	−6.130	0.469	6.008

cients is due to the small sample size. Compare these coefficients with the ones given in Table 2.6 for the original data.

The bootstrap samples give us an indication of the variability of the rule. This would otherwise be difficult to see. It also indicates that we can expect a large optimistic bias for resubstitution.

We can now compute the bootstrap estimate of bias:

$$\omega_{\text{boot}} = \frac{(0.1 - 0.1) + (0.1 - 0.1) + (0.1 - 0) + (0.1 - 0)}{4} \quad \text{or} \quad 0.2/4 = 0.05$$

Since the apparent error rate is zero, the bootstrap estimate of the error rate is also 0.05.

Table 2.6 Linear Discriminant Function Coefficients for the Original Sample

Class Number	Constant Term	Variable No. 1	Variable No. 2
1	−1.493	0.563	−1.726
2	−4.044	0.574	3.653

The e_0 estimate is the average of the four estimates obtained by counting in each bootstrap sample the fraction of the observations that do not appear in the bootstrap sample and that would be misclassified. We see from the results above that these estimates are 0.0, 0.0, 0.5, and 1.0 for bootstrap samples 1, 2, 3, and 4, respectively. This yields an estimated value of 0.375.

Another estimate similar to e_0 but distinctly different is obtained by counting all the observations left out of the bootstrap samples that would have been misclassified by the bootstrap sample rule and dividing by the total number of observations left out of the bootstrap samples. Since only two of the left out observations were misclassified and only a total of eight observations were left out, this would give us an estimate of 0.250. This gives more weight to those bootstrap samples with more observations left out.

For the leave-one-out method, observation 1 would be misclassified as coming from class 2 and observation 8 would be misclassified as coming from class 1. This leads to a leave-one-out estimate of 0.200.

Now the .632 estimator is simply $0.368 \times$ (apparent error rate) $+ 0.632e_0$. Since the apparent error rate is zero, the .632 estimate is 0.237.

Since the data were taken from independent Gaussian distributions, each with variance one and with the mean equal to zero for population 1 and with the mean equal to one for population 2, the expected error rate for the optimal rule based on the distributions being known is easily calculated to be approximately 0.240.

The actual error rate for the classifier based on a training set of size 10 can be expected to be even higher. We note that, in this example, the apparent error rate and the bootstrap both underestimate the true error rate, whereas the e_0 overestimates it.

The .632 estimator comes surprisingly close to the optimal error rate and clearly gives a better estimate of the conditional error rate (0.295, discussed below) than the others. The number of bootstrap replications is so small in this numerical example that it should not be taken too seriously. It is simply one numerical illustration of the computations involved. Many more simulations are required to draw conclusions. Simulation studies, such as the ones already discussed, are what we should rely on.

The true conditional error rate given the training set can be calculated by integrating the appropriate Gaussian densities over the regions defined by the discriminant rule based on the original 10 sample observations. An approximation based on Monte Carlo generation of new observations from the two classes, classified by the given rule, yields for a sample size of 1000 new observations (500 from each class) an estimate of 0.295 for this true conditional error rate.

Since (for equal error rates) this Monte Carlo estimator is based on a binomial distribution with parameters $n = 1000$ and p equal to the true conditional error rate, using $p = .3$, we have that the standard error of this estimate is approximately 0.0145 and an approximate 95% confidence interval for p is [0.266, 0.324]. So our estimate of the true conditional error rate is not very accurate.

   If we are really interested in comparing these estimators to the true conditional error rate, we probably should have taken 50,000 Monte Carlo replications to better approximate it. By increasing the sample size by a factor of 50, we decrease the standard error of the estimate by slightly more than a factor of 7 and hence the standard error of the estimate would be about 0.002 and the confidence interval would be $[p_h - 0.004, p_h + 0.004]$, where p_h is the estimate of the true conditional error rate based on 50,000 Monte Carlo replications. We use 0.004 as the half-width for the interval since a 95% confidence interval requires a half-width of 1.96 standard errors (approximately 2 standard errors).

   The width of the interval would then be less than 0.01 and would be useful for comparison. Again, we should caution the reader that, even if the true conditional error rate were close to the .632 estimate, we could not draw a strong conclusion from it as we would by looking at only one .632 estimate, one e_0 estimate, one apparent error rate estimate, and so on. It really takes simulation studies that account for the variability of the estimates for us to make valid comparisons.

2.1.4. Efron's Patch Data Example

Sometimes in making comparisons we are interested in computing the ratio of two quantities. We are given a set of data that enables us to estimate both quantities and we are interested in estimating the ratio of the two quantities. What is the best way to do it? The natural inclination is to take the ratio of the two estimates. Such estimators are called ratio estimators.

   However, statisticians know quite well that if both estimates are unbiased the ratio estimate will be biased. To see this, suppose X is an unbiased estimate of θ and Y is an unbiased estimate of μ. We want to estimate θ/μ. Since X is unbiased $E(X) = \theta$ and since Y is unbiased $E(Y) = \mu$. But the estimator X/Y has expected value $E(X/Y)$. Now $\theta/\mu = E(X)/E(Y)$, but this is not equal to $E(X/Y)$. In fact, if we further assume that X and Y are independent then

$$E(X/Y) = E(X)E(1/Y) = \theta E(1/Y).$$

The reciprocal function $f(z) = 1/z$ is a convex function and therefore Jensen's inequality (see Ferguson, 1967, pp. 76–78) implies that $f(E(Y)) \leqslant E(f(Y))$ or $f(E(Y)) = f(\mu) = 1/\mu \leqslant E(f(Y)) = E(1/Y)$. Consequently, $E(X/Y) = \theta E(1/Y) \geqslant \theta/\mu$. The only instance in which equality would hold would be if Y were constant. Otherwise $E(X/Y) > \theta/\mu$ and the bias $B = E(X/Y) - \theta/\mu$ is positive. This bias can be large and it is natural to want to improve the estimate by adjusting for the bias. Ratio estimators are also common in survey sampling [see Cochran (1977) for examples].

In Efron and Tibshirani (1993) an example of ratio estimation is given in Section 10.3 on pages 126–133. This was a small clinical trial used to show the FDA that a product produced at a new plant was equivalent to the product produced at the old plant where the agency had previously approved the product. The product is a patch that infuses a certain natural hormone into the bloodstream of the patient. The trial was a crossover trial involving eight subjects. Each subject was given three different patches: one patch manufactured at the old plant and containing the hormone, a second patch manufactured at the new plant and containing the hormone, and a third patch (placebo) that contained no hormone.

The purpose of the placebo is to establish a baseline level for the hormone. Presumably the patients were treated in random order and a waiting time is used to allow the effect of the first patch to wear off before taking measurements with the second patch and similarly for the period between measurements with the second and third patches.

The FDA has a well-defined criterion for determining bioequivalence in such trials. It wants the new patch to produce hormone levels that are within 20% of the amount produced by the old patch relative to the gain in hormone level of the old patch to the placebo. Mathematically we express this as

$$\theta = (E(\text{new}) - E(\text{old}))/(E(\text{old}) - E(\text{placebo}))$$

and require that

$$|\theta| = (|E(\text{new}) - E(\text{old})|)/(E(\text{old}) - E(\text{placebo})) \leqslant 0.20$$

So the FDA wants to test the hypothesis that $|\theta| \leqslant 0.20$ versus the alternative that $|\theta| > 0.20$.

For patient I we define $z_I = $ (old patch blood level − placebo blood level) and $y_I = $ (new patch blood level − placebo blood level). The natural estimate of θ is the plug-in estimate y_b/z_b, where y_b is the average of the eight values y_I for $I = 1, 2,\ldots, 8$ and z_b is the average of the values z_I for $I = 1, 2,\ldots, 8$. As we have already seen such a ratio estimator will be biased.

Table 2.7 shows the y and z values. Based on these data we find that the plug-in estimate for θ is -0.0713, which is considerably less than 0.20 in absolute value. However, the estimate is considerably biased and we might be able to obtain an improved estimate by bias adjustment. The bootstrap can estimate the bias as we have shown previously in the error rate estimation problem.

The real problem is one of confidence interval estimation or hypothesis testing and so the methods of Chapter 3 might better solve this problem. Nevertheless, we can see if the bootstrap can provide a better point estimate. Efron and Tibshirani (1993) generated 400 bootstrap samples and estimated the bias to be 0.0043. They also estimated the standard error of the estimate and showed that the ratio of the bias estimate to the standard error of the

Table 2.7   Patch Data Summary

Subject	Old–Placebo (z)	New–Old (y)
1	8406	−1200
2	2342	2601
3	8187	−2705
4	8459	1982
5	4795	−1290
6	3516	351
7	4796	−638
8	10238	−2719
Average	6342	−452.3

Source: Efron and Tibshirani (1993, p. 373), with permission from CRC Press, LLC.

estimate is only 0.041. This is small enough to indicate that bias adjustment will not matter.

It is important to recognize that although the bootstrap adjustment will reduce the bias of a biased estimator and can do so substantially when the bias is large, it is not clear whether or not it improves the accuracy of the estimate. If we define the accuracy to be the root mean square (rms) error then since rms error is the square root of the square of the bias plus the variance, there is the possibility that although we lower the bias we could be increasing the variance enough so that the rms error increases. This bias versus variance trade-off comes up in a number of statistical problems including kernel density estimation and the error rate estimation problem that we have already encountered. Efron and Tibshirani (1993, p. 138) caution about the hazards of bias correction methods.

2.2. ESTIMATING LOCATION AND DISPERSION

In this section, we consider point estimates of location parameters. For distributions with finite first and second moments, the population mean is a natural location parameter. The sample mean is the "best" estimate and bootstrapping adds nothing to the parametric approach. We shall discuss this briefly.

For distributions without first moments, the median is a more natural parameter to estimate the location or center of a distribution. Again, the bootstrap adds nothing to the point estimation (the sample median is the natural estimate), but we see in Section 2.2.2 that the bootstrap is useful in estimating standard errors and percentiles, which provide measures of the dispersion and measures of the accuracy of the estimates that we are considering.

2.2.1. Means and Medians

For population distributions with finite first moments, the mean is a natural measure of central tendency. If the first moment does not exist, sample estimates can still be calculated but they tend to be unstable and they lose their meaning (i.e., the sample mean no longer converges to a population mean as the sample size increases).

One common example that illustrates this point is the standard Cauchy distribution. Given a sample size n from a standard Cauchy distribution, the sample mean is also standard Cauchy. So no matter how large we take n to be, we cannot reduce the variability of the sample mean.

Unlike the Gaussian or exponential distributions that have finite first and second moments and have sample means that converge in probability to the population mean, the Cauchy has a sample mean that does not converge in probability.

For distributions like the Cauchy, the sample median does converge to the population median as the sample size tends to infinity. Hence, for such cases, the sample median is a more useful estimator of the center of the distribution since the population median of the Cauchy and other heavy-tailed symmetric distributions best represents the "center" of the distribution.

If we know nothing about the population distribution at all, we may want to estimate the median since the population median always exists and is consistently estimated by the sample median regardless of whether or not the mean exists.

How does the bootstrap fit in when estimating a location parameter of a population distribution? In the case of Gaussian or the exponential distributions, the sample mean is the maximum likelihood estimate, is consistent for the population mean, and is the minimum variance unbiased estimate. How can the bootstrap top that?

In fact, it cannot. In these cases the bootstrap could be used to estimate the mean but we would find that the bootstrap estimate is nothing but the sample mean itself and the Monte Carlo estimate is just an approximation to the sample mean. It would be silly to bootstrap in such a case.

Bootstrapping was designed to estimate the accuracy of estimators. This is accomplished by using the bootstrap samples to estimate the standard deviation and possibly the bias of a particular estimator for problems where such estimates are not easily derived from the sample. In general, bootstrapping is not used to produce a better point estimate.

A notable exception was given in Section 2.1, where bias correction to the apparent error rate actually produced a better point estimate of the error rate. This, however, is an exception to the rule.

In the remainder of this book, we will learn about examples for which the estimators are given to us. We will need to estimate their standard errors or construct confidence regions or test hypotheses about the corresponding population parameters.

For the case of distributions with heavy tails, we may be interested in robust estimates of location (the sample median being one such example). The robust estimators are given (e.g., Winsorized mean, trimmed mean, or sample median).

However, the bootstrap is useful to estimate the standard errors and to obtain confidence intervals for the location parameters based on these robust estimators. Texts that deal with robust statistical procedures are Chatterjee and Hadi (1988), Hampel, Ronchetti, Rousseeuw, and Stahel (1986), and Huber (1981).

2.2.2. Standard Errors and Quartiles

The standard deviation of an estimator (also referred to as the standard error for unbiased estimators) is a commonly used estimate of an estimator's variability. This estimate only has meaning if the distribution of the estimator of interest has a finite second moment. In examples for which the estimator's distribution does not have a finite second moment, the interquartile range (the 75th percentile minus the 25th percentile of the estimator's distribution) is often used as a measure of the variability.

Staudte and Sheather (1990, pp. 83–85) provide an exact calculation [originally derived by Maritz and Jarrett (1978)] for the bootstrap estimate of the standard error of the median and compare it to the Monte Carlo approximation for cell lifetime data (obtained as the absolute differences of seven pairs of independent identically distributed exponential random variables).

We shall review Staudte and Sheather's development and present their results here. For the median, they assume for convenience that the sample size n is odd (i.e., $n = 2m + 1$, for m an integer). This makes the exposition easier but is not a requirement.

The Maritz and Jarrett (1978) paper actually provides explicit results for any n. It is just that the median has a different definition for even values of n. For n even, it is defined as the average of the two "middle" values, whereas for n odd it is uniquely the "middle" observation, namely, observation $m + 1$.

The functional representing the median is just $T(F) = F^{-1}(\frac{1}{2})$, where F is the population cumulative distribution and F^{-1} is its inverse function. The sample median is just $X_{(m+1)}$, where $X_{(i)}$ denotes the ith order statistic (i.e., the ith observation when ordered from smallest to largest).

An explicit expression for the variance of the median of the bootstrap distribution can then be derived based on well-known results about order statistics. Let $X^*_{(1)}, \ldots, X^*_{(n)}$ denote the ordered observations from a bootstrap sample taken from $X_1, \ldots, X_n$. Let $x_{(i)}$ denote the ith smallest observation from the original sample. Let $N^*_i = \# \{j: X^*_j = x_{(i)}\}$, $i = 1, \ldots, n$.

Then it can be shown that $\Sigma^k_{i=1} N^*_i$ has the binomial distribution with parameters n and p, where $p = k/n$. Let P^* denote the probability under bootstrap sampling. It follows that

$$P^*\{X^*_{(m+1)} > x_{(k)}\} = P^*\left\{\sum_{i=1}^{k} N^*_i \leqslant n\right\} = \sum_{j=0}^{n} \binom{n}{j}\left(\frac{k}{n}\right)^j\left(\frac{n-k}{n}\right)^{n-j}$$

Using well-known relationships between binomial sums and the incomplete beta function, Staudte and Sheather (1990) find, letting $w_k = P^*\{X^*_{(m)} = x_{(k)}\}$, that

$$w_k = \frac{n!}{(m!)^2} \int_{(k-1)/n}^{k/n} (1-y)^m y^m \, dy$$

and then by simple probability calculations the bootstrap variance of $X^*_{(m+1)}$ is

$$\sum_{k=1}^{n} w_k x_{(k)} - \left(\sum_{k=1}^{n} w_k x_{(k)}\right)^2$$

This result was first obtained by Maritz and Jarrett (1978) and was also presented in Efron (1978). Taking the square root of the above expression, we have explicitly obtained, using properties of the bootstrap distribution for the median, the bootstrap estimate of the standard deviation of the sample median without doing any Monte Carlo approximation.

Table 2.8 from Staudte and Sheather (1990, p. 85) shows the results required to compute the standard error for the "sister cell" data set. In the table, p_k plays the role of w_k and, in the above equation using p_k, gives $\text{SE}_{\text{Boot}} = 0.173$. However, if we replace p_k with $\hat{p}_k$, we get 0.167 for a Monte Carlo approximation based on 500 bootstrap samples.

For other estimation problems, the Monte Carlo approximation to the bootstrap may be required, since we may not be able to provide explicit calculations as we have just done for the median. The Monte Carlo approximation is straightforward.

Let $\hat{\theta}$ be the sample estimate of θ and let $\hat{\theta}_i^*$ be the bootstrap estimate of θ for the ith bootstrap sample. Given k bootstrap samples, the bootstrap estimate of the standard deviation of the estimator $\hat{\theta}$ is, according to Efron (1982a),

$$SD_b = \left\{\frac{1}{k-1} \sum_{j=1}^{k} [\theta_i^* - \bar{\theta}^*]^2\right\}^{1/2}$$

where $\bar{\theta}^*$ is the average of the bootstrap samples. Instead of $\bar{\theta}^*$ one could equally well use $\hat{\theta}$ itself.

Table 2.8 Comparison of Exact and Monte Carlo Bootstrap Distributions for the Ordered Absolute Differences of Sister Cell Lifetimes

k	1	2	3	4	5	6	7
p_k	0.0102	0.0981	0.2386	0.3062	0.2386	0.0981	0.0102
$\hat{p}_k$	0.01	0.128	0.548		0.208	0.098	0.008
$x_{(k)}$	0.3	0.4	0.5	0.5	0.6	0.9	1.7

Source: Staudte and Sheather (1990, p. 85), with permission from John Wiley & Sons, Inc.

The choice of $k - 1$ in the denominator was made as the analog to the unbiased estimate of the standard deviation for a sample. There is no compelling argument for using $k - 1$ instead of k in the formula.

For the interquartile range, one straightforward approach is to order the bootstrap sample estimates from smallest to largest. The bootstrap sample observation that equals the 25th percentile (or an appropriate average of the two bootstrap sample estimates closest to the 25th percentile) is subtracted from the bootstrap sample observation that equals the 75th percentile (or an appropriate average of the two bootstrap sample observations closest to the 75th percentile). Once these bootstrap sample estimates are obtained, bootstrap standard error estimates or other bootstrap measures of spread for the interquartile range can be determined.

Other estimates of percentiles from a bootstrap distribution can be used to obtain bootstrap confidence intervals and test hypotheses, as will be discussed in Chapter 3. Such methods could be applied to get approximate confidence intervals for standard errors, interquartile ranges, or any other parameters that can be estimated from a bootstrap sample (e.g., medians, trimmed means, Winsorized means, or other robust location estimates).

2.3. HISTORICAL NOTES

For the error rate estimation problem, there is a great deal of literature. For developments up to 1974, see the survey article by Kanal (1974) and the extensive bibliography by Toussaint (1974). In addition, for multivariate Gaussian features McLachlan (1976) has derived the asymptotic bias of the apparent error rate (i.e., the resubstitution estimate) and it is not zero!

The bias of plug-in rules under parametric assumptions is discussed in Hills (1966). A collection of articles including some bootstrap work can be found in Choi (1986).

There have been a number of simulation studies showing the superiority of versions of the bootstrap over cross-validation when the training sample size is small. Most of the studies have considered linear discriminant functions [although Jain, Dubes, and Chen (1987) consider quadratic discriminants]. Most consider the two-class problem with two-dimensional feature vectors.

However, Efron (1982a, 1983) and Chernick, Murthy, and Nealy (1985, 1986, 1988a) considered five-dimensional feature vectors as well. Also, in Chernick, Murthy, and Nealy (1985, 1986, 1988a), some three-class problems were considered. Chernick, Murthy, and Nealy (1988a,b) were the first to simulate the performance of these bootstrap estimators for linear discriminant functions when the populations were not Gaussian. Hirst (1996) proposes a smoothed estimator (a generalization of the Snapinn and Knoke approach) for cases with three or more classes and provides detailed simulation studies showing the superiority of his method. He also compares .632 with the smoothed estimator of Snapinn and Knoke (1985) in two-class problems.

Chatterjee and Chatterjee (1983) considered only the two-class problem, doing only one-dimensional Gaussian simulations with equal variance. They were, however, the first to consider a variant of the bootstrap that Efron (1983) later refers to as e_0. They also provided an estimated standard error for their bootstrap error rate estimation.

The smoothed estimators have also been compared with cross-validation by Snapinn and Knoke (1984, 1985a). They show that their estimators have smaller mean square error than cross-validation for small training sample sizes but unfortunately not much has been published comparing the smoothed estimates with the bootstrap estimates. We are aware of one unpublished study — Snapinn and Knoke (1985b) — and some results in Hirst (1996).

In the simulation studies of Efron (1983), Chernick, Murthy, and Nealy (1985, 1986), Chatterjee and Chatterjee (1983), and Jain, Dubes, and Chen (1987), only Gaussian populations were considered.

Only, Jain, Dubes, and Chen (1987) considered classifiers other than linear discriminants. They looked at quadratic and nearest neighbor rules. Performance was measured by mean square error of the conditional expected error rate.

Jain, Dubes, and Chen (1987) and Chatterjee and Chatterjee (1983) also considered confidence intervals and the standard error of the estimators, respectively. Chernick, Murthy, and Nealy (1988a, b), Hirst (1996), and Snapinn and Knoke (1985b) considered certain non-Gaussian populations. The most recent results on the .632 estimator and an enhancement to it are given in Efron and Tibshirani (1997a).

McLachlan has done a lot of research in discriminant analysis and particularly on error rate estimation. His survey article (McLachlan, 1986) provides a good review of the issues and the literature including bootstrap results up to 1986. Some of the developments discussed here appear in McLachlan (1992), where he devotes an entire chapter (Chap. 10) to the estimation of error rates. It includes a section on bootstrap (p. 346–360).

An early account of discriminant analysis methods is given by Lachenbruch (1975). Multivariate simulation methods such as those used in the studies of Chernick, Murthy, and Nealy are covered in detail in Johnson (1987).

The bootstrap distribution for the median is also discussed in Efron (1982a, Chap. 10, pp. 77–78), where the results of Maritz and Jarrett (1978) are again used. Mooney and Duval (1993) discuss the problem of estimating the difference between two medians.

CHAPTER 3

Confidence Sets and Hypothesis Testing

Because of the close relationship between tests of hypotheses and confidence intervals, both are included in this chapter. Section 3.1 deals with "nonparametric" bootstrap confidence intervals (i.e., some or no assumptions are made about the form of the distribution being sampled).

There has also been some work on parametric forms of bootstrap confidence intervals and on methods for reducing or eliminating the Monte Carlo replication. We shall not discuss these methods in this chapter but do include references to the most relevant work in Section 3.4 on historical notes. Also, the parametric bootstrap is discussed briefly in Chapter 6.

Section 3.1.2 considers the simplest technique, the percentile method. This method works well when the statistic used is a pivotal quantity and has a symmetric distribution (see Efron 1981c, 1982a).

The percentile method and various other bootstrap confidence interval estimates require a large number of Monte Carlo replications for the intervals to be both accurate (i.e., be as small as possible for the given confidence level) and nearly exact (i.e., if the procedure were repeated many times the percentage of intervals that would actually include the "true" parameter value is approximately the stated confidence level).

This essentially states for exactness that the actual confidence level of the interval is the stated level. So, for example, if we construct a 95% confidence interval, we would expect that our procedure would produce intervals that contain the true parameter in 95% of the cases. Such is the definition of a confidence interval.

Unfortunately for "nonparametric" intervals, we cannot generally do this. The best we can hope for is to have approximately the stated coverage. Such intervals will be called approximately correct or almost exact. As the sample size increases and the number of bootstrap Monte Carlo replications increases, we can expect the percentile method to be approximately correct and accurate.

Another method, which Hall (1992a) refers to as the percentile method, is also mentioned in Section 3.1.2. Hall refers to Efron's percentile method as the "other" percentile method.

For pivotal quantities that do have symmetric distributions, the intervals can be improved on by a bias adjustment and the use of an acceleration constant. This is the approach taken in Efron (1987) and is the topic of Section 3.1.3.

Another approach that also provides better bootstrap confidence intervals is called bootstrap iteration (or double bootstrap). This approach has been studied in detail by Hall and Martin, among others, and is covered in Section 3.1.4. There we provide a review of research results and the developments from Martin (1990a) and Hall (1992a).

In each of these sections, examples are given to instruct the reader in the proper application of the methods and to illustrate their accuracy and correctness. Important asymptotic results will be mentioned but we shall not delve into the asymptotic theory.

Section 3.1.5 deals with the bootstrap t method for generating bootstrap-type confidence intervals. In some problems the bootstrap t method may be appropriate and have better accuracy and correctness than the percentile method. It is easier to implement than methods involving Efron's corrections. It is not as computer intensive as the iterated bootstrap. Consequently, it is popular in practice. We applied it in the Passive Plus DX trial at Pacesetter. So Section 3.1.5 is intended to provide the definition of it so that the reader may apply it. The bootstrap t goes back at least to the Efron monograph (Efron, 1982a).

In Section 3.2, the reader is shown the connection between confidence intervals and hypothesis tests. This close connection enables the reader to see how a confidence interval for a parameter can be interpreted in terms of the acceptance or rejection of a hypothesis that the parameter is a specified value.

The confidence level is directly related to the significance level of the test. Knowing this, the reader will be able to test hypotheses by constructing bootstrap confidence intervals for the parameter.

In Section 3.3, we provide examples of hypothesis tests to illustrate the usefulness of the bootstrap approach. In some cases, we can compare the bootstrap tests with other nonparametric tests including the permutation tests from Good (1994) or Manly (1991, 1997).

Section 3.4 provides an historical perspective on the literature of confidence intervals and hypothesis testing using the bootstrap approach.

3.1. CONFIDENCE SETS

Before introducing the various bootstrap-type confidence intervals, we will review what a confidence set or region is and then, in Section 3.1.1, present Hartigan's typical value theorem in order to motivate the percentile method of Section 3.1.2. Section 3.1.3 then explains how refinements can be made to handle asymmetric cases where the percentile method does not work well.

Section 3.1.4 presents bootstrap iteration. Bootstrap iteration or double bootstrapping is another approach to confidence intervals that overcomes the deficiencies of the percentile method. In Section 3.1.5, we present the bootstrap t method that also overcomes deficiencies of the percentile method but is simpler and more commonly used in practice than the iterated bootstrap and other bootstrap modifications to the percentile method.

What is a confidence set for a parameter vector? Suppose we have a parameter vector **v** that belongs to an n-dimensional Euclidean space (denoted by R^n). A confidence set with confidence coefficient $1 - \alpha$ is a set in R^n determined on the basis of a random sample and having the property that if the random sampling were repeated infinitely many times with a new region generated each time, then $100(1 - \alpha)\%$ of the time the region will contain **v**.

In the simplest case where the parameter is one dimensional, the confidence region will be an interval or the union of two or more disjoint intervals.

In parametric families of population distributions involving nuisance parameters (parameters required to uniquely specify the distribution but that are not of interest to the investigator) or when very little is specified about the population distribution, it may not be possible to construct confidence sets that have a confidence that is exactly $1 - \alpha$ for all possible **v** and all possible values of the nuisance parameters [see Bahadur and Savage (1956), for example, if you would like to see a mathematical proof of this result]. We shall see that the bootstrap percentile method will at least provide us with confidence intervals that have confidence coefficient approaching $1 - \alpha$ as the sample size becomes very large.

If we only assume that the population distribution is symmetric, then the typical value theorem of Hartigan (1969) tells us that subsampling methods (e.g., random subsampling) can provide confidence intervals that are exact (i.e., have confidence coefficient $1 - \alpha$ for finite sample sizes). We shall now describe these subsampling methods and present the typical value theorems.

3.1.1. Typical Value Theorems for M-Estimates

We shall consider the case of independent identically distributed observations from a symmetric distribution on the real line. We denote the n random variables by $X_1, X_2, \ldots, X_n$ and their distribution by F_θ.

For any set A, let $P_\theta(A)$ denote the probability that a random variable X with distribution F_θ has its value in the set A. As in Efron (1982a, p. 69) we will assume that F_θ has a symmetric density function $f(.)$ so that

$$P_\theta(A) = \int_A f(x - \theta)\, dx$$

where

$$\int_{-\infty}^{+\infty} f(x)\, dx = 1, \quad f(x) \geq 0, \quad \text{and} \quad f(-x) = f(x)$$

An M-estimate $\hat{\theta}(x_1, x_2, \ldots, x_n)$ for θ is any solution to the equation

$$\sum_i \Psi(x_i - \theta) = 0$$

Here we assume that the observed data $X_i = x_i$, for $i = 1, 2, \ldots, n$, are fixed while θ is the variable to solve for.

We note that in general M-estimates need not be unique. The function Ψ is called the kernel, and Ψ is assumed to be antisymmetric and simply increasing [i.e., $\Psi(-z) = -\Psi(z)$ and $\Psi(z + h) > \Psi(z)$ for all z and for $h > 0$]. Examples of M-estimates are given in Efron (1982a). For an appropriately chosen function y, many familiar estimates can be shown to be M-estimates including the sample mean and the sample median.

Consider the set of integers $(1, 2, 3, \ldots, n)$. The number of nonempty subsets of this set is $2^n - 1$. Let S be any one of these nonempty subsets. Let $\hat{\theta}_S$ denote an M-estimate based on only those values x_i for i belonging to S.

Under our assumptions about Ψ these M-estimates will be different for differing choices of S. Now let $I_1, I_2, \ldots, I_{2^n}$ denote the following partition of the real line:

$$I_1 = (-\infty, a_1), \quad I_2 = [a_1, a_2), \quad I_3 = [a_2, a_3), \ldots, I_{2^n - 1} = [a_{2^n - 2}, a_{2^n - 1}),$$

and

$$I_{2^n} = [a_{2^n - 1}, +\infty)$$

where a_1 is the smallest $\hat{\theta}_S$, a_2 is the second smallest $\hat{\theta}_S$, and so on.

We now are able to state the first typical value theorem.

Theorem 3.1.1.1. The Typical Value Theorem (Hartigan, 1969). *The true value of θ has probability $1/2^n$ of being in the interval I_i for $i = 1, 2, \ldots, 2^n$, where I_i is defined as above.*

The proof of this theorem is given in Efron (1982a, pp. 70–71). He attributes the method of proof to the paper by Maritz (1979). The theorem came originally from Hartigan (1969) who attributes it to Tukey and Mallows.

We now define a procedure called random subsampling. Let $S_1, S_2, S_3, \ldots, S_{B-1}$ be $B - 1$ of the $2^n - 1$ nonempty subsets of $\{1, 2, \ldots, n\}$ selected at random without replacement and let $I_1, I_2, \ldots, I_B$ be the partition of the real line obtained by ordering the corresponding $\hat{\theta}_S$ values. We then have the following typical value theorem, which can be viewed as a corollary to the previous theorem.

Theorem 3.1.1.2. *The true value of θ has probability $1/B$ of being in the interval I_i for $i = 1, 2, \ldots, B$, where I_i is defined as above.*

For more details and discussion about these results see Efron (1982a). The important point here is that we know the probability that each interval contains θ.

We can then construct an exact $100(j/B)$ percent confidence region for $1 \leqslant j \leqslant B - 1$, by simply combining any j of the intervals. The most sensible approach would be to paste together the j intervals in the "middle" if a two-sided interval is desired.

3.1.2. Percentile Method

The percentile method is the most obvious way to construct a confidence interval for a parameter based on bootstrap estimates. Suppose that $\hat{\theta}_i^*$ is the ith bootstrap estimate from the ith bootstrap sample, where each bootstrap sample is of size n.

By analogy with the case of random subsampling, we would expect that if we ordered the observations from smallest to largest, the interval containing 90% of the $\hat{\theta}_i^*$ values would be a 90% confidence interval for θ.

A bootstrap confidence interval generated this way is called a percentile method confidence interval. This would be an exact confidence interval if the typical value theorem could be applied to bootstrap sample estimates just as it is applied to random subsample estimates. However, we cannot do this. You should remember that we also had the symmetry condition and the estimator had to be an M-estimator for Hartigan's theorem to apply. Unfortunately, even if the distribution is symmetric and the estimator is an M-estimator, as is the case for the sample median of, say, a Cauchy distribution, the bootstrap percentile method would not be exact (i.e., the parameter is contained in the generated intervals in the advertised proportion of intervals as the number of generated cases becomes large).

Efron (1982a, pp. 80–81) shows that, for the median, the percentile method provides nearly the same confidence interval as the nonparametric interval based on the binomial distribution. So the percentile method works well in some cases even though it is not exact.

Really, the main difference between random subsampling and bootstrapping is that bootstrapping involves sampling with replacement from the original sample whereas random subsampling selects without replacement from the set of all possible subsamples. As the sample size becomes large, the difference in the distribution of the bootstrap estimates and the subsample estimates becomes small. Therefore, we expect the bootstrap percentile interval to be almost the same as the random subsample interval. So the percentile intervals inherit the exactness property of the subsample interval asymptotically (i.e., as the sample size becomes infinitely large).

Unfortunately, in the case of small samples (especially for asymmetric distributions), the percentile method does not work well. Fortunately, there are modifications that will let us overcome these difficulties as we shall see in the next section.

In Chapter 3 of Hall (1992a) several bootstrap confidence intervals are defined. In particular, see Section 3.2 of Hall (1992a). In Hall's notation F_0 denotes the population distribution, F_1 the empirical distribution, and F_2 the distribution of the samples drawn at random and with replacement from F_1.

Let φ_0 be the unknown parameter of interest that is expressible as a functional of the distribution F_0. So $\varphi_0 = \varphi(F_0)$. A theoretical α-level percentile confidence interval for φ_0 (by Hall's definition) is the interval $I_1 = (-\infty, \hat{\psi} + t_0)$, where t_0 is defined so that $P(\varphi_0 \leqslant \hat{\psi} + t_0) = \alpha$.

Alternatively, if we define

$$f_1(F_0, F_1) = I\{\varphi(F_0) \leqslant \varphi(F_1) + t\} - \alpha$$

then t_0 is a value of t such that $f_t(F_0, F_1) = 0$.

By analogy, a bootstrap one-sided percentile interval for φ_0 would be obtained by solving the equation

$$f_t(F_1, F_2) = 0 \tag{3.1}$$

since in bootstrapping, F_1 replaces F_0 and F_2 replaces F_1. If $\hat{t}_0$ is a solution to Eq. 3.1, the interval $(-\infty, \varphi(F_2) + \hat{t}_0)$ is a one-sided bootstrap percentile confidence interval for φ. Here $\varphi(F_2)$ is the bootstrap sample estimate of φ.

This is a natural way to define a percentile confidence interval according to Hall. This percentile estimate can easily be approximated by Monte Carlo but differs from Efron's percentile method. Hall refers to Efron's percentile as the "other" percentile method or the "backwards" percentile method.

3.1.3. Bias Correction and the Acceleration Constant

Efron and Tibshirani (1986, pp. 67–70) describe four methods for constructing approximate confidence intervals for a parameter θ. They provide the assumptions required for each method to work well. In going from the first method to the fourth, the assumptions become less restrictive while the methods become more complicated but more generally applicable.

The first method is referred to as the standard method. It is obtained by taking the estimator $\hat{\theta}$ of θ and an estimate of its standard deviation $\hat{\sigma}$. The interval $[\hat{\theta} - \hat{\sigma}z_a, \hat{\theta} + \hat{\sigma}z_a]$ is then the standard $100(1 - \alpha)\%$ approximate confidence interval for θ. This method works well if $\hat{\theta}$ has an approximate Gaussian distribution with mean θ and standard deviation σ, independent of θ.

The second method is the bootstrap percentile method (Efron's definition) described in Section 3.1.2. It works well when there exists a monotone transformation $\hat{\phi} = g(\theta)$ such that $\hat{\phi} = g(\hat{\theta})$ is approximately Gaussian with mean ϕ and standard deviation τ independent of ϕ.

The third method is the bias-corrected bootstrap interval that we shall discuss in this section. It works well if the transformation $\hat{\phi} = g(\hat{\theta})$ is approximately Gaussian with mean $\phi - z_0\tau$, where z_0 is the bias correction and τ is the standard deviation of $\hat{\phi}$.

The fourth method is the BC_a method that incorporates an acceleration constant a. For it to work well $\hat{\phi}$ is approximately Gaussian with mean $\phi - z_0\tau_\phi$, where z_0 is the bias correction and τ_ϕ is the standard deviation of $\hat{\phi}$ that depends on θ as follows: $\tau_\phi = 1 + a\phi$, where a is the acceleration constant to be defined later in this section. These results are summarized in Table 6 of Efron and Tibshirani (1986) and are reproduced in Table 3.1.

Efron and Tibshirani (1986) claim that the percentile method automatically incorporates normalizing transformations. To illustrate the difficulties that can be encountered with the percentile method, they consider the case where θ is the bivariate correlation coefficient from a two-dimensional Gaussian distribution and the sample size is 15.

In this case, there is no monotone transformation g that maps $\hat{\theta}$ into $\hat{\phi}$ with $\hat{\phi}$ Gaussian with mean ϕ and constant variance τ^2 independent of ϕ. Efron and Tibshirani (1986) show for a set of data referred to as the "law school data" that the sample bivariate correlation is 0.776.

Assuming we have bivariate Gaussian data with a sample of size 15 and a sample correlation estimate equal to 0.776, we would find that, for a bootstrap sample, the probability that the correlation coefficient is less than 0.776 is only 0.431.

For any monotone transformation, this would also be the probability that the transformed value of the bootstrap sample correlation is less than the transformed value of the original sample correlation [i.e., $g(0.776)$]. However, for the transformed values to be Gaussian or at least a good approximation to the Gaussian distribution and centered about $g(0.776)$, this probability would have to be 0.500 and not 0.431. Note that for symmetric distributions like the Gaussian the mean is equal to the median. But we do not see that here for the correlation coefficient.

What we see here is that, at least for some values of θ different from zero, no such transformation will work well. Efron and Tibshirani remedy this problem by making a bias correction to the percentile method. Basically the percentile method works if exactly 50% of the bootstrap distribution for $\hat{\theta}$ is less than $\hat{\theta}$.

By applying the Monte Carlo approximation, we determine an approximation to the bootstrap distribution. We find the 50th percentile of this distribution and call it $\hat{\theta}_{50}^*$. Taking this bias B to be $\hat{\theta} - \hat{\theta}_{50}^*$ we see that $\hat{\theta} - B$ equals $\hat{\theta}_{50}^*$ and so B is called the bias correction.

Another way to look at it, which is explicit but may be somewhat confusing, is to define $z_0 = \Phi^{-1}\{\hat{G}(\hat{\theta})\}$ (where Φ^{-1} is the inverse of the cumulative Gaussian distribution and $\hat{G}$ is the cumulative bootstrap sample distribution for θ). For a central $100(1 - 2\alpha)\%$ confidence interval, we then take the lower endpoint to be $\hat{G}^{-1}(\Phi\{2z_0 + Z^{(\alpha)}\})$ and the upper endpoint to be $\hat{G}^{-1}(\Phi\{2z_0 + Z^{(1-\alpha)}\})$.

This is how the bias correction method is defined in Efron (1982a) and Efron and Tibshirani (1986), where $Z^{(\alpha)}$ satisfies $\Phi(Z^{(\alpha)}) = \alpha$. Note that we use the "hat" notation over the cumulative bootstrap distribution G to indicate that a Monte Carlo estimate of it is used.

Table 3.1 Four Methods of Setting Approximate Confidence Intervals for a Real-Valued Parameter θ

Method	Abbreviation	α-Level Endpoint	Correct if
1. Standard	$\theta_S[\alpha]$	$\hat\theta + \hat\sigma z^{(\alpha)}$	$\hat\theta \approx N(\theta, \sigma^2)$ $\quad$ σ constant
			There exists monotone transformation $\hat\phi = g(\hat\theta),\ \phi - g(\theta)$ such that:
2. Percentile	$\theta_P[\alpha]$	$\hat G^{-1}(\alpha)$	$\hat\phi \approx N(\phi, \tau^2)$ $\quad$ τ constant
3. Bias-corrected	$\theta_{BC}[\alpha]$	$\hat G^{-1}(\Phi\{2z_0 + z^{(\alpha)}\})$	$\hat\phi \approx N(\phi - z_0\tau, \tau^2)$ $\quad$ $z_0,\ \tau$ constant
4. BC_a	$\theta_{BC_a}[\alpha]$	$\hat G^{-1}\left(\Phi\left\{z_0 + \dfrac{(z_0 + z^{(\alpha)})}{1 - a(z_0 + z^{(\alpha)})}\right\}\right)$	$\hat\phi \approx N(\phi - z_0\tau_0, \tau_0^2)$ where $\tau_0 = 1 + a\phi$ $\quad$ $z_0,\ a$ constant

Note: Each method is correct under more general assumptions than its predecessor. Methods 2, 3, and 4 are defined in terms of the percentiles of $\hat G$, the bootstrap distribution.

Source: Efron and Tibshirani (1986, Table 6), with permission from the Institute of Mathematical Statistics.

It turns out that in the case of the law school data (assuming that it is a sample from a bivariate Gaussian distribution) the exact central 90% confidence interval is [0.496, 0.898]. The percentile method gives an interval of [0.536, 0.911] and the bias-corrected method yields [0.488, 0.900]. Since the bias-corrected method comes closer to the exact interval, we can conclude, in this case, that it is better than the percentile method for the correlation coefficient.

What is important here is that this bias-correction method will work no matter what the value of θ really is. This means that, after the adjustment, the monotone transformation leads to a distribution that is approximately Gaussian and whose variance does not depend on the transformed value ϕ. If the variance cannot be made independent of ϕ then a further adjustment, referred to as the acceleration constant a, is required.

Schenker (1985) provides an example for which the bias-corrected percentile method did not work very well. It involves a χ^2 random variable with 19 degrees of freedom. In Efron and Tibshirani (1986) and Efron (1987) it is shown that the use of an acceleration constant overcomes the difficulty. It turns out in examples like Schenker's that there is a monotone transformation that works after a bias correction. The problem is that the resulting Gaussian distribution has a standard deviation τ_ϕ that depends linearly on ϕ (i.e., $\tau_\phi = 1 + a\phi$, where a is the acceleration constant). A difficulty in the application of this modification to the bootstrap is the determination of the acceleration constant, a.

Efron found that a good approximation to the constant is one-sixth of the skewness of the score statistic evaluated at $\hat{\theta}$. See Efron and Tibshirani (1986) for details and examples of the computations involved.

Although this method seems to work in very general cases, it is complicated and may not be necessary. Bootstrap iteration, to be explained in Section 3.1.4, is an alternative, as is the bootstrap t method of Section 3.1.5.

These methods have a drawback that they share with the bootstrap t intervals, namely, that they are not monotone in the assumed level of coverage (i.e., one could decrease the confidence level and not necessarily get a shorter interval that is contained in the interval with the higher confidence level).

3.1.4. Iterated Bootstrap

A number of authors have contributed to the literature on bootstrap iteration and we mention many of these contributors in the historical notes (Section 3.4). Major contributions to the theory of the iterated bootstrap were made by Peter Hall and his graduate student Michael Martin. Martin (1990a) provides a clear and up-to-date summary of these advances (see also Hall, 1992a, Chap. 3).

Under certain regularity conditions on the population distributions, there has developed an asymptotic theory for the degree of closeness of the bootstrap confidence intervals to their stated coverage probability. Details can be found in a number of papers (e.g., Hall, 1988b; Martin, 1990a).

An approximate confidence interval is said to be first-order accurate if its coverage probability differs from its advertised coverage probability by terms that go to zero at a rate of $n^{-1/2}$. The standard intervals discussed in Section 3.1.3 are first-order accurate. The BC_a intervals of Section 3.1.3 and the iterated bootstrap intervals to be discussed in this section are both second-order accurate (i.e., the difference goes to zero at rate n^{-1}).

A more important property for a confidence interval than just being accurate would be for the interval to be as small as possible for the given coverage probability. It may be possible to construct a confidence interval using one method that has coverage probability of 0.95 and yet it may be possible to find another method to use that will also provide a confidence interval with coverage probability 0.95 but the latter interval will actually be shorter!

Confidence intervals that are "optimal" in the sense of being the shortest possible for the given coverage are said to be "correct." Efron (1990b) provides a very good discussion of this issue along with some examples.

A nice property of these bootstrap intervals (i.e., the BC_a and the iterated bootstrap) is that in addition to being second-order accurate, they are also close to the ideal or "correct" interval in a number of problems where it makes sense to talk about "correct" intervals.

In fact, the theory has gone further to show for certain broad classes of parametric families of distributions that corrections can be made to get third-order accurate (i.e., with rate $n^{-3/2}$) intervals (Hall, 1988b; Cox and Reid, 1987a; Welch and Peers, 1963).

Bootstrap iteration provides another way to improve the accuracy of bootstrap confidence intervals. Martin (1990a) discusses the approach of Beran (1987) and shows for one-sided confidence intervals that each bootstrap iteration improves the coverage by a factor of $n^{-1/2}$ and by a factor of n^{-1} for two-sided intervals.

What is a bootstrap iteration? Let us now describe the process. Suppose we have a random sample $\mathbf{X}$ of size n with observations denoted by X_1, X_2, $X_3, \ldots, X_n$. Let $X_1^*, X_2^*, X_3^*, \ldots, X_n^*$ denote a bootstrap sample obtained from this sample.

Let I_0 denote a nominal $(1 - \alpha)$-level confidence interval for a parameter θ of the population from which the original sample was taken. For example, I_0 could be a $(1 - \alpha)$-level confidence interval for θ obtained by Efron's percentile method. To illustrate the dependence of I_0 on the original sample $\mathbf{X}$ and the level $1 - \alpha$, we denote it as $I_0(\alpha|\mathbf{X})$. We then denote the actual coverage of the interval $I_0(\alpha|\mathbf{X})$ by $\pi_0(\alpha)$.

Let β_α be the solution to

$$\pi_0(\beta_\alpha) = P\{\theta \in I_0(\beta_\alpha|\mathbf{X})\} = 1 - z \qquad (3.2)$$

Now let $I_0(\beta_\alpha|\mathbf{X}^*)$ denote the version of I_0 computed using the resample in place of the original sample. The resampling principle of Hall and Martin (1988a) states that to obtain better coverage accuracy than given by the original interval I_0 we use $I_0(\hat{\beta}_\alpha|\mathbf{X}^*)$, where $\hat{\beta}_\alpha$ is the estimate for β_α given in Eq. 3.2 and obtained by replacing θ with $\hat{\theta}$ and $\mathbf{X}$ with $\mathbf{X}^*$. To iterate again, we just use the newly obtained interval in place of I_0 and apply the same procedure to it. When bootstrap iteration involves a single iteration it is also referred to as the double bootstrap.

The algorithm just described is theoretically possible but in practice a Monte Carlo approximation must be used. In the Monte Carlo approximation, B bootstrap samples are generated. Details of the bootstrap iterated confidence interval are given in Martin (1990a, pp. 1113–1114).

Although it is a complicated procedure to describe, the basic idea is that by resampling from the B bootstrap samples, we can estimate the point β_α and use that estimate to correct the percentile intervals. Results for particular examples using simulations are also given in Martin (1990a).

Clearly, the price paid for this added accuracy in the coverage of the confidence interval is an increase in the number of Monte Carlo replications. If we have an original sample size n and each bootstrap resample is of size n then (for a single iteration) the number of replications will be nB_1B_2, where B_1 is the number of bootstrap samples taken from the original sample and B_2 is the number of bootstrap samples taken from each resample.

In his example of two-sided intervals for the studentized mean from a folded normal distribution, Martin (1990a) uses $n = 10$, and $B_1 = B_2 = 299$. The examples do seem to be in agreement with the asymptotic theory in that a single bootstrap iteration does improve the coverage in all cases considered.

Bootstrap iteration can be applied to any bootstrap confidence interval to improve the rate of convergence to the level $1 - \alpha$. Hall (1992a) remarks that although his version of the percentile method may be more appropriate than Efron's, bootstrap iteration works better on Efron's percentile method. The reason is not clear and the observation is based on empirical findings. A single bootstrap iteration provides the same type correction as BC_a does to Efron's percentile method. Using more than one bootstrap iteration is not common in practice. This is probably due to the large increase in complexity and computation compared to the small gain in accuracy of the confidence level.

3.1.5. Bootstrap t Confidence Intervals

The iterated bootstrap and the BC_a intervals both provide improvements over Efron's percentile method but both are complicated and the iterated bootstrap is even more computer intensive than other bootstraps. The idea of the bootstrap t method is found in Efron (1982a). A clearer presentation can be found in Efron and Tibshirani (1993, pp. 160–167).

It is a simple method and has higher order accuracy than Efron's percentile method. To be precise, bootstrap t confidence intervals are second-order accurate (when they are appropriate). See Efron and Tibshirani (1993, pp. 322–325). Consequently, it is popular in practice. We used it in the Passive Plus DX clinical trial.

We shall now describe it briefly. Suppose that we have a parameter θ and an estimate θ_h for θ. Let θ^* be a nonparametric bootstrap estimate for θ based on a bootstrap sample and let S^* be an estimate of the standard deviation for θ based on the bootstrap sample. Define $T^* = (\theta^* - \theta_h)/S^*$. For each of the B bootstrap estimates θ^* there is a corresponding T^*. We find the percentiles of T^*. For an approximate two-sided $100(1 - 2\alpha)\%$ confidence interval for θ, we take the interval $[\theta_h - t^*_{(1-\alpha)}S, \theta_h - t^*_{(\alpha)}S]$ where $t^*_{(1-\alpha)}$ is the $1 - \alpha$ percentile of the T^* values $t^*_{(\alpha)}$ is the α percentile of the T^* values and S is estimated standard deviation for θ_h. This we call the bootstrap t (or percentile t, as Hall refers to it) two-sided $100(1 - 2\alpha)\%$ confidence interval for θ.

A difficulty with the bootstrap t is the need for an estimate of the standard deviation, S, for θ_h and the corresponding bootstrap estimate S^*. In some problems there are obvious estimates, as in the simple case of a sample mean or the difference between two independent sample means (as is the case for Passive Plus DX, where we are interested in the difference between the experimental group and control group means). For more complex parameters (e.g., C_{pk}) S may not be available.

3.2. RELATIONSHIP BETWEEN CONFIDENCE INTERVALS AND TESTS OF HYPOTHESES

In Section 3.1 of Good (1994) hypothesis testing for a single location parameter θ of a univariate distribution is introduced. In this section we show how confidence intervals can be generated based on the hypothesis test. Namely, for a $100(1 - \alpha)\%$ confidence interval, you include the values of θ at which you would not reject the null hypothesis at the significance level α.

To be more precise, suppose we test the null hypothesis that $\theta = \theta_0$ versus the alternative that θ differs from θ_0 at the significance level α. If the test statistic tells us to accept the null hypothesis, then θ_0 is included in the interval and it is excluded otherwise. We apply this to all possible values for θ_0 and the set of θ_0 that are included, form the $100(1 - \alpha)\%$ confidence interval for θ.

Conversely, if we have a $100(1 - \alpha)\%$ confidence interval for θ, we can construct an α-level hypothesis test by simply accepting the null hypothesis that $\theta = \theta_0$ if θ_0 is contained in the $100(1 - \alpha)\%$ confidence interval for θ and rejecting the null hypothesis if θ is outside the interval.

In problems involving nuisance parameters, this procedure becomes more complicated. Consider the case of estimating the mean μ of a normal distribution when the variance σ^2 is unknown. The statistic $(\bar{x} - \mu)/(s/\sqrt{n})$ has Student's t distribution with $n - 1$ degrees of freedom, where

$$\bar{x} = \sum_{i=1}^{n} \frac{x_i}{n}, \qquad s = \left(\frac{\Sigma(x_i - \bar{x})}{n-1} \right)^{1/2}$$

n is the sample size, and x_i is the ith observed value.

What is nice about the t statistic is that its distribution is independent of the nuisance parameter σ^2 and it is a pivotal quantity (to be discussed later). Because its distribution does not depend on σ^2 or any other unknown quantities, we can use the tables of the t distribution to determine probabilities, such as

$$P[a \leqslant t \leqslant b] \quad \text{where} \quad t = \frac{\bar{x} - \mu}{s/\sqrt{n}}$$

Note t is also a pivotal quantity, which means that probability statements like the one above can be converted into confidence statements involving μ. So if

$$P[a \leqslant t \leqslant b] = 1 - \alpha \tag{3.3}$$

then the probability is also $1 - \alpha$ that the random interval

$$\left[\bar{x} - \frac{bs}{\sqrt{n}}, \bar{x} - \frac{as}{\sqrt{n}} \right] \tag{3.4}$$

includes the true value of the parameter μ. This random interval is then a $100(1 - \alpha)\%$ confidence interval for μ.

The interval in Eq. 3.4 is a $100(1 - \alpha)\%$ confidence interval for μ. We can start with Eq. 3.3 and get the interval in Eq. 3.4 or vice versa. If we are testing the hypothesis that $\mu = \mu_0$ versus the alternative that μ differs from μ_0, using Eq. 3.2, we replace μ with μ_0 in the t statistic and reject the hypothesis at the α level if t is less than a or greater than b.

We have learned earlier in this chapter how to construct various bootstrap confidence intervals with confidence level approximately $100(1 - \alpha)\%$. Using these bootstrap confidence intervals, we will be able to construct hypothesis tests by rejecting parameter values if and only if they fall outside the confidence interval. In the case of a translation family of distributions, the power of the test for the translation parameter is connected to the width of the confidence interval.

In the next section we shall illustrate the procedure by using a bootstrap confidence interval for the ratio of two variances in order to test the equality of the variances. This one example should suffice to illustrate how bootstrap tests can be obtained.

3.3. HYPOTHESIS TESTING PROBLEMS

In principle, we can use any bootstrap confidence interval for a parameter to construct a hypothesis test just as we have described it in the previous section (if we have a pivotal or at least an asymptotically pivotal quantity or have no nuisance parameters). Bootstrap iteration and the use of bias correction with an acceleration constant are two ways by which we can provide more accuracy to the advertised level of the test.

Another key point that relates to accuracy is the choice of a test statistic that is asymptotically pivotal. It is pointed out by Fisher and Hall (1990) that tests based on pivotal statistics often result in significance levels that differ from the advertised level by $O(n^{-2})$ as compared to $O(n^{-1})$ for tests based on nonpivotal statistics.

As an example, Fisher and Hall (1990) show that for the one-way analysis of variance, the F ratio is appropriate for testing equality of means when the variances are equal from group to group. For equal (homogeneous) variances the F ratio test is asymptotically pivotal.

However, when the variances differ (i.e., are heterogeneous) the F ratio depends on these variances, which are nuisance parameters. For the heterogeneous case the F ratio is not asymptotically pivotal. They use a statistic first proposed by James (1951), which is asymptotically pivotal. Additional work on this topic can be found in James (1954).

In our example we will be using an F ratio to test for equality of two variances. Under the null hypothesis that the two variances are equal, the F ratio will not depend on the common variance and is therefore pivotal.

In Section 3.3.2 of Good (1994), he points out that permutation tests had not been devised for this problem. On the other hand there is no difficulty in solving this problem by bootstrapping. If we have n_1 samples from one population and n_2 from the second, we can independently resample with sample sizes of n_1 and n_2 from population 1 and population 2, respectively.

We construct a bootstrap value for the F ratio, by using a bootstrap sample of size n_1 from the sample from population 1 to calculate the numerator (a sample variance estimate for population 1) and a bootstrap sample of size n_2 from the sample from population 2 to calculate the denominator (a sample variance estimate for population 2). Since the two variances are equal under the null hypothesis, we expect the ratio to be close to one.

By repeating this many times, we are able to get a Monte Carlo approximation to the bootstrap distribution for the F ratio. This distribution should be centered at one when the null hypothesis is true. The extremes of the bootstrap distribution provide us with information as to how far from one we would need to set our threshold for this test.

Since the F ratio is pivotal under the null hypothesis, we use the percentiles of the Monte Carlo approximation to the bootstrap distribution to get critical points for the hypothesis test. Alternatively, we could use the more sophisticated bootstrap confidence intervals, but in this case it is not crucial.

A simulation study comparing this bootstrap approach to testing the equality of variances to other approaches is given in Good and Chernick (1993). Further simulations for this problem are given in Hall and Padmanabhan (1997).

In the above example under the null hypothesis $\sigma_1^2/\sigma_2^2 = 1$ and we would normally reject the null hypothesis in favor of the alternative $\sigma_1^2/\sigma_2^2 \neq 1$ if the

F ratio differs significantly from 1. However, Hall (1992a, Sec. 3.12) points out that the F ratio for the bootstrap sample should be compared or "centered" at the sample estimate rather than the hypothesized value.

Such an approach is known to generally lead to more powerful tests than the approach based on sampling at the hypothesized value. See Hall (1992a) or Hall and Wilson (1991) for more examples and a more detailed discussion of this point.

3.3.1. Tendril DX Lead Clinical Trial Analysis

In 1995 Pacesetter Inc., a St. Jude Medical Company that produces pacemakers and leads for patients with bradycardia, submitted a protocol to the United States Food and Drug Administration (FDA) for a clinical trial to demonstrate the safety and effectiveness of an active fixation steroid eluting lead. The study called for the comparison of the Tendril DX model 1388T with a concurrent control, the market-released Tendril model 1188T active fixation lead.

The two leads are almost identical with the only differences being the use of titanium nitride on the tip of the 1388T lead and the steroid eluting plug also in the 1388T lead. Both leads were designed for implantation in either the atrial or the ventricular chambers of the heart, to be implanted along with a dual chamber pacemaker (most commonly Pacesetter's Trilogy DR + pulse generator).

It is known from the successful clinical trials of competitor's steroid eluting leads and other research literature that the steroid drug reduces inflammation at the area of implantation. Inflammation results in an increase in the capture threshold for the pulse generator in the acute phase (the acute phase is usually considered to be the first six months postimplant).

Pacesetter statisticians (myself included) proposed, as the primary endpoint for effectiveness, demonstration of a 0.5 volt or greater reduction in the mean capture threshold at the three-month follow-up for patients with 1388T leads implanted in the atrial chamber when compared to similar patients with 1188T leads implanted in the atrial chamber. The same hypothesis test was used for the ventricular chamber.

Patients entering the study were randomized as to whether they received the 1388T steroid lead or the 1188T lead. Since the effectiveness of steroid is well established from other studies in the literature, Pacesetter argued that it would be unfair to the patients in the study to give them only a 50–50 chance of receiving the 1388T lead (which is expected to provide less inflammation and discomfort and lower capture thresholds).

So Pacesetter designed the trial to have reasonable power to detect a 0.5 volt improvement and yet give the patient a 3:1 chance of receiving the 1388T lead. Such an unbalanced design required more patients for statistical confirmation of the hypothesis (i.e., based on Gaussian assumptions, a balanced design required 50 patients in each group, whereas with the 3:1 randomization

99 patients were required in the experimental group and 33 in the control group to achieve the same power, that is, 0.80 probability of rejecting the null hypothesis of no difference when the steroid lead is at least 0.50 volt lower in mean capture threshold, when testing at the 0.05 significance level).

The protocol was approved by the FDA and the trial proceeded. Interim reports and a premarket approval report (PMA) (based on a clinical report published April 9, 1996) were submitted to the FDA and the leads were approved for market release in June 1997.

Capture thresholds take on very discrete values due to the discrete programmed settings. Since the early data at three months were expected to be convincing but the sample size possible was relatively small, nonparametric approaches were taken as alternatives to the standard Gaussian t tests.

The parametric procedures would only be approximately valid for large sample sizes due to the non-Gaussian nature of capture threshold distributions (possibly skewed, discrete, and truncated). The Wilcoxon rank sum test was used as the nonparametric standard for showing improvement in the mean (or median) of the capture threshold distribution and the bootstrap percentile method was also used to test the hypothesis.

Figures 3.1 and 3.3 show the distributions (i.e., histograms) of bipolar capture thresholds for (1188T and 1388T) leads in the atrium and the ventricle, respectively, at the three-month follow-up. The variable "leadloc" refers to the chamber of the heart where the lead was placed.

Figures 3.2 and 3.4 provide the bootstrap histogram of the difference in mean atrial bipolar capture threshold and mean ventricular bipolar capture

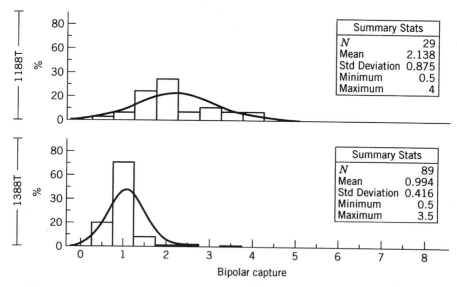

Figure 3.1 Capture threshold distributions for three-month visit (leadloc atrial chamber).

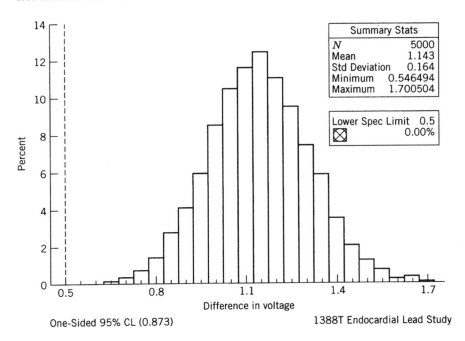

Figure 3.2 Distribution of bootstrapped data sets (atrium): bipolar three-month data as of March 15, 1996.

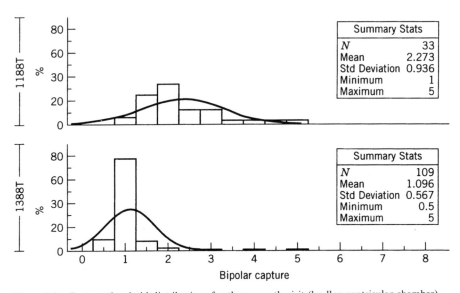

Figure 3.3 Capture threshold distributions for three-month visit (leadloc-ventricular chamber).

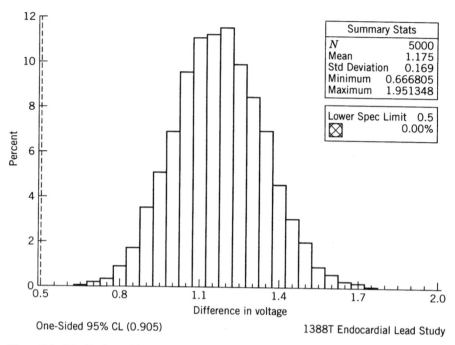

One-Sided 95% CL (0.905)   1388T Endocardial Lead Study

Figure 3.4 Distribution of bootstrapped data sets (ventricle): bipolar three-month data as of March 15, 1996.

threshold, respectively, for the 1388T leads versus the 1188T leads at the three-month follow-up.

The summary statistics in the box are N, the number of bootstrap replications; Mean, the mean of the bootstrap samples; Std Deviation, the standard deviation of the bootstrap samples; Minimum, the smallest value out of the 5000 bootstrap estimates of the mean difference; and Maximum, the largest value of the 5000 bootstrap estimates. Listed on the figures are the respective number of samples for the control (1188T) leads and the number of samples for the investigational (1388T) leads in the original sample for which the comparison is made.

It also shows the mean difference of the original data that should be (and is) close in value to the bootstrap estimate of the sample mean. The estimate of the standard deviation for the mean difference is also given on the figures. We note that this too is very close to the bootstrap estimate for these data.

The histograms are based on 5000 bootstrap replications on the mean differences. Also shown on the graph of the histogram is the lower 5th percentile (used in Efron's percentile method as the lower bound on the true difference for the hypothesis test). The proportion of the bootstrap distribution below zero provides a bootstrap percentile p-value for the hypothesis of no improvement versus a positive improvement in capture threshold.

Due to the slight skewness that can be seen in Figs. 3.1 and 3.3, the Pacesetter statisticians were concerned that the percentile method for determining the bootstrap lower confidence bound on the difference in the mean values might not be sufficiently accurate.

The percentile t was considered but time did not permit the method to be developed for the submission. In a later clinical trial, Pacesetter took the same approach with the comparison of the control and treatment leads for the Passive Plus DX clinical trial.

The percentile t method is a simple method to program and appears to overcome some of the shortcomings of Efron's percentile method without the complications of bias correction and acceleration constants. This technique was first presented by Efron as the bootstrap t (Efron, 1982a, Sec. 10.10). Later, Peter Hall (1986a) developed asymptotic formulas for the coverage error of the percentile t method. This is the method discussed previously in Section 3.1.5.

The Passive Plus DX lead is a passive fixation steroid eluting lead that was compared with a nonsteroid approved version of the lead. The 3:1 randomization was used in the Passive Plus study also.

In the Passive Plus study the capture thresholds behaved similarly to those for the leads in the Tendril DX study. The distributions were similar in shape to the Tendril study. The main difference in the results was that the mean differences were not quite as large (closer to a 0.5 volt improvement for the steroid lead whereas for Tendril the mean difference was observed to be closer to a 1 volt improvement).

The Passive Plus study included calculation of both the percentile method lower 95% confidence bound and the percentile t lower 95% confidence bound.

3.4. HISTORICAL NOTES

Bootstrap confidence intervals were introduced in Efron (1982a, Chap. 10). Efron's percentile method and the bias-corrected percentile method were introduced at that time. Efron (1982a) also introduced the bootstrap t intervals and illustrated these techniques with the median as the parameter.

It was recognized at that time (the early 1980s) that confidence interval estimation was a tougher problem than estimating standard errors and considerably more bootstrap samples would be required (i.e., 1000 bootstrap samples were recommended for confidence intervals whereas only 100 would be required for estimates of standard errors). For more discussion of this issue see Section 7.1.

In an important work, Schenker (1985) shows that bias adjustment to Efron's percentile method does not always provide a sufficient improvement to yield "good" confidence intervals. Nat Schenker's examples motivated Efron to come up with the use of an acceleration constant as well as a bias correction in the modification of the percentile estimate of the interval endpoints. This led to significant improvements in the bootstrap confidence intervals and overcame Schenker's objections.

The idea of bootstrap iteration to improve confidence interval estimation appears in Hall (1986a), Beran (1987), Loh (1987), Hall and Martin (1988a), and DiCiccio and Romano (1988). The methods of Hall, Beran, and Loh all differed in the way they corrected the critical point. Loh referred to his approach as a bootstrap calibration.

Hall (1986b) deals with sample size requirements. Specific application to the confidence interval estimation for the correlation coefficient is given in Hall, Martin, and Schucany (1989). For recent developments in bootstrap iteration see Martin (1990a), Hall (1992a), or Davison and Hinkley (1997).

Some of the asymptotic theory is based on formal Edgeworth expansions, which were rigorously developed in Bhattacharya and Ghosh (1978) [see Hall (1992a) for a detailed account with applications to the bootstrap]. Other asymptotic expansions such as saddlepoint approximations may provide comparable confidence intervals without the need for Monte Carlo [see the monograph by Field and Ronchetti (1990) and the papers by Davison and Hinkley (1988) and Tingley and Field (1990)].

DiCiccio and Efron (1992) also obtain very good confidence intervals without Monte Carlo for data from an exponential family of distributions. DiCiccio and Romano (1989a) also produce accurate confidence limits by making some parametric assumptions.

Much of the latest research on the bootstrap suggests that sometimes the Monte Carlo approximation may not be necessary (see Section 7.3 and the references above) or that the number of Monte Carlo replications can be considerably reduced by variance reduction techniques [see Section 7.2 and Davison, Hinkley, and Schechtman (1986), Therneau (1983), Hesterberg (1988), Johns (1988), and Hinkley and Shi (1989)]. The most recent developments can be found in Hesterberg (1995a, b, 1996, 1997).

Discussions of bootstrap hypothesis tests appear in an early paper by Efron (1979a) and some work can be found in Beran (1988c), Hinkley (1988), Fisher and Hall (1990), and Hall and Wilson (1991). Specific applications and Monte Carlo studies of bootstrap hypothesis testing problems are given in Dielman and Pfaffenberger (1988), Rayner (1990b), and Rayner and Dielman (1990).

Fisher and Hall (1990) point out that even though there are close connections between bootstrap hypothesis tests and confidence intervals there are also important differences, which lead to specialized treatment. They recommend the use of asymptotic pivotal quantities in order to maintain a close approximation to the advertised significance level for the test.

Ideas are illustrated using the analysis of variance problem with both real and simulated data sets. Results based on Edgeworth expansions and Cornish–Fisher expansions clearly demonstrate the advantage of bootstrapping pivotal statistics for both hypothesis testing and confidence intervals (see Hall, 1992a). Lehmann (1986) is the classic reference on hypothesis testing.

The first application of Edgeworth expansions to derive properties for the bootstrap is Singh (1981). The work by Bickel and Freedman (1981) is similar to Singh (1981) and also uses Edgeworth expansions. Their work shows how bootstrap methods correct for skewness.

Both papers applied one-term Edgeworth expansion corrections. Much of the development of Edgeworth expansions goes back to the determination of particular cumulants, as in James (1955, 1958).

The importance of asymptotically pivotal quantities was not brought out in the early papers because the authors considered a nonstudentized sample mean and assumed the population variance is known. Rather, this result regarding pivotal quantities was first shown by Babu and Singh in a series of papers (Babu and Singh, 1983, 1984a, 1985). Another key paper on the use of Edgeworth expansions for hypothesis testing is Abramovitch and Singh (1985).

Hall (1986a, 1988b) wrote two key papers that demonstrate the value of asymptotically pivotal quantities in the accuracy of bootstrap confidence intervals.

Hall (1986a) derives asymptotic formulas for coverage error of percentile t (i.e. bootstrap t) confidence intervals. Theoretical comparisons of variations on bootstrap t confidence intervals are given in Bickel (1992). Hall (1988b) gives a general theory for bootstrap confidence intervals. Other papers that support the use of pivotal statistics are Beran (1987) and Liu and Singh (1987).

Methods based on symmetric bootstrap confidence intervals are introduced in Hall (1988a). Hall (1988b) also defines "short" bootstrap confidence intervals [see also Hall (1992a) for some discussions]. The idea for the "short" bootstrap confidence intervals goes back to Buckland (1980, 1983).

Efron first proposed his version of the percentile method in 1979 (Efron, 1979a) [see also Efron (1982a) for detailed discussions]. The BC_a intervals were first given in Efron (1987). Several papers by Buckland (1983, 1984, 1985) provide applications for Efron's bias correction intervals along with algorithms for their construction.

Bootstrap iteration in the context of confidence intervals is introduced in Hall (1986a) and Beran (1987). Hall and Martin (1988a) develop a general framework for bootstrap iteration. Loh (1987) introduced the notion of bootstrap calibration. When applied to bootstrap confidence intervals, calibration is equivalent to bootstrap iteration.

Other important works related to confidence intervals and hypothesis testing include Beran (1986, 1990a, b).

CHAPTER 4

Regression Analysis

This chapter is divided into three parts along with an historical notes section. Section 4.1 deals with linear regression and Section 4.2 deals with the nonlinear regression problems. Section 4.3 deals with nonparametric regression models. In Section 4.4, we provide historical notes regarding the development of the bootstrap procedures in both the linear and nonlinear cases.

In Section 4.1.1, we will briefly review the well-known Gauss–Markov theory that applies to least squares estimation in the linear regression problem. A natural question for the practitioner to ask is: "Why bootstrap in the linear regression case? Isn't least squares a well-established approach that has served us well in countless applications?" The answer is that for many problems least squares regression has served us well and is always useful as a first approach but is problematic when residuals come from heavy-tailed distributions or even when only a few outliers are present in the data.

The difficulty is that in some applications certain key assumptions may be violated. These assumptions are: (1) the error term in the model has a probability distribution that is the same for each observation and does not depend on the predictor variables (i.e., independence and homoscedasticity); (2) the predictor variables are observed without error; and (3) the error term has a finite variance.

Under these three assumptions, the least squares procedure provides the best linear unbiased estimate of the regression parameters. However, if assumption (1) is violated because the variance of the residuals varies as the predictor variables change, a weighted least squares approach may be more appropriate.

The strongest case for least squares estimation can be made when the error term has a Gaussian or approximately Gaussian distribution. Then the theory of maximum likelihood also applies and confidence intervals and hypothesis tests for the parameters can be applied using the standard theory and the standard statistical packages.

However, if the error distribution is non-Gaussian and particularly if the error distribution is heavy tailed, least squares estimation may not be suitable (robust regression methods may be better). When the error distribution is non-Gaussian, regardless what estimation procedure is used, it is difficult to

70

determine confidence intervals for the parameters or to obtain prediction intervals for the response variable.

This is where the bootstrap can help and we will illustrate it for both the linear and nonlinear cases. In the nonlinear case, even the standard errors for the estimates are not easily obtained, but bootstrap estimates are fairly straightforward.

There are two basic approaches to bootstrapping in the regression problem. One is to first fit the model and apply the bootstrap to the residuals. This requires that the residuals be independent and identically distributed (or at least exchangeable).

In a quasi-optical experiment (Shimabukuro, Lazar, Dyson, and Chernick 1984), I used the bootstrap to estimate the standard errors for two of the parameters in the nonlinear regression model. Results are discussed in Section 4.2.2. The residuals appear to be correlated with the incident angle of the measurement. This invalidates the exchangeability assumption, but how does it affect the standard errors of the parameters?

Our suspicion is that bootstrapping the residuals makes the bootstrap samples vary more than in actuality and consequently it biases the estimated standard errors on the high side. This, however, remains an open question. Clearly from the intuitive point of view the bootstrapping is not properly mimicking the variation in the actual residuals and the procedure can be brought into question.

A second method with more general applicability is to apply the bootstrap to the vector of the observed response variable and the associated predictor variables. This only requires that the vectors be exchangeable and does not place explicit requirements on the residuals from the model.

However, some statisticians, particularly from the British school, view this philosophically as an inappropriate approach. To them the regression problem requires that the predictor variables be fixed for the experiment and not selected at random from a probability distribution. The bootstrapping of the vector of the response variables and the predictor variables implicitly assumes a joint probability distribution for the vector of predictors and response. From their point of view this is an inappropriate model and hence the vector approach is not an option.

However, from the practical point of view, if the approach of bootstrapping the vector has nice robustness properties related to model specification, it should be justified. This was suggested by Efron and Tibshirani (1993, p. 113) for the case of one predictor variable. Then it really doesn't matter that we are not mimicking the assumed regression model. Presumably their observation extends to the case of more than one predictor variable.

On the other hand, some might argue that bootstrapping the residuals is only appropriate when the predictor variables are not fixed. This comes down to another philosophical issue that only statisticians would care about. The question is one of whether conditional inference is valid when the experiment really involves an unconditional joint distribution for the predictor and response variables.

This is a familiar technical debate for statisticians as it is the same issue regarding the appropriateness of conditioning on marginal totals in a 2×2 contingency table. Conditioning on (ancillary) information in the observed data is something Sir Ronald A. Fisher advocated as a general approach to statistical inference and he clearly applied it in his permutation test for independence in a 2×2 contingency table (Fisher's exact test).

For the practitioner, I would repeat what my wise friend and colleague, V. K. Murthy, always told me: "the proof of the pudding is in the eating." This applies here to these bootstrap regression methods as it does in the comparison of variants of the bootstrap in discriminant analysis. If we simulate the process under accepted modeling assumptions, the method that performs best in the simulation is the one to use regardless of how much you believe or like some particular theory.

These two methods for bootstrapping in regression are given by Efron (1982a, pp. 35–36). These methods are very general. They apply to linear and nonlinear regression models and can be used for least squares or for any other estimation procedure. We shall now describe these bootstrap methods.

A general regression model can be given by

$$Y_i = g_i(\boldsymbol{\beta}) + \varepsilon_i \quad \text{for } i = 1, 2, \dots, n$$

The functions g_i are of known form and may depend on a fixed vector of covariates c_i. The vector $\boldsymbol{\beta}$ is a $p \times 1$ vector of unknown parameters and the ε_i are independent and identically distributed with some distribution F.

We assume that F is "centered" at zero. Usually this means that the expected or average value of ε_i is zero. However, in cases where the expected value does not exist we may use the criterion that $P(\varepsilon < 0) = 0.50$.

Given the observed vector

$$\mathbf{y} = \begin{bmatrix} y_1 \\ y_2 \\ \vdots \\ y_a \end{bmatrix}$$

where the ith component y_i is the observed valued of the random variable Y_i, we find the estimate of $\boldsymbol{\beta}$, which minimizes the distance measure between $\mathbf{y}$ and $\lambda(\boldsymbol{\beta})$, where

$$\lambda(\boldsymbol{\beta}) = \begin{bmatrix} g_1(\boldsymbol{\beta}) \\ g_2(\boldsymbol{\beta}) \\ \vdots \\ g_n(\boldsymbol{\beta}) \end{bmatrix}$$

Denote the distance measure by $D(\mathbf{y}, \lambda(\boldsymbol{\beta}))$. If

$$D(\mathbf{y}, \lambda(\boldsymbol{\beta})) = \sum_{i=1}^{n} [y_i - g_i(\boldsymbol{\beta})]^2$$

we get the usual least squares estimates. For least absolute deviations, we would choose

$$D(\mathbf{y}, \lambda(\boldsymbol{\beta})) = \sum_{i=1}^{n} \left| y_i - g_i(\boldsymbol{\beta}) \right|$$

Now by taking $\hat{\boldsymbol{\beta}} = \min D(\mathbf{y}^*, \lambda(\boldsymbol{\beta}))$ we have our parameter estimate of $\boldsymbol{\beta}$. The residuals are then obtained as $\hat{\varepsilon}_i = y_i - g_i(\boldsymbol{\beta})$.

The first bootstrap approach is simply to bootstrap the residuals. This is accomplished by constructing the distribution F_n that places probability $1/n$ on each $\hat{\varepsilon}_i$. We then generate bootstrap residuals ε_i^* for $i = 1, 2, \ldots, n$, where the ε_i^* are obtained by sampling independently from F_n (i.e., we sample with replacement from $\hat{\varepsilon}_1, \hat{\varepsilon}_2, \ldots, \hat{\varepsilon}_n$).

We then have a bootstrap sample data set:

$$y_i^* = g_i(\hat{\boldsymbol{\beta}}) + \varepsilon_i^* \quad \text{for } i = 1, 2, \ldots, n$$

For each such bootstrap data set y^*, we obtain $\hat{\boldsymbol{\beta}}^* = \min(\mathbf{y}^*, \lambda(\boldsymbol{\beta}))$. The procedure is repeated B times and the covariance matrix for $\hat{\boldsymbol{\beta}}$ is estimated as

$$\hat{\Sigma} = \frac{1}{B-1} \sum_{j=1}^{B} (\hat{\boldsymbol{\beta}}_j^* - \hat{\boldsymbol{\beta}}^*)(\hat{\boldsymbol{\beta}}_j^* - \hat{\boldsymbol{\beta}}^*)^T$$

where $\hat{\boldsymbol{\beta}}_j^*$ is the bootstrap estimate from the jth bootstrap sample and $\hat{\boldsymbol{\beta}}^* = (1/B)\Sigma_{j=1}^{B}\hat{\boldsymbol{\beta}}_j^*$. This is the covariance estimate suggested by Efron (1982a, p. 36).

We note that bootstrap theory suggests simply using $\hat{\beta}$ in place of $\hat{\boldsymbol{\beta}}^*$. The resulting covariance matrix estimate should then be close to that suggested by Efron. Confidence intervals for β can be obtained by the methods described in Chapter 3, but with the bootstrap samples for the $\hat{\beta}$ values.

The second approach is to bootstrap the vector

$$\mathbf{Z}_i = \begin{pmatrix} y_i \\ \mathbf{c}_i \end{pmatrix}$$

of the observations y_i and the covariates or predictor variables $\mathbf{c}_i$ for $i = 1, 2, \ldots, n$. The bootstrap samples are then $\mathbf{z}_i^*$ for $i = 1, 2, \ldots, n$, obtained by giving probability of selection $1/n$ to each $\mathbf{z}_i$. Taking

$$\mathbf{z}_i^* = \begin{pmatrix} y_i^* \\ \mathbf{c}_i^* \end{pmatrix}$$

we use y_i^* to obtain the estimate $\hat{\beta}^*$ just as before.

Efron claims that although the two approaches are asymptotically equivalent for the given model, the second approach is less sensitive to model misspecification. It also appears that since we do not bootstrap the residuals, the second approach may be less sensitive to the assumptions concerning independence or exchangeability of the error terms.

4.1. LINEAR MODELS

In the case of the linear regression model, if the least squares estimation procedure is used, there is nothing to be gained by bootstrapping. As long as the error terms are independent and identically distributed with mean zero and common variance σ^2, the least squares estimates of the regression parameters will be the best among all linear unbiased estimators.

The covariance matrix corresponding to the least squares estimate $\hat{\beta}$ of the parameter vector $\boldsymbol{\beta}$ is given by

$$\Sigma = \sigma^2 (X^T X)^{-1}$$

where X is called the design matrix and $(X^T X)^{-1}$ is well defined if X is a full rank matrix. If $\hat{\sigma}^2$ is the least squares estimate of the residual variance σ^2, then

$$\hat{\Sigma} = \hat{\sigma}^2 (X^T X)^{-1}$$

is the commonly used estimate of the parameter covariance matrix.

For more details see Draper and Smith (1981). These least squares estimates are the standard estimates, which can be found in all the standard statistical computer programs.

If, in addition, the error terms are Gaussian, or approximately Gaussian, the least squares estimates are also the maximum likelihood estimates. Also, the confidence intervals for the regression parameters, hypothesis tests about the parameters, and prediction intervals for a new observation, based on known values of the regression variables, can be determined in a straightforward way.

In the non-Gaussian case, even though we can estimate the parameter covariance matrix, we will not know the probability distribution for $\hat{\beta}$ and so we cannot determine the confidence intervals or prediction intervals or perform hypothesis tests using the standard methods. The bootstrap approach does, however, provide a method for approximating the distribution of $\hat{\beta}$ through bootstrap sample estimates $\hat{\beta}^*$.

First, we review the Gauss–Markov theory of least squares estimation in Section 4.1.1. In Section 4.1.2, we discuss, in more detail, situations where we might prefer to use other estimates of β such as the least absolute deviation estimates or the M estimates.

In Section 4.1.3, we discuss bootstrap residuals and the possible problems that can arise. If we bootstrap the vector of response and predictor variables, we can avoid some of the problems of bootstrapping residuals.

4.1.1. Gauss–Markov Theory

The least squares estimators of the regression parameters are maximum likelihood when the error terms are assumed to be Gaussian. Consequently, the least squares estimates have the usual optimal properties under the Gaussian model. They are unbiased and asymptotically efficient. In fact, they have the minimum variance among unbiased estimators. The Gauss–Markov theorem is a more general result in that it applies to linear regression models with general error distributions. All that is assumed is that the error distribution has mean zero and variance σ^2. The theorem states that among all estimators, which are both unbiased and a linear function of the responses y_i for $i = 1, 2, 3, \ldots, n$, the least squares estimate has the smallest possible variance.

The result was first shown by Carl Friedrich Gauss in 1821. For more details about the theory, see the *Encyclopedia of Statistical Science*, Volume 3, pages 314–316.

4.1.2. Why Not Just Use Least Squares?

In the face of all these optimal properties, one should ask why least squares shouldn't always be the method of choice. The basic answer is that the least squares estimates are very sensitive to violations in the modeling assumptions. If the error distribution has heavy tails or the data contain a few "outliers" (observations that depart significantly from most of the data), the least squares estimates will not be very good.

This is particularly true if these outliers are at high leverage points (i.e., points that will have a large influence on the slope parameters). High leverage points occur at or near the extreme values of the predictor variables. In cases of heavy tails or outliers, the method of least absolute deviations or other robust regression procedures such as M-estimation or the method of repeated medians provide better solutions although analytically they are more complex.

Regardless of the procedure used, we may be interested in confidence regions for the regression parameters or prediction intervals for future cases. Under the Gaussian theory for least squares, this is possible. However, if the error distribution is non-Gaussian and unknown, the bootstrap provides a method for computing standard errors for the regression parameters or prediction intervals for future values, regardless of the method of estimation.

There are many other complications to the regression problem that can be handled by bootstrapping. These include the problem of heteroscedasticity of the variance of the error term (variance is dependent on the values of the predictor variables), nonlinearity in the model terms, and bias adjustment when transformation of variables is used.

For a bootstrap-type approach to the problem of retransformation bias, see Duan (1983). Bootstrap approaches to the problem of heteroscedasticity are covered in Carroll and Ruppert (1988).

An application of bootstrapping residuals for a nonlinear regression problem is given in Shimbukuro, Lazar, Dyson, and Chernick (1984) and will be discussed later. When procedures other than least squares are used, confidence intervals and prediction intervals are still available by bootstrapping.

Two editions of a book by Miller (1986, 1997) deal with linear models. They are very excellent references for understanding the importance of modeling assumptions. They also demonstrate when and why the methods are robust to departures from the basic assumptions. These texts also point out when robust statistical procedures and bootstrap procedures are more appropriate.

4.1.3. Should I Bootstrap the Residuals from the Fit?

From Efron (1979a, Sec. 7), the bootstrap estimate of the covariance matrix for the coefficients in a linear regression model is shown to be

$$\hat{\Sigma} = \hat{\sigma}^2 \left(\sum_{i-1}^{n} c_i c_i \right)^{-1}$$

where

$$\hat{\sigma}^2 = \frac{1}{n} \sum_{i=1}^{n} \hat{\varepsilon}_i^2$$

The model is given by $y_i = c_i \beta + \varepsilon_i$ for $i = 1, 2, \ldots, n$ and $\hat{\varepsilon}_i$ is the residual estimate obtained by least squares. The only difference between this estimate and the standard one from the Gauss–Markov theory is the use of n in the denominator of the estimate for σ^2.

The standard theory would use $n - p$, where p is the number of covariates in the model (i.e., the dimension of the vector β of slopes for predictor variables). So we see that at least when the linear least squares model is an appropriate method, bootstrapping the residuals gives nearly the same answer as the Gauss–Markov theory when n is large. Of course, in such a case, we do not need to bootstrap since we already have an adequate model.

It is important to ask how well this approach to bootstrapping residuals works when there is not an adequate theory for estimating the covariance matrix for the regression parameters. There are many situations involving violations of the standard assumptions that we would like to consider: (1)

heteroscedasticity in the residual variance, (2) correlation structure in the residuals, (3) nonlinear models, (4) non-Gaussian error distributions, and (5) more complex econometric and time series models.

Unfortunately, the theory has not quite reached the level of maturity to give complete answers in these cases. There are still many open research questions to be answered. In this section and in Section 4.2, we will try to give partial answers to (1) through (4) with (5) being deferred to Chapter 5, where we cover time series methods.

A second approach to bootstrapping in a regression problem is to bootstrap the entire vector

$$\mathbf{Z}_i = \begin{pmatrix} y_i \\ \mathbf{c}_i \end{pmatrix}$$

which is a $(p + 1)$-dimensional vector of the response variable and the covariate values. A bootstrap sample is obtained by choosing integers at random with replacement from the set $1, 2, 3, \ldots, n$ until n integers have been chosen. If, on the first selection, say, integer j is chosen, then the bootstrap observation $Z_i^* = Z_j$.

After a bootstrap sample has been chosen, the regression model is fit to the bootstrap samples producing an estimate $\boldsymbol{\beta}$. By repeating this B times, we get $\beta_1^*, \beta_2^*, \ldots, \beta_B^*$ the bootstrap sample estimates of β. The usual sample estimates of variance and covariance can then be applied to $\beta_1^*, \beta_2^*, \ldots, \beta_B^*$.

Efron and Tibshirani (1986) claim that the two approaches are asymptotically equivalent (presumably when the covariates are assumed to be chosen from a probability distribution) but can perform differently in small sample situations.

The latter method does not take full advantage of the special structure of the regression problem. Whereas bootstrapping the residuals leads to the estimates $\hat{\Sigma}$ and $\hat{\sigma}^2$ as defined earlier when $B \to \infty$, this latter procedure does not.

The advantage is that it provides better estimates of the variability in the regression parameters when the model is not correct. We recommend it over bootstrapping the residuals when there is (1) heteroscedasticity in the residual variance, (2) correlation structure in the residuals, or (3) if we suspect that there may be other important parameters missing from the model.

Wu (1986) discusses the use of a jackknife approach in regression analysis, which he views to be superior to the bootstrap approaches that we have mentioned. His approach works particularly well in the case of heteroscedasticity of residual variances.

There are several discussants to Wu's paper. Some strongly support the bootstrap approach and point out modifications for heteroscedastic models. Wu claims that even such modifications to the bootstrap will not work for nonlinear and binary regression problems. The issues are far from settled.

The two bootstrap methods described in this section apply equally to nonlinear (homoscedastic, i.e., constant variance) models as well as the linear (homoscedastic) models. In the next section, we will give some examples of nonlinear models. We will then consider a particular experiment where we bootstrap the residuals.

4.2. NONLINEAR MODELS

The theory of nonlinear regression models advanced greatly in the 1970s and 1980s. Much of this development has been well documented in recent textbooks devoted strictly to nonlinear models. Two such books are Bates and Watts (1988) and Gallant (1987).

The nonlinear models can be broken up into two categories. In the first category, local linear approximations can be made using a Taylor series, for example. When this can be done, approximate confidence or prediction intervals can be generated based on asymptotic theory.

Much of this theory is covered in Gallant (1987). In the aerospace industry, there has been great success applying local linearization methods in the construction of Kalman filters for missiles, satellites, and other orbiting objects.

The second category is the highly nonlinear model for which the linear approximation will not work. Bates and Watts (1988) provide methods for diagnosing the severity of the nonlinearity.

The bootstrap method can be applied to any type of nonlinear model. The two methods as described in Efron (1982a) can be applied to fairly general problems. To bootstrap, we do not need to have a differentiable functional form. The nonlinear model could even be a computer algorithm rather than an analytical expression. We do not need to restrict the residual variance to have a Gaussian distribution. The only requirements are that the residuals should be independent and identically distributed (exchangeability may be sufficient) and their distribution should have a finite variance.

The distribution of the residuals should not change as the predictor variables are changed. This requirement imposes homoscedasticity on the residual variance. For models with heteroscedastic variance, modifications to the bootstrap are available. We shall not discuss these modifications here. To learn more about it, look at the discussion in Wu (1986).

4.2.1. Examples of Nonlinear Models

In Section 4.2.2, we discuss a quasi-optical experiment that was performed to determine the accuracy of a new measurement technique for the estimation of optical properties of materials used to transmit and/or receive millimeter wavelength signals. This experiment was conducted at the Aerospace Laboratory.

As a statistician in the engineering group, I was asked to determine the standard errors of their estimates. The statistical model was nonlinear and I chose to use the bootstrap to estimate the standard error. Details on the model and the results of the analysis are given in Section 4.2.2.

Many problems that arise in practice can be solved by approximate models that are linear in the parameters (remember in statistical models the distinction between linear and nonlinear is in the parameters and not in the predictor variables). The scope of applicability of linear models can, at times, be extended to other cases by including transformations of the variables.

However, there are limits to what can adequately be approximated by linear models. In many practical scientific endeavors, the model may arise from a solution to a differential equation. A nonlinear model that could arise as the solution of a simple differential equation might be the function

$$f(x, \boldsymbol{\sigma}) = \sigma_1 + \sigma_2 \exp(x\sigma_3)$$

where x is a predictor variable and

$$\boldsymbol{\sigma} = \begin{bmatrix} \sigma_1 \\ \sigma_2 \\ \sigma_3 \end{bmatrix}$$

is a three-dimensional parameter vector.

A common problem in time series analysis is the so-called harmonic regression problem. We may know that the response function is periodic or the sum of a few periodic functions, but we do not know the amplitude or the frequency of the periodic components.

Here, it is the fact that the frequencies are included among the unknown parameters that makes the model nonlinear. The simple case of a single periodic function can be described by the following function:

$$f(t, \boldsymbol{\varphi}) = \varphi_0 + \varphi_1 \sin(\varphi_2 t + \varphi_3)$$

where t is the time since a specific epoch and

$$\boldsymbol{\varphi} = \begin{bmatrix} \varphi_0 \\ \varphi_1 \\ \varphi_2 \\ \varphi_3 \end{bmatrix}$$

is a vector of unknown parameters. The parameters φ_1, φ_2, and φ_3 all have physical interpretations. φ_1 is called the amplitude, φ_2 is the frequency, and φ_3 is the phase delay.

Because of the trigonometric identity

$$\sin(A + B) = \sin A \cos B + \cos A \sin B$$

we can reexpress $\varphi_1 \sin(\varphi_2 t + \varphi_3)$ as $\varphi_1 \cos \varphi_3 \sin \varphi_2 t + \varphi_1 \sin \varphi_3 \cos \varphi_2 t$. The problem can then be reparameterized as

$$f(t, A) = A_0 + A_1 \sin A_2 t + A_3 \cos A_2 t$$

where

$$A = \begin{bmatrix} A_0 \\ A_1 \\ A_2 \\ A_3 \end{bmatrix}$$

and $A_0 = \varphi_0$, $A_1 = \varphi_1 \cos \varphi_3$, $A_2 = \varphi_2$, and $A_3 = \varphi_1 \sin \varphi_3$.

This reparameterized form of the model is the form given by Gallant (1987, p. 3) with slightly different notation.

There are many other examples where nonlinear models are solutions to differential equations or systems of differential equations. Even in the case of linear differential equations or systems of linear differential equations, the solutions involve exponential functions (both real and complex valued). The solutions are then real-valued functions that are periodic or exponential or a combination of both.

If constants involved in the differential equation are unknown, then their estimates will be obtained through the solution of a nonlinear model. As a simple example consider the equation

$$\frac{d}{dx} y(x) = -\sigma_1 y(x)$$

subject to the initial condition $y(0) = 1$. The solution is then

$$y(x) = e^{-\varphi_1 x}$$

Since φ_1 is an unknown parameter, the function $y(x)$ is nonlinear in φ_1.

For a commonly used linear system of differential equations whose solution involves a nonlinear model, see Gallant (1987 pp. 5–8). Such systems of differential equations arise in compartmental models commonly used in chemical kinetics problems.

4.2.2. A Quasi-optical Experiment

In this experiment, I was asked as consulting statistician to determine estimates of two parameters that were of interest to the experimenters. More importantly, they needed a "good" estimate of the standard errors of these estimates since they were proposing a new measurement technique that they believed would be more accurate than previous methods.

Since the model was nonlinear and I was given a computer program rather than an analytic expression, I chose to bootstrap the residuals. The results were published in Shimabukuro, Lazar, Dyson, and Chernick (1984).

The experimenters were interested in the relative permittivity and the loss tangent (two material properties related to the transmission of signals at millimeter wavelengths through a dielectric slab). The experimental setup is graphically depicted in Fig. 4.1. Measurements are taken to compute $|T|^2$, where T is a complex number called the transmission coefficient. An expression for T is given by

$$T = \frac{(1 - r^2)e^{-(\beta_1 - \beta_0)di}}{1 - r^2 e^{-2\beta_1 di}}$$

where

$$\beta_1 = \frac{2\pi}{\lambda_0}\sqrt{\varepsilon_1/\varepsilon_0 - \sin^2 \varphi}$$

$$\beta_0 = \frac{2\pi}{\lambda_0}\cos \varphi$$

$$\varepsilon_1 = \varepsilon_r \varepsilon_0 \left(1 - \frac{i\sigma}{\omega \varepsilon_r \varepsilon_0}\right)$$

Figure 4.1 Photograph of experimental setup. The dielectric sample is mounted in the Teflon holder. (From Shimabukuro, Lazar, Dyson, and Chernick, 1984.)

and

ε_0 = permittivity of free space

ε_r = relative permittivity

σ = conductivity

λ_0 = free-space wavelength

$\dfrac{\sigma}{\omega \varepsilon_r \varepsilon_0}$ = $\tan \delta$ = loss tangent

d = thickness of the slab

r = reflection coefficient of a plane wave incident on a dielectric boundary

ω = free-space frequency

i = $\sqrt{-1}$

For more details on the various conditions of the experiment see Shimbukuro, Lazar, Dyson, and Chernick (1984).

We applied the bootstrap to the residuals using the nonlinear model

$$y_i = g_i(\mathbf{v}) + \varepsilon_i \quad \text{for } i = 1, 2, \ldots, N$$

where y_i is the power transmission measurement at incident angle φ_i with $\varphi_i = i - 1$ degrees. The nonlinear function $g_i(\mathbf{v})$ is $|T|^2$ and $\mathbf{v}$ is a vector of two parameters — ε_r (relative permittivity) and $\tan \delta$ (loss tangent).

For simplicity, the wavelength λ, the slab thickness d, and the angle of incidence φ_i are all assumed to be known for each observation. The experimenters believe that measurement error in these variables would be relatively small and have little effect on the parameter estimates. Some checking of these assumptions was made.

For most of the materials, 51 observations were taken. We chose to do 20 bootstrap replications for each model. Results were given for eight materials and are shown in Table 4.1.

The actual least squares fit to the eight materials is shown in Fig. 4.2. We notice that the fit is generally better at the higher incidence angles.

This suggests a violation of the assumption of independent and identically distributed residuals. There may be a bias at the low incidence angles indicative of either model inadequacy or poorer measurements.

Looking back on the experiment there are several possible ways we might have improved the bootstrap procedure. Since bootstrapping residuals is more sensitive to the correctness of the model, it may have been better to bootstrap the vector.

Recent advances in bootstrapping in heteroscedastic models may also have helped. A rule of thumb for estimating standard errors is to take 100 to 200 bootstrap replications, whereas we did only 20 replications in this research.

From a data analytic point of view, it may have been helpful to delete the low-angle observations and see the effect on the fit. We might then have

Table 4.1   Estimates of Permittivities and Loss Tangents ($f = 93.788$ GHz)

Material	Least Squares Estimate		Bootstrap Estimates with Standard Error	
	ε_r	tan δ	ε_r	tan δ
Teflon	2.065	0.0002	2.065 ± 0.004	0.00021 ± 0.00003
Rexolite	2.556	0.0003	2.556 ± 0.005	0.00026 ± 0.00006
TPX	2.150	0.0010	2.149 ± 0.005	0.0009 ± 0.0001
Herasil	3.510	0.0010	3.511 ± 0.005	0.0010 ± 0.0001
(fused quartz)				
36D	2.485	0.0012	2.487 ± 0.008	0.0011 ± 0.0002
	(2.45)	(<0.0007)		
36DA	3.980	0.0012	3.980 ± 0.009	0.0014 ± 0.0001
	(3.7)	(<0.0007)		
36DK	5.685	0.0040	5.685 ± 0.009	0.0042 ± 0.0001
	(5.4)	(<0.0008)		
36DS	1.765	0.0042	1.766 ± 0.006	0.0041 ± 0.0001
	(1.9)	(<0.001)		

Source: Shimabukuro, Lazar, Dyson, and Chernick (1984).

decided to fit the parameters and bootstrap only for angles greater than, say, 15 degrees.

By bootstrapping the residuals, the large residuals at the low angles would be added at the higher angles for some of the bootstrap samples. We believe that this would tend to increase the variability in the parameter estimates of the bootstrap sample and hence lead to an overestimate of their standard errors.

Since the estimated standard errors were judged to be good enough by the experimenters, we felt that our approach was adequate. The difficulty with the residual assumptions was recognized at the time.

4.3. NONPARAMETRIC MODELS

Given a vector $\mathbf{X}$, the regression function $E(y|X)$ is often a smooth function in $\mathbf{X}$. In Sections 4.1 and 4.2, we considered specific linear and nonlinear forms for the regression function. Nonparametric regression is an approach that allows for more general smooth functions as possibilities for the regression function. The nonparametric regression model for an observed data set $(y_i, \mathbf{x}_i)$ for $1 \leqslant i \leqslant n$ is

$$y_i = g(\mathbf{x}_i) + \varepsilon_i, \qquad 1 \leqslant i \leqslant n$$

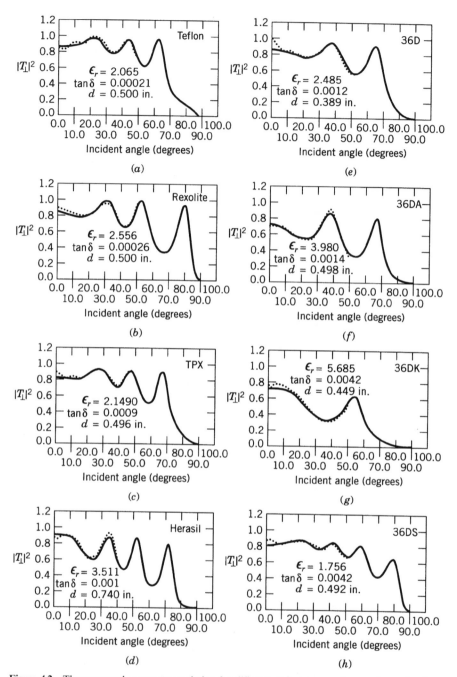

Figure 4.2 The measured power transmission for different dielectric samples are shown by dots. The line curves are the calculated $|T_\perp|^2$ using the best-fit estimates of ε_r and tan δ. (From Shimabukuro, Lazar, Dyson, and Chernick, 1984.)

where $g(\mathbf{x}) = E(y|\mathbf{x})$ is the function we wish to estimate. We assume that the ε_i values are independent and identically distributed with mean zero and variance σ^2.

In the regression model, $\mathbf{x}$ is assumed to be specified (i.e., numerical values are known exactly), as in a designed experiment. One approach to the estimation of the function g is kernel smoothing (see Hardle, 1990a, b or Hall, 1992a, pp. 257–269). The bootstrap is used to help determine the degree of smoothing (i.e., determine the trade-off between variance and bias analogous to its use in nonparametric density estimation).

Cox's proportional hazards model is a standard regression method for dealing with censored data (see Cox, 1972). The hazard function $h(t|\mathbf{x})$ is the derivative of $-\ln[S(t|\mathbf{x})]$ where $S(t|\mathbf{x})$ is the survival function (i.e. the probability of surviving t or more time units given predictor variables $\mathbf{x}$). In Cox'x model $h(t|\mathbf{x}) = h_0(t)e^{\beta \mathbf{x}}$, where $h_0(t)$ is an arbitrary unspecified function assumed to depend solely on t.

Through the use of the "partial likelihood" function, the regression parameters β can be estimated independently of the function $h_0(t)$. Because of the form of $h(t|\mathbf{x})$, the method is sometimes referred to as semiparametric.

Efron and Tibshirani (1986) apply the bootstrap to leukemia data for mice in order to assess the effectiveness of a treatment. They use a Cox proportional hazards model. See their article for more details.

Without going into the details, we mention projection pursuit regression and alternating conditional expectation (ACE) as two other "nonparametric" regression techniques that have been studied recently. Efron and Tibshirani (1986) provide examples of applications of both methods and show how the bootstrap can be applied when using these techniques.

The interested reader can consult Friedman and Stuetzle (1981) for the original source of project pursuit. The original work describing ACE is Brieman and Friedman (1985).

Briefly, projection pursuit searches for linear combinations of the predictor variables and takes smooth functions of those linear combinations to form the prediction equation. ACE generalizes the Box–Cox power transformation regression model by transforming the response variable with an unspecified smooth function as opposed to a simple power transformation.

4.4. HISTORICAL NOTES

Although regression analysis is one of the most widely used statistical techniques, application of the bootstrap to regression problems has only appeared fairly recently. The many fine books on regression analysis including Draper and Smith (1981) for linear regression and Gallant (1987) and Bates and Watts (1988) do not mention or pay much attention to bootstrap methods. A recent exception is Sen and Srivastava (1990).

Draper and Smith (1998) also incorporate a discussion of the bootstrap.

Early discussion of the two methods of bootstrapping in the nonlinear regression model with homoscedastic errors can be found in Efron (1982a). Carroll, Ruppert, and Stefanski (1995) deal with the nonlinear calibration problem (measurement error models and other nonlinear regression problems, pp. 273–279, Appendix A.6) Order determination problems in regression models are treated in Miller (1990).

Efron and Tibshirani (1986) provide a variety of interesting applications and some insightful discussion of bootstrap applications in regression problems. They go on to discuss nonparametric regression applications including projection pursuit regression and methods for deciding on transformations for the response variable such as the alternating conditional expectation (ACE) method of Brieman and Friedman (1985). Texts devoted to nonparametric regression and smoothing methods include Hardle (1990a, b), Hart (1997), and Simonoff (1996). Belsley, Kuh, and Welsch (1980) deal with multicollinearity and other regression diagnostics.

Bootstrapping the residuals is an approach that also can be applied to time series models. We shall discuss time series applications in the next chapter. An example of a time series application to the famous Wolfer sunspot numbers is given in Efron and Tibshirani (1986, p. 65).

Shimabukuro, Lazar, Dyson, and Chernick (1984) present an early example of a practical application of a nonlinear regression problem. The first major study of the bootstrap as applied to the problem of estimating the standard errors of the regression coefficients by constrained least squares with an unknown, but estimated, residual covariance matrix can be found in Freedman and Peters (1984a). Similar analysis for econometric models can be found in Freedman and Peters (1984b).

Peters and Freedman (1984) also deal with issues regarding bootstrapping in regression problems. Their study is very interesting because it shows that the conventional asymptotic formulas, which are correct for very large samples, do not work well in small to moderate sample size problems. They show that these standard errors can be too small by a factor of nearly three! On the other hand, the bootstrap method gives accurate answers. The motivating example is an econometric equation for the energy demand by industry.

In Freedman and Peters (1984b) the bootstrap is applied to a more complex econometric model. Here the authors show that the three-stage least squares estimates and the conventional estimated standard errors of the coefficients are good. However, conventional prediction intervals based on the model are too small due to forecast bias and underestimation of the forecast variance.

The bootstrap approach given by Freedman and Peters (1984b) seems to provide better prediction intervals in their example. The authors point out that there is unfortunately no good rule of thumb to apply to determine when the conventional formulas will work or when it may be necessary to resort to the bootstrap. They suggest that the development of such a rule of thumb could be the result of additional research. Even the bootstrap procedure has problems in this context.

Theoretical work on the use of bootstrap in regression is given in Freedman (1981), Bickel and Freedman (1983), Weber (1984), Wu (1986) and Shao (1988a, b). Another application to an econometric model appears in Daggett and Freedman (1985).

Theoretical work related to robust regression can be found in Shorack (1982). Rouseeuw (1984) applies the bootstrap to the least median of squares algorithm. Efron (1992a) discusses the application of the bootstrap to estimating regression percentiles.

Jeong and Maddala (1993) review various developments in resampling tests for econometric models. Hall (1989c) shows that the bootstrap applied to regression problems can give unusually accurate confidence intervals.

Various recent regression applications include Breiman (1992) for model selection related to x-fixed prediction, Brownstone (1992) regarding admissibility of linear model selection procedures, Bollen and Stine (1993) regarding fitting of structural equation models, Cao-Abad (1991) regarding rates of convergence for a type of bootstrap called "wild bootstrap," which is useful in nonparametric regression [see also Mammen (1993) who applies the wild bootstrap in linear models], DeAngelis, Hall, and Young (1992a) related to L^1 regression, Lahiri (1992c) for M-estimation in multiple linear regression problems, Ditka (1990) for nearest neighbor regression, and Green, Hahn, and Rocke (1987) for an economic application to estimation of elasticities.

Wu (1986) gives a detailed theoretical treatment of jackknife methods applied to regression problems. He deals mainly with the problem of heteroscedastic errors. He is openly critical of the blind application of bootstrap methods and illustrates that certain bootstrap approaches will give incorrect results when applied to data for which heteroscedastic models are appropriate. A number of the discussants, including Beran, Efron, Freedman, and Tibshirani, defend the appropriate use of the "right" bootstrap in this context. The issue is a complex one that even today is not completely settled.

It is fair to say that Jeff Wu's criticism of the bootstrap in regression problems was a reaction to the "euphoria" expressed for the bootstrap in some of the earlier works such as Efron and Gong (1983, Sec. 1) or Diaconis and Efron (1983).

Although enthusiasm for the bootstrap approach is justified, some statements could leave naive users of statistical methods with the idea that it is easy just to apply the bootstrap to any problem they might have. I think every bootstrap researcher would agree that careful analysis of the problem is a necessary step in any applied problem and that if bootstrap methods are appropriate for the application, one must be careful to choose the "right" bootstrap method from the many possible bootstraps.

Stine (1985) deals with bootstrapping for prediction intervals and Bai and Olshen as discussants to Hall (1988b) provide some elementary asymptotic theory for prediction intervals. See Beran (1992) for further theoretical developments on prediction intervals. Olshen, Biden, Wyatt, and Sutherland (1989) provide a very interesting practical application in gait analysis.

A theoretical treatment of nonparametric kernel methods in regression problems is given in Hall (1992a). His development is based on asymptotic expansions (i.e., Edgeworth expansions). Other key articles related to bootstrap applications in nonparametric regression include Hardle and Bowman (1988) and Hardle and Marron (1991).

The reader may first want to consult Silverman (1986) for a treatment of kernel density methods and some applications of the bootstrap in density estimation. Devorye and Gyorfi (1985) also deal with kernel density methods as does Hand (1982) and for multivariate densities, see Scott (1995). Hardle (1990a) provides a recent account of nonparametric regression techniques.

Hayes, Perl, and Efron (1989) have extended bootstrap methods to the case of several unrelated samples with application to estimating contrasts in particle physics problems. Hastie and Tibshirani (1990) treat a general class of models called generalized additive models. These include both the linear models discussed in this chapter and the generalized linear models as special cases. It can be viewed as a form of curve fitting but is not quite as general as nonparametric regression.

Bailer and Oris (1994) provide regression examples for toxicity testing and compare bootstrap methods with likelihood methods for Poisson regression models (generalized linear models). One of their examples was used for practical number 6, pages 383–384 of Davison and Hinkley (1997).

CHAPTER 5

Forecasting and Time
Series Analysis

5.1. METHODS OF FORECASTING

One of the most common problems in the "real world" is forecasting. We try to forecast tomorrow's weather or when the next big earthquake will hit. When historical data are available and models can be developed that fit the historical data well, we may be able to produce accurate forecasts. For certain problems (e.g., earthquake predictions or the Dow Jones Industrial Average) the lack of a good statistical model makes forecasting problematic (i.e., no better than crystal ball gazing).

Among the most commonly used forecasting techniques are exponential smoothing and autoregressive integrated moving average (ARIMA) modeling. The ARIMA models are often referred to as the Box–Jenkins models after George Box and Gwilym Jenkins who popularized the approach in Box and Jenkins (1970, 1976). The autoregressive models that are a subset of the ARIMA models actually go back to Yule (1927).

Exponential smoothing is an approach that provides forecasts of future values using exponentially decreasing weights on the past values. The weights are determined by smoothing constants that are estimated from the data. The simplest form — single exponential smoothing — is also a special case of the ARIMA models, namely, the IMA (1, 1) model. The smoothing constant can be determined from the moving average parameter of the IMA (1, 1) model.

5.2. TIME SERIES MODELS

ARIMA models are attractive because they provide good empirical approximations to a large class of time series. There is a body of statistical theory showing that "most" stationary stochastic processes can be well approximated by high-order autoregressive processes.

The term stationary stochastic process generally means strictly stationary. A stochastic process is said to be strictly stationary if the joint probability distribution of k consecutive observations does not depend on the time t for all $k = 1, 2, \ldots, \infty$. Informally, this means that if we are looking at the first k observations in a time series, the statistical properties would be the same if we look at any other set of k consecutive observations.

A weaker form of stationarity is second-order or weak stationarity. Second-order stationarity requires only that the second-order moments exist and that the first- and second-order moments, the mean, and the autocorrelation function, respectively, do not depend on time (i.e., are constant over time).

Strict stationarity implies weak stationarity but there can be second-order stationary processes that are not strictly stationary. Second-order stationary Gaussian processes are strictly stationary because their joint distributions depend only on their second moments.

Box and Jenkins used the mixed autoregressive moving average model to provide a parsimonious representation for these high-order autoregressive processes (i.e., by including just a few moving average terms an equivalent model is found with only a small number of parameters to estimate). To generalize this further to handle trends and seasonal variations (i.e., non-stationarity), Box and Jenkins (1976) include differencing and seasonal differences of the series. Using mathematical operator notation, let

$$W_t = \Delta^d Y_t$$

where Y_t is the original observation at time t and the operation Δ^d applies the difference operation, Δ, d times, where Δ is defined by $\Delta y_t = y_t - y_{t-1}$. So

$$\Delta^2 y_t = \Delta(y_t - y_{t-1}) = \Delta y_t - \Delta y_{t-1} = (y_t - y_{t-1}) - (y_{t-1} - y_{t-2})$$
$$= y_t - 2y_{t-1} + y_{t-2}$$

In general,

$$\Delta^d y_t = \Delta^{d-1}(\Delta y_t) = \Delta^{d-1}(y_t - y_{t-1}) = \Delta^{d-1} y_t - \Delta^{d-1} y_{t-1}$$

After differencing the time series, W_t is a stationary ARMA (p, q) process given by

$$W_t = b_1 W_{t-1} + b_2 W_{t-2} + \cdots + b_p W_{t-p} + e_t + a_0 e_{t-1} + \cdots + a_q e_{t-q}$$

where $e_t, e_{t-1}, \ldots, e_{t-q}$ are the assumed random innovations and $W_{t-1}, W_{t-2}, \ldots, W_{t-p}$ are past values of the dth difference of the Y_t series.

These ARIMA models can handle polynomial trends in the time series. Additional seasonal components can be handled by seasonal differences [see Box and Jenkins (1976) for details].

Although the Box–Jenkins models cover a large class of time series and provide very useful forecasts and prediction intervals, they have drawbacks in some cases. The models are linear and the least squares or maximum likelihood parameter estimates are good only if the innovation series e_t is nearly Gaussian.

If the innovation series e_t has heavy tails or there are a few spurious observations in the data, the estimates can be distorted and the prediction intervals are not valid. In fact, the Box–Jenkins methodology for choosing the order of the model (i.e., deciding on the values for p, d, and q) will not work if outliers are present. This is due to the fact that estimates for the autocorrelation and partial autocorrelation functions are very sensitive to outliers (e.g., see, Chernick, Downing, and Pike, 1982; or Martin, 1980).

One approach that overcomes the difficulty is to detect and remove the outliers and then fit the Box–Jenkins model with some missing observations. Another approach is to use robust estimation procedures for parameters (see Rousseeuw and Leroy, 1987).

In the past decade there have also been a number of interesting theoretical developments in bilinear and other nonlinear time series models, which may help to extend the applicability of statistical time series modeling (see Tong, 1983, 1990).

Even if an ARIMA model is appropriate and the innovations e_t are uncorrelated, but not Gaussian, it may be appropriate to bootstrap the residuals to obtain appropriate standard errors for the model parameters and the predictions. Bootstrap prediction intervals may also be appropriate.

The approach is the same as we have discussed in Chapter 4 on regression analysis. The confidence interval methods of Chapter 3 may be appropriate for the prediction intervals. We shall discuss this further in the next section.

5.3. WHEN DOES BOOTSTRAPPING HELP WITH PREDICTION INTERVALS?

Some results are available on the practical application of the bootstrap to time series models. These results apply to stationary autoregressive (AR) processes, a subset of the stationary autoregressive moving average (ARMA) models discussed in the previous section.

To illustrate how the bootstrap can be applied to an autoregressive model, we will illustrate the approach with a simple first-order autoregressive process. This model is sufficient to illustrate the key points. For the first-order autoregression (AR(1) model), the model is given by

$$y_t = b_1 y_{t-1} + e_t$$

where y_t is the observation at time t (possibly centered to have zero mean) and e_t are the innovations.

If the average of the observed series is not zero, a sample estimate of the mean is subtracted from each observation in order to center the data. In practice, if the original series appears to be nonstationary, differencing methods or other forms of trend removal would be applied first.

For Gaussian processes, least squares or maximum likelihood estimates for b_1 are computed along with standard errors for the estimates. If y_{1m} is the last observation, a one-step-ahead prediction is obtained at $t_m + 1$ using $\hat{b}_1 y_{t_m}$ as the prediction where $\hat{b}_1$ is the estimate of b_1. Statistical software packages (e.g., SAS/ETS, BMDP, and IMSL) provide such estimates of parameters and also produce forecast intervals.

These procedures work well when the e_t have approximately a Gaussian distribution with mean zero. Stine (1987) provides forecasts and prediction intervals with the classical Gaussian model but uses a bootstrap approach. He shows that, although the bootstrap is not as efficient as the classical estimate when the Gaussian approximation is valid, it provides much better prediction intervals for non-Gaussian cases.

In order to apply the bootstrap to the AR(1) model, we need to generate a bootstrap sample. First we need an estimate $\hat{b}_1$. We may take the Gaussian maximum likelihood estimate generated by a software program such as Proc ARIMA from SAS. We then generate the estimated residuals, namely,

$$\hat{e}_t = y_t - \hat{b}y_{t-1} \quad \text{for } t = 2, 3, \ldots, t_m$$

Note that we cannot compute a residual $\hat{e}_1$ since y_0 not available to us. A bootstrap sample, $y_1^*, y_2^*, \ldots, y_{t_m}^*$ is then generated by bootstrapping the residuals.

We simply generate $e_2^*, e_3^*, \ldots, e_{t_m}^*$ by sampling with replacement from $\hat{e}_2$ $\hat{e}_3, \ldots, \hat{e}_{t_m}$ and defining by recursion

$$y_2^* = \hat{b}_1 y_1^* + e_2^*, \qquad y_3^* = \hat{b}_1 y_2^* + e_3^*, \ldots, y_{t_m}^* = b_1 y_{t_m-1}^* + e_{t_m-1}^*$$

Efron and Tibshirani (1986) take $y_1^* = y_1$ for each bootstrap sample. With autoregressive processes, since we have a first time point, which we denote as $t = 1$, we need initial values. In the AR(1) example, we see that we need a single initial value to start the process. In this case we let $y_1^* = y_1$.

In general, for the pth-order autoregression, we will need p initial values. Stine (1987) and Thombs and Schucany (1990) provide alternative methods for obtaining starting values for the bootstrap samples.

Now, for each bootstrap sample, an estimate $\hat{b}_1^*$ is obtained by applying the estimation procedure to $y_1^*, y_2^*, \ldots, y_{t_m}^*$. Efron and Tibshirani (1986) illustrate this on the Wolfer sunspot data. They obtain the standard errors for $\hat{b}_1$ by this procedure. They then go on to fit an AR(2) (second-order autoregressive model) to the sunspot data and obtain bootstrap estimates of the standard errors for the two parameters in the AR(2) model. They did not go on to consider prediction intervals.

For the Gaussian case, the theory has been developed to obtain the minimum mean square error predictions based on "known" autoregressive parameters. Formulas for the predictions and their mean square errors can be found in Box and Jenkins (1976) or Fuller (1976). Stine (1987) shows the well-known result that when the autoregressive parameter b_1 is replaced by the estimate $\hat{b}_1$ in the forecasting equations the prediction mean square error increases.

Stine (1987) provides a Taylor series expansion to estimate the mean square error of the prediction that works well for Gaussian data. The bootstrap estimates of mean square error that he obtains are biased but his bootstrap approach does provide good prediction intervals. We shall describe this approach, which we recommend when the residuals do not fit well to the Gaussian model.

Stine (1987) assumes that the innovations have a continuous and strictly increasing distribution with finite moments. He also assumes that the distribution is symmetric about zero. The key difference between Stine's approach and that of Efron and Tibshirani is the introduction of the symmetric error distribution. Instead of sampling with replacement from the empirical distribution for the estimated residuals (the method of Efron and Tibshirani previously described) Stine does the following.

Let

$$F_T(x) = \tfrac{1}{2} + (L(x)/[2(T-p)]), \; x \geq 0, \qquad t = p+1, \ldots, T$$
$$= 1 - F_T(-x), \; x < 0$$

where $L(x)$ = number of t such that $k|\hat{\varepsilon}_t| \leq x$, and

$$k = [(T-p)/(T-2p)]^{1/2}$$

This choice of F_T produces bootstrap residuals that are symmetric about zero and have a variance that is the same as the original set of residuals.

A bootstrap approximation to the prediction error distribution is easily obtained given the bootstrap estimates of the autoregressive parameters and the bootstrap observations $y_1^*, y_2^*, \ldots, y_{t_m}^*$. The prediction formulas are used to obtain a bootstrap prediction $\hat{y}_{t_m+f}^*$ for the time point $t_m + f$, f time steps in the future. The variable $\hat{y}_{t_m+f}^* - \hat{y}_{t_m+f}$ provides the bootstrap sample estimate of prediction error f steps ahead, where $\hat{y}_{t_m+f}$ is the original prediction based on the original estimates of the autoregressive parameters and the observations $y_1, y_2, \ldots, y_{t_m}$. Actually Stine uses a more sophisticated approach based on the structure of the forecast equation [see Stine (1987) for details].

Another difference between Stine's approach and that of Efron and Tibshirani is that Efron and Tibshirani fix the first p values of the process in generating the bootstrap sample whereas Stine chooses a block of p consecutive observations at random to initiate the bootstrap sample.

In practice, we will know the last p observations when making future predictions. Autoregressive forecasts for $1, 2, \ldots, f$ steps ahead depend only on the autoregressive parameters and the last p observations. Consequently, it makes sense to condition on the last p observations when generating the bootstrap predictions.

Thombs and Schucany (1990) use a time-reversal property for autoregressive processes to fix the last p observations and generate bootstrap samples for the earlier observations. They apply the backward representation (Box and Jenkins, 1976, pp. 197–200) to express values of the process at time t as a function of future values. This representation is based on generating the process backward in time, which is precisely what we want to do with the bootstrap samples. The correlation structure for the reversed process is the same as for the forward process.

For Gaussian processes this means that the two series are distributionally equivalent. Weiss (1975) has shown that for linear processes (including autoregressions) the time-reversed version is distributionally equivalent to the original only if the process is Gaussian.

Chernick, Daley, and Littlejohn (1988) provide an example of a first-order autoregression with exponential marginal distributions whose reversed version also has exponential marginals, is first-order Markov, and has a special structure. The process is not time reversible (in the strict sense where reversibility means distributional equivalence of the two stochastic processes, original and time reversed) as can be seen by looking at sample paths.

Thombs and Schucany (1990) also present simulation results that show that their method has promise. They did not use the symmetrized distribution for the residuals. In small samples, they concede that some refinements such as the bias-corrected percentile method might be helpful.

Unfortunately, it is still too early to recommend a particular bootstrap procedure as a "best" approach to bootstrapping time series even for generating prediction intervals for autoregressive time series. The method of Stine (1987) is recommended for use when the distributions are non-Gaussian. For nearly Gaussian time series, the standard methods available in most statistical time series programs are more efficient. Alternative bootstrap approaches to time series problems are described in Sections 5.4 and 5.5.

5.4. MODEL-BASED RESAMPLING VERSUS BLOCK RESAMPLING

The methods described thus far all fall under the category of model-based resampling methods, because the residuals are generated and resampled based on a time series model [i.e., the AR(1) model in the earlier illustration]. Refinements to the above approach are described in Davison and Hinkley (1997, pp. 389–391).

There, they center the residuals by subtracting the average of the residuals. They then use a prescription just as we have described above. However, they point out that the generated series is not stationary. This is due to the initial values. This could be remedied by starting the series in equilibrium or more practically by allowing a "burn-in" period of k observations that are discarded. We choose k so that the series has "reached" stationarity.

To use the model-based approach, we need to know the parameters and the structure of the model and this is not always easy to discern from the data. If we choose an incorrect structure, the resampled series will have a different structure from the original data and hence will have different statistical properties. So if we know we have a stationary series but we do not know the structure, we would like a bootstrap resampling procedure that doesn't depend on this unknown structure.

Bose (1988) showed that if an autoregressive process is a correct model (or for the practitioner at least approximately correct) there is an advantage to using the model-based resampling approach, namely, good higher-order asymptotic properties for a wide variety of statistics that can be derived from the model. On the other hand, we could pay a heavy price, in that the estimates could be biased and/or grossly inaccurate if the model structure is wrong.

A remedy, introduced first by Kunsch (1989), is to resample the time series in blocks (possibly overlapping blocks). For uncorrelated exchangeable sequences, the original nonparametric bootstrap that resamples the individual observations is appropriate. For stationary time series, observations removed sufficiently far in time are uncorrelated (or at least nearly so).

An idea in block resampling is that, for stationary series, individual blocks of observations that are separated far enough in time will be approximately uncorrelated and can be treated as exchangeable. So suppose the time series has length $n = bl$. We can generate b nonoverlapping blocks each of length l.

The key idea that underlies this approach is that if the blocks are sufficiently long, each block preserves, in the resampled series, the dependence present in the original data sequence. The resampling scheme here is to sample with replacement from the b blocks.

There are several variants on this idea. One is to allow the blocks to overlap. This was one of Kunsch's proposals and it allows for more blocks than if they are required not to overlap.

Suppose we take the first block to be (y_1, y_2, y_3, y_4) and the second to (y_2, y_3, y_4, y_5), and so on. The effect of this approach is that the first $l - 1$ observations from the original series appear in fewer blocks than the rest.

Note that observation y_1 appears in only one block and y_2 in only two blocks, and so on. This effect can be overcome by wrapping the data around in a circle (i.e., the last observation in the series is followed again by the first, etc.).

The block resampling approach is currently the subject of much additional research. In addition to the properties described above, it does have some drawbacks. In particular, resampled blocks do not quite mimic the behavior of the time series. They have a tendency to weaken the dependency in the series.

Two methods, postblackening and resampling blocks of blocks, help to remedy this problem. The interested reader should consult Davison and Hinkley (1997 pp. 397–398) for some discussion of these methods.

Another simple way to overcome this difficulty is what is called the stationary bootstrap. The stationary bootstrap is a block resampling scheme, but instead of having fixed-length blocks, the blocks are chosen to have a length determined at random. The block length is a random variable L, where

$$\Pr(L = j) = (1 - p)^{j-1}p, \quad \text{for } j = 1, 2, 3, \ldots, \infty$$

This length distribution is the geometric distribution with parameter p. The mean block length for L is $\lambda = p^{-1}$. We may choose λ as one might choose the length for fixed block length. Since $\lambda = 1/p$, determining λ also determines p. The stationary bootstrap was devised by Politis and Romano (1994a).

It appears that the block resampling method has desirable properties of robustness to model specification in that it applies to a broad class of stationary series. Other variations and some theory related to block resampling can be found in Davison and Hinkley (1997, pp. 401–403 for choice of block length and pp. 405–408 for the underlying theory). Hall (1998) provides an overview of the subject.

Davison and Hinkley (1997) illustrate the application of block resampling using data on the river heights over time for the Rio Negro. A concern of the study was that there is a trend for the heights of the river near Manaus to increase over time due to deforestation. A test for trend was applied and there is some evidence that a trend may be present but the test was inconclusive. The trend test was based on a test statistic that is a linear combination of the observations namely,

$$T = \Sigma a_i Y_i$$

where Y_i for $i = 1, 2, 3, \ldots, n$ is the sequence of measurements of the river level at Manaus and

$$a_i = \{(-1)[1 - ((i - 1)/(n + 1))]\}^{1/2} - \{i[1 - (i/(n + 1)]\}^{1/2}$$
$$\text{for } i = 1, 2, 3, \ldots, n$$

The test based on this statistic is optimal for detecting a monotonic trend when the observations are independent (and identically distributed under the null hypothesis). However, the time series data show clear autocorrelation. A smoothed version of the Rio Negro (a centered ten-year moving average) is shown in Fig. 5.1 taken from Davison and Hinkley (1997).

The test statistic T above is still used and its value in the example turns out to be 7.908. But is this significantly large based on the null hypothesis? Instead of using the distribution of the test statistic under the null hypothesis, Davison and Hinkley choose to estimate its null distribution using block resampling. This is more realistic for the Rio Negro data. They compare the stationary

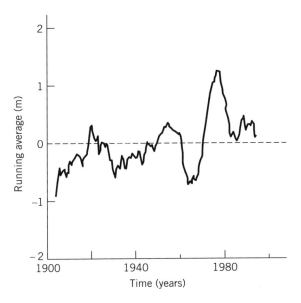

Figure 5.1 Ten-year running average of Manaus data. [From Davidson and Hinkley (1997, Fig. 8.9, p. 403), with permission from Cambridge University Press.]

bootstrap with a fixed block length method. The purpose is to use the bootstrap to estimate the variance of T under the null hypothesis that the series is stationary but correlated. The asymptotic normality of T is used to make the inference. Many estimates were obtained by the two methods depending on the chosen block length (in the fixed case) or the average block length in the case of the stationary bootstrap. The bottom line is that the variance for T is about 25 based on the first 120 time points but the lowest reasonable estimate for the variance based on the full series is approximately 45. This gives a p-value of 0.12 for the test statistic, indicating a lack of strong evidence for a trend.

5.5. FREQUENCY DOMAIN APPROACHES

As we have mentioned before, second-order stationary Gaussian processes are strictly stationary and are characterized by their mean value and their autocorrelation function. The Fourier transformation of the autocorrelation function is a function of frequency called the spectral density function.

Since a mean zero stationary Gaussian process is characterized by its autocorrelation function and the Fourier transform of the autocorrelation is invertible, the spectral density function also characterizes zero mean stationary Gaussian processes. This helps explain the importance of the autocorrelation function and the spectral density function in the theory of stationary time series. Brillinger (1981) gives a nice theoretical account of the frequency domain approach to time series.

The periodogram, the sample analog to the spectral density, and smoothed versions that are estimates of the spectral density have many interesting properties, which are discussed in Brillinger (1981). The Fourier transform of the time series data itself is a complex-valued function called the empirical Fourier transform.

If a stationary process has a spectral density function and admits an infinite moving average representation, then as the series length $n \to \infty$, the real and imaginary parts of this empirical Fourier transform at the Fourier frequencies $\omega_k = 2\pi k/n$ are approximately independent and normally distributed with mean zero and variance $ng(\omega_k)/2$, where $g(\omega_k)$ is the true spectral density function for the time series at the frequency ω_k.

This asymptotic result is important and practically useful. The empirical Fourier transform is easy to compute thanks to a technique known as the fast Fourier transform. Independent identically distributed normal random variables are much easier to deal with.

Instead of bootstrapping the time series itself, we could consider bootstrapping the empirical Fourier transform by parametric bootstrapping. The series is then generated through the inverse Fourier transform. This idea is exploited in the phase scrambling algorithm. Although the concept is simple, the exact algorithm is somewhat complicated. The interested reader is referred to Davison and Hinkley (1997, pp. 408–409) for details. Davison and Hinkley then apply the phase resampling method to the Rio Negro data [assuming an AR(2) model under the null hypothesis] to again get an estimate of the variance of the test statistic T for the monotonic trend test. Again they find the variance for T based on the full series to be around 51 (a result comparable to the one obtained with the stationary bootstrap).

Now under the conditions described above, the periodogram has its values at the Fourier frequencies approximately independent and exponentially distributed. If one is interested only in confidence intervals for the spectral density at a certain frequency or in assessing the variability of estimates that are based on the periodogram values, it is only necessary to resample periodogram values and not the time series or its transform. Details of this method called periodogram resampling, with an application to inference about the spectral density function can be found in Davison and Hinkley (1997, pp. 412–414).

5.6. HISTORICAL NOTES

The use of ARIMA and seasonal ARIMA models for forecasting and control problems was first popularized by Box and Jenkins (1970, 1976). A very recent update of this classic book is Box, Jenkins, and Reinsel (1994).

A popular common theoretical account of time series analysis both in the time and frequency domains can be found in Brockwell and Davis (1991).

Fuller (1976) is another excellent theoretical book that covers both domains well. Works by Tong (1983, 1990) deal with nonlinear time series models.

Bloomfield (1976), Brillinger (1981), and Priestley (1981) present the spectral analysis of time series (i.e., frequency domain analysis) only. Another recent text on time series methods is Hamilton (1994). Braun and Kulperger (1997) did work on the Fourier bootstrap method.

The idea of bootstrapping residuals was described in Efron (1982a) in the context of regression. It is not clear who was the first to make the obvious extension of this to ARMA time series models. Findley (1986) was probably the first to point out some difficulties with the bootstrap approach, particularly regarding the estimation of mean square error.

Efron and Tibshirani (1986) showed how bootstrapping residuals provided improved standard error estimates for the autoregressive parameter estimates for the Wolfer sunspot data. Stine (1987) and Thombs and Schucany (1990) provide refinements to obtain better prediction intervals. Other empirical studies are Chatterjee (1986) and Holbert and Son (1986). McCullough (1994) provides an application of bootstrapping forecast intervals for AR(p) models.

Results for nonstationary autoregressions appear in Basawa, Mallik, McCormick, and Taylor (1989) and Basawa, Mallik, McCormick, Reeves, and Taylor (1991a, b). Theoretical developments are given in Bose (1988) and Kunsch (1989).

Basawa, Green, McCormick, and Taylor (1990) show that the bootstrap is consistent for finite Markov chains. Other results for Markov chains can be found in Kulperger and Prakasa Rao (1989) for the finite-state Markov chain and in Athreya and Fu (1992a, b), Datta and McCormick (1992), and Datta and McCormick (1995) for the special case of a first-order autoregression with positive innovations.

Kunsch (1989) attempts to develop a general theory for bootstrapping stationary time series. Bose (1988) showed good asymptotic higher-order properties when applying model-based resampling to a wide class of statistics in autoregressive processes. Hall and Jing (1996) apply resampling methods in general dependent data situations. Shao and Yu (1993) apply the bootstrap for the sample mean for a general class of time series (stationary mixing processes).

Model-based resampling for time series was discussed by Freedman (1984), Freedman and Peters (1984a, b), Swanepoel and van Wyk (1986), and Efron and Tibshirani (1986). Li and Maddala (1996) provide a survey of the related time domain literature on bootstrapping with emphasis on econometric applications.

Peters and Freedman (1985) deal with bootstrapping for the purpose of comparing competing forecasting equations. Tsay (1992) provides an applied account of parametric bootstrapping of time series.

Kabaila (1993a) discusses prediction in time series. Stoffer and Wall (1991) apply the bootstrap to state space models for time series. Chen, Davis, Brockwell, and Bai (1993) use model-based resampling to determine the appropriate order for an autoregressive model.

Good higher-order asymptotic properties for block resampling [similar to Bose (1988)] were demonstrated by Lahiri (1991) and Gotze and Kunsch (1996). Davison and Hall (1993) show that the good asymptotic properties depend crucially on the choice of a variance estimate. Lahiri (1992b) applies an Edgeworth correction using the moving block bootstrap for both stationary and nonstationary time series.

The block resampling approach was introduced by Carlstein (1986) and Kunsch (1989) followed with important theoretical developments including ideas about overlapping blocks. The stationary bootstrap was introduced by Politis and Romano (1994a). In an earlier work Politis and Romano (1992a) proposed a circular block resampling method for stationary data. Liu and Singh (1992b) obtain general results on moving block jackknife and bootstrap approaches to general types of weak dependence and Liu (1988) and Liu and Singh (1995) deal with bootstrap approaches to general data sets that are not independent and identically distributed.

Theoretical developments for general block resampling schemes followed Kunsch in Politis and Romano (1993a, 1994b), Buhlmann and Kunsch (1995), and Lahiri (1995). Issues of block length are addressed in Hall, Horowitz, and Jing (1995).

Fan and Hung (1997) use balanced resampling (a variance reduction technique that we describe in Chapter 7) to bootstrap finite Markov chains. Liu and Tang (1996) use bootstrap methods for control charting in both independent and dependent situations.

Frequency domain resampling has been discussed by Franke and Hardle (1992) with an analogy to nonparametric regression. Janas (1993) and Dahlhaus and Janas (1996) extended these results. Politis, Romano, and Lai (1992) provide bootstrap confidence bands for spectra and cross-spectra (frequency domain analog, respectively, to autocorrelation and cross-correlation functions in the time domain).

CHAPTER 6

Which Resampling Method Should You Use?

Through the first five chapters of this book we have discussed the bootstrap and several variations in many different contexts, including point estimation, confidence intervals, hypothesis tests, regression problems, and time series predictions. In addition to considering whether or not to use the empirical distribution, a smoothed version or a parametric version of it, we also considered improvements on bootstrap confidence intervals through the use of bias correction and an acceleration constant or by applying bootstrap iteration or bootstrap t intervals.

In some applications, related resampling methods such as the jackknife, cross-validation, and the delta method were considered. These three other resampling methods have been traditional and are the older solutions to the problem of estimating the standard error of an estimate.

In the case of linear regression with heteroscedastic errors, Wu (1986) pointed out problems with the standard bootstrap approach and offered more effective jackknife estimators. Other authors have proposed variants to the bootstrap, which they claim work just as well as in the heteroscedastic case.

In the case of error rate estimation in discriminant analysis, Efron (1983) showed convincingly that the bootstrap and some variants (particularly the .632 estimator) are superior to cross-validation. Other simulation studies supported and extended his work.

With regard to bootstrap sampling, in Chapter 7 we shall illustrate certain variance reduction techniques that help to reduce the number of bootstrap resamples needed to adequately approximate the bootstrap estimate. Sometimes variants of the bootstrap are merely applications of a different variance reduction method.

With all these variants of the bootstrap and related resampling techniques, the practitioner may naturally wonder which of these various techniques should be applied to a particular problem. The answer may or may not be clear-cut depending on the application. Also, because the research work on the bootstrap is somewhat immature, the jury is still out for many real world problems.

Nevertheless, the purpose of this chapter is to sort out the various resampling techniques and describe them for the practitioner. We discuss their similarities and differences and, where possible, recommend the preferred techniques.

The title of the chapter is intended to be provocative. This chapter will not provide a complete answer to the question. In fact, in most situations there is not a definitive answer at this time.

6.1. RELATED METHODS

6.1.1. Jackknife

As pointed out in Section 2.1, the jackknife goes back to Quenouille (1949), whose goal was to improve an estimate by correcting for bias. Later it was discovered to be more useful as a way to estimate variances or standard errors of estimators.

In general, we consider an estimate $\hat{\varphi}$ based on a sample $x_1, x_2, \ldots, x_n$ of observations that are independently drawn from a common distribution F. Suppose that $\hat{\varphi}$ can be represented as a functional of F_n, the empirical distribution [i.e., $\varphi = \varphi(F)$ and $\hat{\varphi} = \varphi(F_n)$]. Now we define $\hat{\varphi}_{(i)} = \varphi(F_{n(i)})$, where $F_{n(i)}$ is the distribution that places probability mass $1/(n-1)$ on the observations $x_1, x_2, \ldots, x_{i-1}, x_{i+1}, \ldots, x_n$ and no mass on x_i. The jackknife estimate of variance is then defined as

$$\hat{\sigma}^2_{\text{jack}} = \left(\frac{n-1}{n}\right) \sum_{i=1}^{n} [\hat{\varphi}_{(i)} - \hat{\varphi}_*]^2$$

where

$$\hat{\varphi}_* = \frac{1}{n} \sum_{i=1}^{n} [\hat{\varphi}_{(i)}.$$

The jackknife estimate of the standard error of $\hat{\varphi}$ is just the square root of $\hat{\sigma}^2_{\text{jack}}$.

Tukey defined a quantity

$$\hat{\varphi}_i = \hat{\varphi} + (n-1)(\hat{\varphi} - \hat{\varphi}_{(i)})$$

and called it the ith pseudovalue. The reason for this is that for general statistics $\hat{\varphi}$, the jackknife estimate of φ, is

$$\hat{\varphi} = \frac{\sum_{i=1}^{n} \hat{\varphi}_i}{n}$$

and

$$\hat{\sigma}^2_{\text{jack}} = \frac{\sum_{i-1}^{n} (\hat{\varphi}_i - \hat{\varphi})^2}{n(n-1)}.$$

This is the standard estimate used for the variance of a sample mean (in this case the mean is the mean of the pseudovalues).

The jackknife has proved to be a very useful tool in the estimation of variance of complicated estimators such as robust estimators of location like trimmed and Winsorized means [see Efron (1982a, pp. 14–16) for a discussion]. Simulation studies by Efron have generally shown the bootstrap estimate of standard deviation to be superior to the jackknife [for the trimmed means example see Efron (1982a, pp. 15–16) and for an adaptive trimmed mean see Efron (1982a, pp. 28–29)].

In Section 2.2.2, we provided a bootstrap estimate of the standard error for the sample median. The jackknife prescription provides an estimate that is not even consistent [see Efron (1982a, p. 16 and Chap. 6) for details].

All this empirical and theoretical evidence leads us to a recommendation to use the bootstrap instead of the jackknife when determining a standard error for an estimator. Efron (1982a, Theorem 6.1), shows that the jackknife estimate of a standard error is a bootstrap estimate with $\hat{\varphi}$ replaced by a linear approximation [up to a factor of $\sqrt{n/(n-1)}$]. This result suggests that the jackknife estimate is an approximation to the bootstrap and some researchers use this point to argue in favor of the bootstrap. Beran (1984a) determines jackknife approximations to bootstrap estimates exploiting some of these ideas of Efron.

6.1.2. Delta Method, Infinitesimal Jackknife, and Influence Functions

Many times we may be interested in the moments of an estimator (for variance estimation, the second moment). In such cases, it may be difficult to derive the exact moments. Nevertheless, the estimator may be represented as a function of other estimators whose first moments are known. As an example, the correlation coefficient ρ between random variables X and Y is defined as

$$\rho = \frac{\mathrm{Cov}(X,\ Y)}{\sqrt{\mathrm{Var}(X)\,\mathrm{Var}(Y)}}$$

where $\mathrm{Cov}(X,\ Y)$ is the covariance between X and Y and $\mathrm{Var}(X)$ and $\mathrm{Var}(Y)$ are the respective variances of X and Y. The method described here, known as the delta method, is often used in such situations and particularly in simple cases where we want the variance of a transformed variable such as X^p or $\log(X)$.

To illustrate, assume φ is $f\ (\alpha)$, where φ and α are one-dimensional and f is differentiable with respect to α. The procedure to be described can be generalized to multidimensional φ and α with f a vector-valued function.

Viewing α as a random variable with expected value α_0, we produce a first-order Taylor series expansion of φ about $f(\alpha_0)$:

$$\varphi = f(\alpha) = f(\alpha_0) + (\alpha - \alpha_0)f'(\alpha_0) + \text{remainder terms}$$

After dropping the remainder terms we have

$$f(\alpha) - f(\alpha_0) \approx (\alpha - \alpha_0)f'(\alpha_0).$$

Upon squaring both sides of the above equation and taking expectations, we have

$$E((f(\alpha) - f(a_0))^2 \approx E(\alpha - \alpha_0)^2[f'(\alpha_0)]^2. \tag{6.1}$$

Now $E(\alpha - \alpha_0)^2$ is the variance of α and $f'(\alpha_0)$ is known. The left-hand side of Eq. 6.1 is approximately the variance of $f(\alpha)$ or φ.

In the case where the variance of α is unknown, Efron (1982a, p. 43) suggests the nonparametric delta method where formulas like Eq. 6.1 are derived but applied to the empirical distribution for α [i.e., in the case of Eq. 6.1 the sample estimate of variance for α replaces $E(\alpha - \alpha_0)^2$].

Using geometrical ideas, Efron (1982a) shows that various estimates of standard error are related to bootstrap estimates. Estimates can also be obtained based on influence function estimates. Using the empirical influence function, Efron shows that the influence function estimate of standard error is the same as the one obtained using Jaeckel's infinitesimal jackknife (Efron, 1982a, p. 42). Amari (1985) provides a thorough treatment of related differential geometric methods.

Following Efron (1982a, pp. 39–42), the infinitesimal jackknife estimate of the standard error is defined as follows:

$$\text{SD}_{\text{IJ}}(\theta_e) = \left(\frac{\Sigma U_i^2}{n^2}\right)^{1/2}$$

where θ_e is the estimate of the parameter θ and U_i is a directional derivative in the direction of the ith coordinate centered at the empirical distribution function.

Slight differences in the choice of the influence function estimate can lead to different estimates (e.g., the ordinary jackknife and the positive jackknife). See Efron (1982a, p. 42) for details.

The key result relating these jackknife and influence function estimates to the delta method is Theorem 6.2 of Efron (1982a, p. 43), which states that the nonparametric delta method and the infinitesimal jackknife give identical estimates of the standard error of an estimator in cases where the nonparametric delta method is defined.

We have seen that in the context of estimating standard errors, the jackknife, the bootstrap, and the delta methods are closely related and, in fact, are asymptotically equivalent. For the practitioner, however, their differences in small to moderate sample sizes are important.

General conclusions are that the bootstrap tends to be superior, but the bootstrap usually requires the use of Monte Carlo replications. The ordinary jackknife is second best. Too often the nonparametric delta method (or equivalently the infinitesimal jackknife) badly underestimates the true standard errors.

Hall (1992a, pp. 86–88) discusses a slightly more general version of the delta method. He assumes that S_n and T_n are two asymptotically normal statistics that admit an Edgeworth expansion. If the two statistics differ by an amount that goes to zero in probability at a rate of $n^{-j/2}$ for $j \geq 1$, then the Edgeworth expansions of their distribution functions will differ by no more than $n^{-j/2}$. The standard delta method that we have described in this section amounts to the special case where S_n is a linear approximation to T_n. That is what we get by truncating the Taylor series expansion of φ with only the linear term.

Hall (1992a) goes on to point out the usefulness of this more general delta method. It may be easier to derive the low-order terms of the Edgeworth expansion for S_n rather than for T_n. Because S_n and T_n are "close," their Edgeworth expansions can only differ in the terms of order $n^{-k/2}$, where $k \geq j$.

It may be sufficient to obtain the expansion only up to $n^{-(j-1)/2}$, in which case the delta method is a convenient tool. Note that here we are using the delta method as an analytical device for obtaining Edgeworth expansion terms and not as an estimator per se.

6.1.3. Cross-validation

Cross-validation is a general procedure used in statistical model building. It can be used to decide on the order of a statistical model (including time series models, regression models, mixture distribution models, and discrimination models). It also has been generalized to estimate smoothing parameters in nonparametric density estimation and to construct spline functions. As such, it is a very useful tool. We have also seen its use in the estimation of the error rate of a prediction rule. The bootstrap provides a competitor to cross-validation in all such problems. The research on the bootstrap has not developed to the point where clear guidelines can be given for each of these problems.

The basic idea behind cross-validation is to take two random subsets of the data. Models are fit or various statistical procedures are applied to the first subset and then are tested on the second subset. The extreme case of fitting to all but one observation and then testing on the remaining one is sometimes referred to as leave-one-out and has also been called cross-validation by Efron.

Since leaving only one observation out does not provide an adequate test, the procedure actually is to fit the model n times, each time leaving out a different observation and testing the model on the observation left out each time. This provides a fair test by always testing on observations not used in the fit. It also is efficient in the use of the data for fitting the model since $n - 1$ observations are always used in the fit.

In the context of estimating the error rate of discriminant functions (Section 2.1.2), we found that the bootstrap and variants (particularly the .632 estimate) were superior to the leave-one-out estimator in terms of mean square estimation error.

For classification trees (i.e., discrimination rules based on a series of binary decisions graphically represented in a tree structure), Breiman, Friedman, Olshen, and Stone (1984) use cross-validation to "prune" (i.e., remove branches or shorten) classification tree algorithms. They also discuss a bootstrap approach (pp. 311–313). They refer to Efron (1983) for the discriminant analysis example of advantages of the bootstrap over cross-validation but did not have any theory or simulation to support its use in the case of classification trees. Further work has been done since then but nothing has shown strong superiority for bootstrap.

6.2. BOOTSTRAP VARIANTS

In previous chapters, we have introduced some modifications to the "nonparametric" bootstrap (i.e., sampling with replacement from the empirical distribution). These modifications were found sometimes to provide improvements over the nonparametric bootstrap when the sample size is small.

Recall that for the error rate estimation problem the .632 estimator, the double bootstrap (a form of bootstrap iteration), and the "convex" bootstrap (a form of smoothing the bootstrap distribution) were variations that proved to be superior to the original nonparametric bootstrap in a variety of small sample simulation studies.

For confidence intervals, Hall has shown that the accuracy of both kinds of bootstrap percentile methods can be improved by bootstrap iteration. Efron's bias correction and acceleration constant also provide a way to improve on the accuracy of his version of the bootstrap percentile confidence intervals.

When the problem indicates that the observed data should be modeled as coming from a distribution that is continuous and has a probability density, it may be reasonable to replace the empirical distribution function with a smoothed version (possibly based on kernel methods). This is referred to as the smoothed bootstrap.

Although it is desirable to smooth the distribution, particularly when the sample size is small, there is a catch. Kernel methods generally require large samples, particularly to estimate the tails of the density.

There is also the question of determining the width of the kernel (i.e., the degree to smoothing). Generally kernel width's have been determined by cross-validation. It is therefore not clear whether or not there will be a payoff to using a smoothed bootstrap even when we know the density exists. This issue is clearly addressed by Silverman and Young (1987).

In fact, we may also use the bootstrap as another approach to deciding on the width of a kernel in density estimation (as a competitor to cross-validation). At this time I have not seen much work that provides a comparison of bootstrap with cross-validation in kernel width determination.

Another variation on the bootstrap is Rubin's Bayesian bootstrap (see Rubin, 1981). The Bayesian bootstrap can be viewed as a Bayesian's justification for using bootstrap methods as Efron and others have interpreted it. On the other hand, Rubin used it to point out weaknesses in the original nonparametric version of the bootstrap.

6.2.1. Bayesian Bootstrap

Consider the case where $x_1, x_2, \ldots, x_n$ can be viewed as n independent identically distributed realizations of a random variable X with distribution F and denote the empirical distribution by $\hat{F}$. Recall that the nonparametric bootstrap, samples with replacement from $\hat{F}$. Let ϕ be a parameter of the distribution F. For simplicity we may think of x_i as one dimensional and ϕ as a single parameter, but both could be multidimensional as well. Let $\hat{\phi}$ be an estimate of ϕ based on $x_1, x_2, \ldots, x_n$. As we know, the nonparametric bootstrap can be used to approximate the distribution of $\hat{\phi}$.

Instead of sampling each x_i with replacement and probability $1/n$, the Bayesian bootstrap uses a posterior probability distribution for the x_i. This posterior probability distribution is centered at $1/n$ for each x_i but varies from one Bayesian bootstrap replication to another.

Specifically, the Bayesian bootstrap replications are defined as follows: Draw $n - 1$ uniform random variables from the interval $[0, 1]$. Let $u_{(1)}, u_{(2)}, \ldots, u_{(n-1)}$ denote their values in increasing order. Let $u_{(0)} = 0$ and $u_{(n)} = 1$. Then define $g_i = u_{(i)} - u_{(i-1)}$ for $i = 1, 2, 3, \ldots, n$. Then the g_i are called the gaps between uniform order statistics. The vector

$$g = \begin{bmatrix} g_1 \\ g_2 \\ \vdots \\ g_n \end{bmatrix}$$

is used to assign probabilities to the Bayesian bootstrap sample. Namely, n observations are selected by sampling with replacement from $x_1, x_2, \ldots, x_n$ but instead of each x_i having exactly probability $1/n$ of being selected each time, x_1 is selected with probability g_1, x_2 with probability g_2, and so on. A second Bayesian bootstrap replication is generated in the same way but with a new set of $n - 1$ uniform random numbers and hence a new set of g_i.

It is Rubin's point that the bootstrap and the Bayesian bootstrap are very similar and have common properties. Consequently, he suggests that any limitations attributable to the Bayesian bootstrap may be viewed as limitations of the nonparametric bootstrap as well.

An advantage of this form of bootstrapping to a Bayesian is that it can be used to make the usual Bayesian type inferences about the parameter ϕ based on ϕ's estimated posterior distribution, whereas, strictly speaking, the nonparametric bootstrap has only the usual frequentist interpretation about the distribution of the statistic $\hat{\phi}$.

If we let $g_i^{(1)}$ be the value of g_i in the first Bayesian bootstrap replication and let $g_i^{(2)}$ be the value of g_i in the second replication, we find, based on elementary results for uniform order statistics (e.g., see David, 1982), that

$$E(g_i^{(1)}) = E(g_i^{(2)}) = 1/n,$$
$$\mathrm{Var}(g_i^{(1)}) = \mathrm{Var}(g_i^{(2)}) = (n-1)/n^3$$
$$C(g_i^{(1)}, g_j^{(2)}) = C(g_i^{(2)}, g_j^{(1)}) = -1/(n-1)$$

where $E(\cdot)$, $\mathrm{Var}(\cdot)$, and $C(\cdot,\cdot)$ denote expectation, variance, and correlation over the respective replications. Because of the above properties, the bootstrap distribution for $\hat{\phi}$ and the Bayesian bootstrap posterior distribution for ϕ will be very similar in many applications.

Rubin (1981) provides some examples and shows that the Bayesian bootstrap procedure leads to a posterior distribution for ϕ that is Dirichlet and is based on a conjugate Dirichlet prior distribution. Rubin then goes on to criticize the Bayesian bootstrap because of the odd prior distribution that is implied.

He sees the Bayesian bootstrap as being appropriate in some problems but views the prior distribution as restrictive and hence does not recommend it as a general inference tool. Where he is uncomfortable with the Bayesian bootstrap he is equally uncomfortable with the ordinary nonparametric bootstrap. His main point is that, through the analogy he makes with the nonparametric bootstrap, the nonparametric bootstrap also should not be oversold as a general inference tool. Much of the criticism is directed at the lack of smoothness of the empirical distribution. Versions such as the parametric bootstrap and the smoothed bootstrap overcome some of these objections. See Rubin (1981) for a more detailed discussion along with some examples.

Stacy David has recently applied the Bayesian bootstrap in a missing data application at Eli Lilly. This application is mentioned briefly in Chapter 8.

The Bayesian bootstrap can be generalized by not restricting the prior distribution to the Dirichlet. The generalized version can be viewed as a Monte Carlo approximation to a posterior distribution for ϕ. In recent years there have been a number of papers written on the Bayesian bootstrap. Consult the bibliography for more references (particularly Rubin and Schenker, 1998).

6.2.2. The Smoothed Bootstrap

One motivation for the nonparametric bootstrap is that $\hat{F}$ (the empirical distribution) is the maximum likelihood estimator of F when no assumptions are made about F. Consequently, we can view the bootstrap estimates of parameters of F as nonparametric maximum likelihood estimates of those parameters.

However, in many applications, it is quite sensible to consider replacing $\hat{F}$ by a smooth distribution based on, say, a kernel density estimate of F' (the derivative of F with respect to x in the case of a univariate distribution). The Bayesian version of this is given by Banks (1988).

Efron (1982a) illustrates the application of smoothed bootstrap versions for the correlation coefficient using Gaussian and uniform kernel functions. The observations in the simulation study were Gaussian and the results show that the smoothed bootstrap does a little better than the original nonparametric bootstrap in estimating the standard error of the correlation coefficient.

Although smoothed versions of the bootstrap were considered early on in the history of the bootstrap, some researchers have recently proposed a Monte Carlo approximation based on sampling from a kernel estimate or a parametric estimate of F and have called it a generalized bootstrap (e.g., Dudewicz, 1992).

In his proposed generalized bootstrap, Dudewicz suggests fitting the data to a broad class of distributions and then doing the resampling from the fitted distribution. One such family that he suggests is called the *generalized lambda distribution* (see Dudewicz, 1992, page 35).

This is a four-parameter family of distributions, which can be specified by the mean, variance, skewness, and kurtosis. The method of moments is suggested as a way to determine the parameter estimates and hence the particular distribution from which to resample.

Comparisons of generalized bootstrap with the nonparametric bootstrap in a particular application of studying beaver dispersal patterns is given by Sun and Muller-Schwarze (1996). They apply the generalized lambda distribution for the generalized bootstrap.

It appears that the generalized bootstrap might be a promising alternative to the nonparametric bootstrap since it has the advantage of taking account of the fact that the data are continuous but it does not seem to suffer the drawbacks of the smoothed bootstrap. Another technique also referred to as a generalized bootstrap is presented in Bedrick and Hill (1992).

The value of a smoothed bootstrap is not altogether clear. It depends on the context of the problem and the sample size. See Silverman and Young (1987) for more discussion of this issue.

6.2.3. The Parametric Bootstrap

Efron (1982a) views the original bootstrap as a nonparametric maximum likelihood approach. As such, it can be viewed as a generalization of Fisher's maximum likelihood approach to the nonparametric framework. When looked at this way, $\hat{F}$ is the nonparametric estimate of F. If we make no further assumptions, the ordinary nonparametric bootstrap estimates are "maximum likelihood."

If we assume further that F is absolutely continuous, then smoothed distributions are natural and we are led to the smoothed bootstrap. Taking this a step further, if we assume that F has a parametric form such as, say, the Gaussian distribution, then the appropriate estimator for F would be a Gaussian distribution with the maximum likelihood estimates of μ and σ^2 used for these respective unknown parameters.

Sampling with replacement from such a parametric estimate of F leads to bootstrap estimates that are maximum likelihood estimates in accordance with Fisher's theory. The Monte Carlo approximation to the parametric bootstrap is simply an approximation to the maximum likelihood estimate.

The parametric bootstrap is discussed briefly in Efron (1982a, pp. 29–30). It is interesting to note that a parametric form of bootstrapping is equivalent to maximum likelihood.

However, in parametric problems, the existing theory on maximum likelihood estimation is adequate and the bootstrap adds little or nothing to the theory. Consequently, it is uncommon to see the parametric bootstrap used in real problems.

In more complex problems there may be semiparametric approaches, which might be called a parametric bootstrap. Davison and Hinkley (1997) justify the nonparametric bootstrap in parametric problems as a test of robustness of validity for the parametric method. They introduce the parametric bootstrap through an example of an exponential distribution and describe the implementation in a section on parametric simulation (pp. 15–21). They justify the use of the parametric bootstrap in cases where the estimator of interest has a distribution that is difficult to derive analytically or has an asymptotic distribution that does not provide a good small sample approximation, particularly for the variance, which is where the bootstrap is often useful.

6.2.4. Double Bootstrap

The double bootstrap was a method originally suggested in Efron (1983) as a way to improve on the bootstrap bias correction of the apparent error rate of a linear discriminant rule. As such, it is the first application of bootstrap iteration (i.e., taking resamples from each bootstrap resample). We briefly discussed this application in Chapter 2.

Normally, bootstrap iteration requires a total of B^2 bootstrap samples, where B is both the number of bootstrap replications from the original sample and the number of bootstrap samples taken from each bootstrap replication. In Efron (1983), a Monte Carlo swindle (variance reduction method) is used to obtain the accuracy of the B^2 bootstrap samples with just $2B$ samples.

Bootstrap iteration has been particularly useful in improving the accuracy of confidence intervals. The theory of bootstrap iteration for confidence intervals was developed by Hall, Beran, and Martin and is nicely summarized in Hall (1992a). See Chapter 3, Section 3.1.4 for more detail.

CHAPTER 7

Efficient and Effective Simulation

In Chapter 1, we introduced the notion of a Monte Carlo approximation to the bootstrap estimate of a parameter θ. We also mentioned that the bootstrap folklore suggests that the number of Monte Carlo iterations should be on the order of 100 to 200 for estimates such as standard errors and bias but 1000 or more for confidence intervals. These rules of thumb are based mostly on simulation studies and experience with a wide variety of applications.

Efron (1987) presented an argument that showed, based on calculations for the coefficient of variation, that 100 bootstrap iterations are all that is really necessary for standard error estimation and sometimes samples as small as 25 would suffice. He also argued for 1000 iterations to get good estimates of the endpoints for confidence intervals.

In a recent work, Booth and Sarkar (1998) challenge Efron's argument. They claim that the number of iterations should be based on the conditional distribution of the coefficient of variation rather than the unconditional distribution of the coefficient of variation. They believe that the number of iterations should be sufficiently large that the Monte Carlo error would not be allowed to affect the statistical inference. They show, using their conditioning argument, that 800 iterations are needed for standard errors as compared to the 100 recommended by Efron. Section 7.1 deals with this topic in detail.

A somewhat theoretical basis for the number of iterations has been developed by Hall using Edgeworth expansions. He also has results suggesting the potential gain from various variance reduction schemes, including the use of antithetic variates, importance sampling, linear approximations, and balanced sampling. Details can be found in Hall (1992a, Appendix II).

In this chapter (with Section 7.1 dealing with uniform resampling or ordinary Monte Carlo), we summarize Hall's finding and provide guidelines for practitioners based on current developments. In addition to Hall, a detailed account of various approaches can be found in Davison and Hinkley (1997, Chap. 9).

7.1. HOW MANY REPLICATIONS?

The usual Monte Carlo method, sampling with probability $1/n$ for each observation and with replacement from the original sample of size n, is referred to as a uniform resampling in Hall (1992a). We shall adopt that terminology here.

Let B be the number of bootstrap replications in a uniform resampling. Let σ_B^2 be the variance of a single bootstrap resample estimate of the parameter. Since the Monte Carlo approximation to the bootstrap estimate is an average of B such estimates independently drawn, the variance of the Monte Carlo approximation is just $B^{-1}\sigma_B^2$.

Of course, this basic result is well known and has been applied for many years to judge how many replications to take in a simulation. There is nothing new here with the bootstrap. For the bootstrap, the particular distribution being sampled is the empirical distribution but otherwise nothing is different.

If the parameter $\sigma = F(x)$, where F is the population distribution and x is a specified value, then $\sigma_B^2 = \sigma(1 - \sigma)$. By substituting $\sigma(1 - \sigma)$ for σ_B^2 in the result above, we obtain the inequality $B^{-1}\sigma(1 - \sigma) < (4B)^{-1}$ since $\sigma(1 - \sigma)$ must be less than or equal to $\frac{1}{4}$.

This result can be generalized slightly. Hall (1992a) points out that if the estimate $\hat{\theta}$ is for a parameter θ that represents a distribution function evaluated at a point x or is a quantile of a distribution function, then the variance of the uniform bootstrap approximation is CB^{-1} for large n and B. The constant C does not depend on B or n but is a function of unknown parameters. Often C can be usefully bounded above such as with the value $\frac{1}{4}$ given previously.

The practitioner then chooses B to make the variance sufficiently small, ensuring that the bootstrap approximation is close to the actual bootstrap estimate. Note that the accuracy of this approximation depends on B and not n. It only expresses how close the approximation is to the bootstrap estimate and does not express how close the bootstrap estimate is to the true parameter value!

If the constant C cannot be easily estimated or bounded, consider the following practical guideline. The practitioner can take, say, 100 bootstrap resamples and then double it to 200 to see how much the bootstrap approximation changes. One continues this process until the change is small enough. With the speed now available from modern computers, this approach is practical and commonly used.

When the parameter is a smooth function of a population mean, the variance is approximately $CB^{-1}n^{-1}$. Variance reduction methods can be used to reduce this variance, either by reducing C or the factor involving n (e.g., changing it from n^{-1} to n^{-2}).

The rules of thumb described in Efron (1987) and Booth and Sarkar (1998) are based on mathematical results, which indicate that after a particular number of iterations the error in the bootstrap estimate is dominated by the error due to the use of the empirical distribution mimicking the true distribution with the error in the Monte Carlo approximation being relatively small. In the 1990s it seems silly to me to argue over the choice of 100 iterations versus 800 iterations. For most simple problems (with the speed of modern computers) it is easy to complete the bootstrapping using 5000 to 10,000 iterations. I see little value in reduced cost from fewer than 5000 iterations (for many common simple problems).

When applicable, Hall's result provides specific accuracy bounds based on the specific problem and the desired accuracy. So I would prefer using it as opposed to either the Efron or the Booth–Sakar general rule of thumb.

7.2. VARIANCE REDUCTION METHODS

Variance reduction methods or swindles (as they are sometimes referred to in the statistics literature) are tricks that adjust the sampling procedure with the goal of reducing the variance for a fixed number of iterations. It is an old idea that goes back to nuclear applications in the 1950s.

Historically, one of the earliest examples is the method of antithetic variates, which can be attributed to Hammersley and Morton (1956). A good survey of these early methods is given by Hammersley and Handscomb (1964).

A principle, which is used to reduce the variance, is to split the computation of the estimate into deterministic and stochastic components and then to apply the Monte Carlo approximation only to the stochastic part. This approach is applicable in special isolated cases and does not have a name associated with it. It was used, for example, in the famous Princeton robustness study (Andrews, Bickel, Hampel, Huber, Rogers, and Tukey, 1972).

7.2.1. Linear Approximation

The linear approximation is a special case of the idea expressed in the preceding paragraph. The estimator is expressed as the expected value of a Taylor series expansion. Since the linear term in the series is known to have zero expectation, the Monte Carlo method is applied only to the estimation of the higher order terms.

As an example, consider the bias estimation problem described in Section 2.1. Recall that the bootstrap estimation of bias is $E(\theta^* - \hat{\theta})$, where θ^* is an estimate of θ based on a bootstrap sample. We further assume that $\hat{\sigma} = g(\bar{x})$, where $\bar{x}$ is a sample mean and g is a smooth function (i.e., g has first-order and higher-order derivatives in its argument).

A Taylor series expansion for $\theta^* - \hat{\theta}$ is

$$U^* = \theta^* - \hat{\theta} = (\bar{x}^* - \bar{x})g'(\bar{x}) + \tfrac{1}{2}(\bar{x}^* - \bar{x})^2 g''(\bar{x}) + \cdots \qquad (7.1)$$

Now $E(U^*|\bar{x})$ from Eq. 7.1 is

$$E(U^*|\bar{x}) = E(\bar{x}^* - \bar{x}|\bar{x})g'(\bar{x}) + \tfrac{1}{2}E((\bar{x}^* - \bar{x})^2 \,|\, \bar{x})g''(\bar{x}) + \cdots \qquad (7.2)$$

$E(U^*|\bar{x})$ is the bootstrap estimate of bias. Uniform sampling would estimate this directly without regard to the expansion.

But $E(\bar{x}^*|\bar{x}) = \bar{x}$, since bootstrap sampling is sampling with replacement from the original sample. Therefore $E(\bar{x}^* - \bar{x}|\bar{x}) = 0$ and the first term [i.e., the linear term in the expansion of $E(U^*|\bar{x})$] can be omitted.

So Eq. 7.2 reduces to

$$E(U^*|\bar{x}) = \tfrac{1}{2}E((\bar{x}^* - \bar{x})^2 |\bar{x})g''(\bar{x}) + \cdots \qquad (7.3)$$

To take advantage of Eq. 7.3 we define $V^* = U^* - (\bar{x}^* - \bar{x})g'(\bar{x})$ and apply uniform sampling to V^* instead. In view of Eq. 7.3, $E(V^*|\bar{x}) = E(U^*|\bar{x})$. So averaging V^* approaches the bootstrap estimate as $B \to \infty$ just as U^* does, but we have removed a deterministic part $E(\bar{x}^* - \bar{x})g'(\bar{x})$, which we know equals zero.

In Hall (1992a, Appendix II), this result is shown for the more general case where $\bar{x}$ is a d-dimensional vector. He shows that $\text{Var}(V^*|\bar{x})$ is of the order $B^{-1}n^{-2}$ as compared to $B^{-1}n^{-1}$ for $\text{Var}(U^*|\bar{x})$.

In principle, if higher-order derivatives exist, we may remove these terms and compute an estimate that approaches the bootstrap at a rate $B^{-1}n^{-k}$, where k is the highest order of derivatives removed. It does, however, require computation of a linear form in the first k central sample moments.

This principle can also be applied through the use of what are called control variates. An estimator T^* is decomposed using the following identity: $T^* = C + (T^* - C)$, where C is a "control variate." Obviously any variable can be chosen to satisfy the identity but C should be picked (1) to have high positive correlation with T^* and (2) so that its statistical properties are known analytically. Then to determine the statistical properties of T^*, we can apply the Monte Carlo approximation to $T^* - C$ instead of T^*.

Because of the high positive correlation between T^* and C, the variable $T^* - C$ will have a much smaller variance than T^* itself. So this device (i.e. the use of a control variate C) enables us to get more precision in estimating T^* by only applying the Monte Carlo approximation to $T^* - C$ and using knowledge of the statistical properties of C. More details with specific application to bootstrap estimates of bias and variance can be found in Davison and Hinkley (1997, pp. 446–450).

7.2.2. Balanced Resampling

Balanced resampling was introduced by Davison, Hinkley, and Schechtman (1986). It is also covered with a number of illustrative examples in Davison and Hinkley (1997, pp. 438–446).

The idea is to control the number of times observations occur in the bootstrap samples so that in the B bootstrap samples, each observation occurs the same number of times (namely, B). Of course for the bootstrap to work, some observations must be missing in certain bootstrap samples, while others may occur two or more times.

Balanced resampling does not force each observation to occur once in each sample but equalizes the number of occurrences of each observation over the set of bootstrap samples. If an observation occurs twice in one bootstrap sample, there must be another bootstrap sample where it is missing from the sample. This is reminiscent of the kind of balancing constraints used in statistical experimental designs (e.g., balanced incomplete block designs).

A simple way to achieve balanced resampling is to create a string of the observations $X_1, X_2, \ldots, X_n$ repeated B times (i.e., we have the sequence Y_1, $Y_2, \ldots, Y_{nB}$, where $Y_i = X_j$ with j being the remainder when dividing i by n). Then take a random permutation π of the integers from 1 to nB. Take $Y_{\pi_{(1)}}$, $Y_{\pi_{(2)}}, \ldots, Y_{\pi_{(n)}}$ as the first bootstrap sample $Y_{\pi_{(n+1)}}, Y_{\pi_{(n+2)}}, \ldots, Y_{\pi_{(2n)}}$ as the second bootstrap sample, and so on until $Y_{\pi_{((B-1)n+1)}}, Y_{\pi_{((B-1)n+2)}}, \ldots, Y_{\pi_{(Bn)}}$ is the Bth bootstrap sample.

Hall (1992a, Appendix II) shows that balanced resampling produces an estimate with conditional variance on the order of $B^{-1}n^{-2}$. His result applies to smooth functions of a sample mean.

Balanced resampling can be applied in much greater generality including the estimation of distributions and quantiles. In such cases there is still an improvement in the variance but not as dramatic an improvement. Unfortunately for distribution functions, the order is still Cn^{-1} and only the constant C is reduced. See Hall (1992a, pp. 333–335) for details.

The MC estimator of Chernick, Murthy, and Nealy (1985), discussed in Section 2.2.2, is a form of controlled selection where an attempt is made to sample with the limiting repetition frequencies of the bootstrap distribution. As such it is similar to variance reduction methods like balanced resampling.

7.2.3. Antithetic Variates

As mentioned earlier, the concept of antithetic variates dates back to Hammersley and Morton (1956). The idea is to introduce negative correlation between pairs of Monte Carlo samples to reduce the variance.

The basis for the idea is as follows. Suppose $\hat{\varphi}_1$ and $\hat{\varphi}_2$ are two unbiased estimates for the parameter φ. We can then compute a third unbiased estimate $\hat{\varphi}_3 = \frac{1}{2}(\hat{\varphi}_1 + \hat{\varphi}_2)$. Then

$$\mathrm{Var}(\hat{\varphi}_3) = \tfrac{1}{4}[(\mathrm{Var}\,\hat{\varphi}_2) + 2\,\mathrm{Cov}(\hat{\varphi}_1, \hat{\varphi}_2) + \mathrm{Var}(\hat{\varphi}_1)]$$

Assume without loss of generality that $\mathrm{Var}(\hat{\varphi}_2) > \mathrm{Var}(\hat{\varphi}_1)$. Then if

$\text{Cov}(\hat{\varphi}_1, \hat{\varphi}_2) < 0$, we have the following inequality:

$$\text{Var}(\hat{\varphi}_3) \leqslant \tfrac{1}{2}(\text{Var } \hat{\varphi}_2)$$

So $\hat{\varphi}_3$ has a variance that is smaller than half of the larger of the variances of the two estimates. If $\text{Var}(\hat{\varphi}_1)$ and $\text{Var}(\hat{\varphi}_2)$ are nearly equal then, roughly, we have guaranteed a reduction by about a factor of two for the variance of the estimate. The larger the negative correlation between $\hat{\varphi}_1$ and $\hat{\varphi}_2$, the greater is the reduction in variance.

One way to do antithetic resampling is to consider the permutation that maps the largest X_i to the smallest, the second largest to the second smallest, and so on. Let the odd bootstrap samples be generated by uniform bootstrap resampling. The even bootstrap samples takes $X_i^* = X_{\pi(k)}$, where $X_i^* = X_k$ for the preceding odd bootstrap sample.

The pairs of bootstrap samples generated in this way are negatively correlated because the permutation symbol π maps indices for larger values to indices for smaller values. So if the first bootstrap sample tends to be higher than average, the second will tend to be lower than average. This provides the negative correlation in each bootstrap pair that creates a variance reduction in the Monte Carlo average as we have shown above for antithetic variates in general.

We compute an estimate for the odd bootstrap samples and also for the even bootstrap samples. Call these estimates U_1^* and U_2^*. The antithetic resampling estimate is then

$$U^* = (U_1^* + U_2^*)/2$$

Unfortunately, Hall (1992a) shows that antithetic resampling only reduces the variance by a constant factor and hence is not as good as balanced resampling or the linear approximation.

7.2.4. Importance Sampling

Importance sampling is an old variance reduction technique. Reference to it can be found in Hammersley and Handscomb (1964). One of the first to suggest its use in bootstrapping is Johns (1988).

Importance sampling (or resampling) is a useful tool when estimating the tails of the distribution function or for quantile estimation. It has limited value when estimating bias and variance. It is therefore especially applicable to hypothesis testing problems where the estimation of a p-value for a test statistic is important.

The idea is to control the sampling, so as to take more samples from the part of the distribution that is important to the particular estimation problem. For example, when estimating the extreme tails of a distribution [i.e., $1 - F(x)$ for very large x or $F(x)$ for very small x], we need to observe values larger than x.

However, if the probability $1 - F(x)$ is very small, n must be extremely large to even observe any values greater than x. Even for extremely large n and $1 - F_n(x) > 0$, where F_n is the empirical distribution, the number of observations greater than x in the sample will be small.

Importance resampling is an idea used to improve such estimates by including these observations more frequently in the bootstrap samples. Of course, any time the sampling distribution is distorted by such a procedure, an appropriate weighting scheme is required to ensure that the estimate is converging to the bootstrap estimate as B gets large.

Basically it exploits the identity that for a parameter μ defined as $\int m(y)\, dG(y)$, we have $\int m(y)\, dG(y) = \int m(y)\{dG(y)/dH(y)\}\, dH(y)$. This suggests sampling from $H(y)$ instead of $G(y)$ but using the weight $dG(y)/dH(y)$ for each value of $m(y)$ that is sampled.

This works if the support of H includes the support of G (i.e., the support of a distribution is the set of values y such that the density function associated with that distribution is nonzero). A detailed description along with examples can be found in Davison and Hinkley (1997, pp. 450–466).

One can view importance resampling as a generalization of uniform resampling. In uniform resampling each X_i has probability $1/n$. In general, we can define an importance resample by assigning probability p_i to X_i, where the only restriction on the p_i is that $p_i \geq 0$ for each i and $\Sigma_{i=1}^n p_i = 1$.

When $p_i = 1/n$, the jth bootstrap sample mean

$$\bar{X}_j^* = \frac{1}{n} \sum_{i=1}^n \bar{X}_i^*$$

is an unbiased estimate of $\bar{X}$ and the Monte Carlo approximation

$$\bar{X}_B^* = \frac{1}{B} \sum_{i=1}^B \bar{X}_j^*$$

approaches $\bar{X}$ as $B \to \infty$. This is a desirable property that is lost if, for some values of i, $p_i \neq 1/n$. However, since

$$\bar{X} = \frac{1}{n} \sum_{i=1}^n X_i$$

define $\bar{X}_j^* = \Sigma_{i=1}^n \alpha_i \bar{X}_i^*$, where α_i is chosen so that if $\bar{X}_i^* = X_k$, then $\alpha_i = 1/np_k$ for $k = 1, 2, \ldots, n$.

This weighting guarantees that conditional on X_1, X_2, $X_3, \ldots,$ X_n, $E(\bar{X}_j^*) = \bar{X}$. One can then look for values for the p_k so that the variance of the estimator (in this case $\bar{X}_j^*$) is minimized.

We shall not go into the details of deriving optimal p_k values for various estimation problems. Our advice to the practitioner is to only consider importance sampling for cases where the distribution function, a p-value, or a quantile is to be estimated or in the special case of bootstrap recycling to be described later.

Hall (1992a, Appendix II), derives the appropriate importance sample for minimizing the variance of an estimate of a distribution function for a studentized asymptotically normal statistic. The interested reader should look there for more details. Other references on importance resampling are Johns (1988), Hinkley and Shi (1989), Hall (1991a), and Do and Hall (1990, 1991a).

A clever application of importance sampling is referred to as bootstrap recycling. This can be applied to the iterated bootstrap by repeated use of the importance sampling identity. It has advantages when the statistic of interest is complicated and costly to estimate, as in the case of a difficult optimization problem such as one that requires Markov chain Monte Carlo methods for its estimation. Details along with applications to bootstrap iteration can be found in Davison and Hinkley (1997, pp. 463–466).

7.2.5. Centering

Recall the linear approximation V^* to the bias estimation problem described in Section 7.2.1 along with the smooth function g defined in that section. Let

$$\bar{X}_*^* = B^{-1} \sum_{j=1}^{B} \bar{X}_j^*$$

Let

$$\hat{X}_B^* = B^{-1} \sum_{j=1}^{B} g(\bar{X}_j^*) - g(\bar{X}_*^*)$$

Now

$$U^* = B^{-1} \sum_{j=1}^{B} g(\bar{X}_j^*) - g(\bar{X})$$

We choose to center at $g(\bar{X}_*^*)$ instead of $g(\bar{X})$. This idea was introduced in Efron (1988). Hall (1992a, Appendix II) shows that it is essentially equivalent to linear approximation in its closeness to the bootstrap estimate.

These variance reduction techniques are particularly useful for complex problems where the estimates themselves might require intensive computing as in some of the examples in Chapter 8. For simple problems that do not require double bootstrapping, it should be easy enough to generate 5000 to 10,000 bootstrap samples using uniform resampling.

7.3. WHEN CAN MONTE CARLO BE AVOIDED?

In the nonparametric setting, we have shown in Chapter 3 several ways to obtain confidence intervals. Most approaches to bootstrap confidence intervals require adjustments for bias and skewness including Efron's BC_a intervals and Hall's bootstrap iteration technique. Each requires many bootstrap replications.

Of course, in parametric formulations without nuisance parameters, classical methods provide exact confidence intervals without any need for Monte Carlo. This is because pivotal quantities can be constructed whose probability distributions are known or can be derived. Because of the duality between hypothesis testing and confidence intervals, the same statement applies to hypothesis tests.

In a "semiparametric" setting where weak distributional assumptions are made (e.g., the existence of a few moments of the distribution) asymptotic expansions of the distribution of these quanitites (which may be asymptotically pivotal) can be used to obtain confidence intervals.

We have seen, for example, in Section 3.1.4, that the asymptotic properties of bootstrap iteration can be derived from Cornish–Fisher expansions. These results suggest that some approaches are more accurate because of their faster rate of convergence.

DiCiccio and Efron (1990, 1992), using properties of exponential families, construct confidence intervals with the accuracy of the bootstrap BC_a intervals but without any Monte Carlo. Edgeworth and Cornish–Fisher expansions can be used in certain problems. The difficulty, in practice, is that they sometimes require large samples to be sufficiently accurate, particularly when estimating the tails of the distribution.

The idea of recentering and then expanding in the neighborhood of a saddlepoint was first suggested by Daniels (1954) to provide good approximations to the distribution of the test quantity in the neighborhood of a point of interest (e.g., the tails of the distribution). Field and Ronchetti (1990) apply this approach, which they refer to as small sample asymptotics in a number of cases.

They claim that their approach works well in small samples and that it obtains the accuracy of the bootstrap confidence intervals without resampling. Detailed discussion of saddlepoint approximations can be found in Davison and Hinkley (1997, pp. 466–485).

Another similar approach due to Hampel (1973) is also discussed in Field and Ronchetti (1990). An expository paper on saddlepoint approximations is Reid (1988).

Applications in Field and Ronchetti (1990) include estimation of a mean using the sample mean, robust location estimators including L estimators, and multivariate M estimators. Confidence intervals in regression problems and connections with the bootstrap in the nonparametric setting are also considered by Field and Ronchetti (1990).

Also, the greatest promise with small sample asymptotics is the ability of high-speed computers to generate the estimates and their apparent high accuracy in small samples. Nevertheless, none of the major statistical packages include small sample asymptotic methods to date and the theory to back up the empirical evidence of small sample accuracy requires further development.

7.4. HISTORICAL NOTES

Variance reduction methods for parametric simulation have a long history and the information is scattered throughout the literature of many disciplines. Some of the pioneering work came out of the nuclear industry of the 1940s and 1950s when computational methods were a real challenge.

There were no fast computers then! In fact, the first vacuum tube computers were developed in the mid-1940s at Princeton, New Jersey, and Aberdeen, Maryland, to aid in the war effort.

Some discussions of these variance reduction methods can be found in various texts on Monte Carlo methods, such as Hammersley and Handscomb (1964), Bratley, Fox, and Schrage (1987), Ripley (1987), Devorye (1986), Mooney (1997), and Niedereiter (1992). My own account can be found in a chapter on Monte Carlo methods in a compendium on risk analysis techniques written for employees at the Army Materiel Systems Analysis Activity (Atzinger, Brooks, Chernick, Elsner, and Foster 1972). An early clever application to the comparison of statistical estimators was the Princeton robustness study (see Andrews, Bickel, Hampel, Huber, Rogers, and Tukey, 1972).

Balanced bootstrap simulation was first introduced by Davison, Hinkley, and Schechtman (1986). Ogbonmwan (1985) proposes a slightly different method for achieving first-order balance. Graham, Hinkley, John, and Shi (1990) discuss ways to achieve second-order balance and they provide connections to the classical experimental designs. A very recent overview of balanced resampling based on the use of orthogonal multiarrays is Sitter (1998). Nigam and Rao (1996) develop balanced resampling for finite populations when applying simple random sampling or stratified sampling with equal samples in each strata. Do (1992) compares balanced and antithetic resampling methods in a simulation study.

The theoretical aspects of balanced resampling were investigated by Do and Hall (1991b). There are mathematical connections to number-theoretical methods of integration (Fang and Wang, 1994) and to Latin hypercube sampling (McKay, Conover, and Beckman, 1979; Stein, 1987; Owen, 1992).

Importance resampling was first suggested in Johns (1988). Hinkley and Shi (1989) applied it to iterated bootstrap confidence intervals. Gigli (1994a) outlines its use in parametric simulation for regression and time series.

The large sample performance of importance resampling has been investigated by Do and Hall (1991). Booth, Hall, and Wood (1993) describe algorithms. Gigli (1994b) provides an overview on resampling simulation techniques.

Linear approximations were used as control variates in bootstrap sampling by Davison, Hinkley, and Schechtman (1986). Efron (1990) took a different approach using the recentered bias estimate and control variates in quantile estimation. Therneau (1983) and Hesterberg (1988) provide further discussion on control methods.

The technique of bootstrap recycling originated with Davison, Hinkley, and Worton (1992) and was derived independently by Newton and Geyer (1994). Properties of bootstrap recycling are discussed for a variety of applications in Ventura (1997).

Another approach to variance reduction is Richardson extrapolation which was suggested by Bickel and Yahav (1988). Davison and Hinkley (1997) and Hall (1992a) both provide sections discussing variance reduction methods. Davison and Hinkley (1997) briefly refer to Richardson extrapolation in Problem 22 in Section 9.7, page 494. Hall (1992a) mentions it only in passing as another approach.

General discussion of variance reduction techniques for bootstrapping appear in Hall (1989a, 1992c). Hall (1989b) deals with bootstrap applications of antithetic resampling.

Saddlepoint methods originated with Daniels (1954). Reid (1988) reviews their use in statistical inference. Longer accounts can be found in Jensen (1992). Recent applications to bootstrapping the studentized mean are given in Daniels and Young (1991). Field and Ronchetti (1990) and Barndorff-Nielsen and Cox (1989) also deal with saddlepoint methods. Other related asymptotic results can be found in Barndorff-Nielsen and Cox (1994).

CHAPTER 8

Special Topics

This chapter deals with a variety of statistical problems. The common theme is the complex nature of the problems. In many cases, classical approaches require special assumptions or they provide incomplete or inadequate answers.

Although the bootstrap theory has not advanced to the stage of explaining how well it works on these complex problems, many researchers and practitioners see the bootstrap as a valuable tool when dealing with these difficult, but practical problems.

8.1. KRIGING

When monitoring air pollution in a given region, the government wants to control the levels of certain particulates. By setting up monitoring stations in particular locations, the levels of these particulates can be measured at various times.

An important practical question is: Given measured levels of a particulate at a set of monitoring stations, what is the level of the particulate in the communities located between the monitoring stations? Spatial stochastic models are constructed to try to realistically represent the way the pollution levels change smoothly as we move from one place to another. The results are graphically represented with contour plots such as in Fig. 8.1.

Kriging is one technique for generating such contour plots. It has optimality properties under certain conditions. Due to the statistical nature of the procedure, there is, of course, uncertainty (at least partly due to measurement uncertainty) about the particulate level and hence also the constructed contours. The bootstrap provides one approach for observing the variability. As Diaconis and Efron (1983) show, bootstrap samples of kriging contours can be used as a visual tool to illustrate this variability.

Diaconis and Efron (1983) considered another important pollution problem that was studied extensively in the 1980s, namely, the question of acidity levels in the rainfall, using 2000 measurements of pH levels for rainfall at nine weather stations in the northeastern United States over a two-year period from September 1978 through August 1980. Figure 8.1 shows the original map generated in Eynon and Switzer (1983) along with five bootstrap replications.

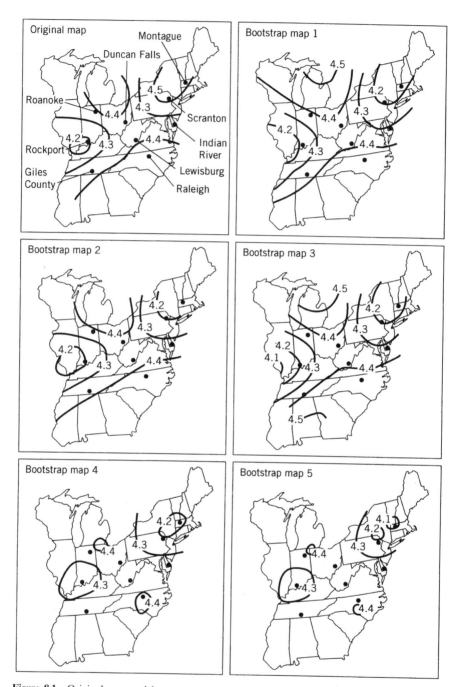

Figure 8.1 Original map and bootstrap maps. [From Diaconis and Efron (1983, p. 117), with permission from Scientific American.]

The weather stations are illustrated with the nine dots and the city names are labeled on the original map; there are also important noticeable differences. Although there is no generally accepted measure of the variability of contour lines on a map, visual inspection of the bootstrap replications provides a sense of this variability and points out that the original map must be interpreted cautiously.

For point estimates, we have confidence intervals to express our degree of uncertainty in the estimates. Similarly, we have uncertainty in the kriging contours. The problem is much more complex, however. The bootstrap procedure provides one way to assess this uncertainty.

A naive approach to bootstrapping spatial data is not appropriate because of spatial correlation. Cressie (1991, pp. 489–497) discusses resampling procedures applied to spatial data including the bootstrap and the jackknife.

We shall not describe the kriging method here [see Cressie (1991) for details]. Diaconis and Efron (1983) do not describe kriging or the particular way the bootstrap was applied by Eynon and Switzer to obtain the bootstrap sample maps. Presumably, they bootstrapped the residuals from the model fit. This must have been done in a way so that the geographic relationship among the weather stations was preserved over the bootstrap replications.

Early work on bootstrapping spatial data that takes account of the local (in space) correlation is due to Hall (1985). He proposed two methods for bootstrapping. Both ideas start with the division of the entire space D into k congruent subregions $D_1, D_2, \ldots, D_k$.

Let $\mathbf{Z}_1, \mathbf{Z}_2, \ldots, \mathbf{Z}_k$ be the k vectors of observations associated with the corresponding subregions. The first scheme assigns the bootstrap data $\mathbf{Z}_i^*$ for $i = 1, 2, \ldots, k$ by sampling with replacement from the original data and if $\mathbf{Z}_i^* = \mathbf{Z}_j$, then the observation is assigned to region D_j.

The second approach is to sample with replacement from the possible subregions and then a chosen subregion D_j^* has its data assigned to region D_j. Note that region D_j^* could be a region D_i that is congruent to D_j but i is not necessarily equal to j. So the data $\mathbf{Z}_j$ is assigned to the region D_i.

This latter scheme is analogous to one proposed by Kunsch (1989) to bootstrap time series taking the time correlation into account. Other semiparametric approaches involve regression equations relating the spatial data to explanatory variables. These methods are described in Cressie (1991, pp. 493–494). The methods of Freedman and Peters (1984a) or Solow (1985) can then be exploited in the spatial data context.

Cressie (1991, p. 496) describes a form of parametric bootstrap. He points out that the Monte Carlo based hypothesis testing approaches of Hope (1968) and Besag and Diggle (1977) predated Efron's celebrated *Annals of Statistics* paper (Efron, 1979a) and each was based on a suggestion from Barnard (1963). Cressie does not offer any suggestion as to a preferred method for resampling in the case of spatial data.

8.2. SUBSET SELECTION

In both regression and classification problems, we may have a large set of variables that we think can help to predict the outcome. In the case of regression, we are trying to predict an "average" response (usually a conditional expectation) and in discriminant analysis we are trying to predict the appropriate category to which the observation belongs.

In either case, a subset of the candidate variables may provide most of the key information. Often, when there are too many candidate variables, it will actually be better for prediction purposes to use a subset. This is true because (1) some of the variables may be highly correlated or (2) we do not have a sufficiently large sample to estimate all the coefficients accurately.

To overcome these difficulties, there have been a number of procedures developed to help pick the "best" subset. For both discrimination and regression there are forward, backward, and stepwise selection procedures based on the use of criteria such as an F ratio, which compares two hypotheses at each stage.

Although there are many useful criteria for optimizing the choice of the subset, in problems with a large number of variables, the search for the "optimal" subset is unfeasible. That is why suboptimal selection procedures such as forward, backward, and stepwise selection are used. For a given data set forward, backward, and stepwise selection often lead to different answers.

This should suggest to the practitioner that there is no unique answer to the problem (i.e., different sets of variables may work equally well). Unfortunately, there is a great temptation for the practitioner to interpret the variables selected as being useful and those left out as not being useful. Such an interpretation can be a big mistake.

Diaconis and Efron (1983) relate an example of a study by Gong, which addresses this issue through the bootstrap [see Efron and Gong (1983, Sec. 10) or Gong (1986) for more details]. In the example, a group of 155 people with acute and chronic hepatitis were initially studied by Dr. Gregory at the Stanford University School of Medicine. Of the 155 patients, 33 died from the disease and 122 survived.

Gregory wanted to develop a model to predict a patient's chance of survival. He had 19 variables available to help with the prediction. The question is which variables to use. With only 155 observations, all 19 are probably too many. Gregory, using his medical judgment, eliminated six of the variables. Forward logistic regression was then used to pick 4 variables from the remaining 13.

Gong's approach was to bootstrap the entire data analysis including the preliminary screening process. The interesting results were in the variability of the final variables chosen by the forward logistic regression procedure. Gong generated 500 bootstrap replications of the data for the 155 patients. In some replications only one predictor variable emerged. It could be ascites (the presence of a fluid in the abdomen), the concentration of bilirubin in the liver, or the physician's prognosis.

Other bootstrap samples led to the selection of as many as six of the variables. None of the variables emerged as very important by appearing in 60% or more of the cases. In addition to showing that caution must be exercised when interpreting the results of a variable selection procedure, this example illustrates the potential of the bootstrap as a tool for assessing the effects of preliminary "data mining" (or exploratory analysis) on the final results. In the past, such effects have been ignored because they are difficult or impossible to assess mathematically.

McQuarrie and Tsai (1998) consider the use of cross-validation and bootstrap for model selection in regression and time series models. They devote Chapter 6, pages 251–291, to this subject. They found little difference between cross-validation and bootstrap for model selection. For the regression problem they determined through simulation that there was little difference whether you bootstrap residuals or vectors in terms of model choice. Consequently, since they find it easier to bootstrap residuals, they recommend that approach for model selection applications.

Data mining is becoming very popular for businesses with large data sets. People believe that intensive exploratory analysis could reveal patterns in the data that can be used to gain a competitive advantage. Consequently, it is now a growing discipline among computer scientists. Statistical methods in general, and possibly the bootstrap in particular, could provide a helpful role in understanding and evaluating data mining procedures, particularly apparent patterns that could have occurred merely by chance and are therefore not really significant.

8.3. DETERMINING THE NUMBER OF DISTRIBUTIONS IN A MIXTURE MODEL

Mixture distribution problems arise in a variety of circumstances. In the case of Gaussian mixtures, they have been used to identify outliers (see Aitkin and Tunnicliffe Wilson, 1980) and to investigate the robustness of certain statistics for departures from normality [e.g., the sample correlation coefficient studied by Srivastava and Lee (1984)].

Increasing use of mixture models can be found in the field of cluster analysis. Various examples of such applications can be found in the papers by Basford and McLachlan (1985a–d).

In my experience, working on defense problems, targets and decoys appeared to have feature distributions that fit well to a mixture of multivariate normal distributions. This multimodal shape to the distribution could occur because the enemy might use different types of decoys or apply coating materials to the targets to try to hide their signatures.

Discrimination algorithms would work well once the appropriate mixture distributions are determined. The most difficult part of the problem is to determine the number of distributions in the mixture for the targets and the decoys.

McLachlan and Basford (1988) apply a mixture likelihood approach to the clustering problem. As one approach to deciding on the number of distributions in the mixture, they apply a likelihood ratio test employing bootstrap sampling.

The bootstrap is used because in most parametric mixture problems, if we define λ to be the likelihood ratio statistic, then $-2 \log \lambda$ fails to have an asymptotic chi-square distribution because the regularity conditions necessary for this standard asymptotic result fail to hold. The bootstrap is used to approximate the distribution of $-2 \log \lambda$ under the null hypothesis.

We now formulate the problem of likelihood estimation for a mixture model and then define the maximum likelihood approach. This will then be followed by the bootstrap test for the number of distributions.

Let $\mathbf{x}_1, \ldots, \mathbf{x}_n$ be p-dimensional random vectors. Each $\mathbf{x}_j$ comes from a multivariate distribution G, which is a mixture of a finite number of probability distributions, say, k, where $F_1, F_2, \ldots, F_k$ represent these distributions and $\pi_1, \pi_2, \ldots, \pi_k$ represent their proportions, where π_j is the probability that population j is selected. Consequently, we require

$$\sum_{i=1}^{k} \pi_i = 1 \quad \text{and} \quad \pi_i \geq 0 \quad \text{for } i = 1, 2, \ldots, k$$

From this definition we also see that the cumulative distribution G is related to the F_i by

$$G(\mathbf{X}) = P[\mathbf{x}_j \leq \mathbf{x}] = \sum_{i=1}^{k} \pi_i F_i(\mathbf{x})$$

where by $\mathbf{u} \leq \mathbf{v}$ we mean that each component of the p-dimensional vector $\mathbf{u}$ is less than or equal to the corresponding component of the p-dimensional vector $\mathbf{v}$.

Assuming that G and the F_i are all differentiable functions (i.e., each has a well-defined density function), then if $g_\phi(\mathbf{x})$ is the density for G and $f_{i,\phi}(\mathbf{x})$ is the density for F_i, then

$$g_\phi(\mathbf{x}) = \sum_{i=1}^{k} \pi_i f_{i,\phi}(\mathbf{x})$$

Here ϕ denotes the vector $\begin{pmatrix} \pi \\ \theta \end{pmatrix}$ where

$$\pi = \begin{bmatrix} \pi_1 \\ \pi_2 \\ \vdots \\ \pi_k \end{bmatrix}$$

and θ is a vector of unknown parameters defining the distributions F_i for $i = 1$, $2, \ldots, k$. If we assume that all the f_i belong to the same parametric family of distributions, there will be an identifiability problem with the mixtures, since any rearrangement of the indices will not change the likelihood.

One way to overcome this problem is to define the ordering so that $\pi_1 \geqslant \pi_2 \geqslant \pi_3 \geqslant \cdots \geqslant \pi_k$. Once this ambiguity is overcome, maximum likelihood estimates of the parameters can be obtained by solving the system of equations

$$\partial L\phi / \partial \phi = 0 \qquad (8.1)$$

where L is the log of the likelihood function and ϕ is the vector of parameters.

The method of solution is the EM algorithm of Dempster, Laird, and Rubin (1977), which was applied earlier to specific mixture models by Hasselblad (1966, 1969), Wolfe (1967, 1970), and Day (1969). They recognized through manipulation of Eq. 8.1 that if we define the a posteriori probabilities

$$\tau_{ij}(\phi) = \tau_i(x_j; \phi) = \text{probability that } x_j \text{ comes from } F_i$$

$$= \pi_i f_{i,\theta}(x_j) \Big/ \sum_{l=1}^{k} \pi_l f_{l,\theta}(x_j) \quad \text{for } i = 1, 2, \ldots, k$$

then

$$\hat{\pi}_i = \sum_{j=1}^{n} \frac{\hat{\tau}_{ij}}{n} \quad \text{for } i = 1, 2, \ldots, k$$

and

$$\sum_{i=1}^{k} \sum_{j=1}^{n} \hat{\tau}_{ij} \frac{\partial \log f_{i,\theta}(x_j)}{\partial \hat{\theta}} = 0$$

There are many issues related to the successful application of the EM algorithm in various parametric problems including the choice of starting values. For details, the reader should consult McLachlan and Basford (1988) or the recent book by McLachlan and Krishnan (1997), dedicated solely to the EM algorithm.

The approach taken in McLachlan and Basford (1988), when choosing the number of distributions, is to compute the generalized likelihood ratio statistics, which compare the likelihood under the null and alternative hypotheses. The likelihood ratio λ is then used to accept or reject the null hypothesis. As in classical theory, the quantity $-2 \log \lambda$ is used for the test, but since the usual regularity conditions are not satisfied, its asymptotic null distribution is not chi-square.

As a simple example, consider the test that the observations come from a single normal distribution versus the alternative that they come from the mixture of two normal distributions. In this case, the mixture model is

$$f_\theta(\mathbf{x}) = \pi_1 f_{1,\theta}(\mathbf{x}) + (1 - \pi_1) f_{2,\theta}(\mathbf{x}).$$

Under the null hypothesis $\pi_1 = 1$ and this solution is on the boundary of the parameter space. This is precisely the reason why the regularity conditions fail.

In certain special cases the asymptotic distribution of $-2 \log \lambda$ has been derived. See McLachlan and Basford (1988, pp. 22–24) for examples. In the general formulation, we assume a nested model (i.e., under H_0, the number of distributions is k, whereas under the alternative H_1, there are $m > k$ distributions).

The bootstrap prescription (assuming n observations in the original sample) is as follows: (1) compute the maximum likelihood estimates of the parameters under H_0; (2) generate a bootstrap sample of size n from the mixture of k distributions defined by the maximum likelihood estimates in step (1); (3) calculate $-2 \log \lambda$ for the bootstrap sample; (4) repeat steps (2) and (3) many times.

The bootstrap distribution of $-2 \log \lambda$ is then used to approximate its distribution under H_0. Since critical values are required for the test, estimation of the tails of the distribution are required. McLachlan and Basford recommend at least 350 bootstrap samples.

In order to get an approximate α level test, suppose we have m bootstrap replications with

$$\alpha = 1 - j/(m + 1)$$

then we reject H_0 if $-2 \log \lambda$ for the original data exceeds the jth smallest value from the m bootstrap replications.

For the simple case of choosing between one normal distribution and the mixture of two normals, McLachlan (1981) performed simulations to show the improvement in the power of the test as m increased from 19 to 99. A number of applications of the above test to clustering problems can be found in McLachlan and Basford (1988).

8.4. CENSORED DATA

The bootstrap has also been applied to some problems involving censored data. Efron (1981a) considers the determination of standard error and confidence interval estimates for parameters of an unknown distribution when the data are subject to right censoring. He applies the bootstrap to the Channing House data first analyzed by Hyde (1980). Additional discussion of the problem can be found in Efron and Tibshirani (1986).

The Channing House in Palo Alto, California, is a retirement center. The data consist of 97 men who lived in the Channing House during the period from 1964 (when it opened) until July 1, 1975 (when the data were collected). During this period 46 residents died while in Channing House while the remaining 51 were still alive at the censoring time, July 1, 1975.

The Kaplan–Meier survival curve (see Kaplan and Meier, 1958) is the standard method for estimating the survival distribution and Greenwood's formula is the standard way to estimate the standard error of the survival curve.

Efron (1981a) compared a bootstrap estimate of the standard error of the Kaplan–Meier curve with Greenwood's formula when applied to the Channing House data. He found very close agreement between the two methods. He also considered bootstrap estimates of the median survival time and percentile method confidence intervals for the median survival time.

Censored data such as the Channing House data consists of a bivariate vector (x_i, d_i), where x_i is the age of the patient at the time of death or at the censoring date and d_i is an indicator variable that is zero if x_i is censored and one if it is not. In the case of the Channing House data, x_i is recorded in months. So, for example, the pair (790, 1) would represent a man who died in the house at age 790 months, while (820, 0) would represent a man living on July 1, 1975, who was 820 months old at that time.

A statistical model for the survival time and the censoring time as described in Efron and Tibshirani (1986, pp. 64–65) is used to describe the mechanism for generating the bootstrap replications. A somewhat simpler approach treats the pair (x_i, d_i) as an observation from a bivariate distribution F.

Simple sampling with replacement from F leads to the same results as the more complex approach, which takes account of the censoring mechanism. The conclusion is that the bootstrap provides appropriate estimates of standard errors when the usual assumptions about the censoring mechanism fail to hold. This result is very reminiscent of the result for bootstrapping paired data in simple regression as compared with bootstrapping residuals from the model.

8.5. *p*-VALUE ADJUSTMENT

In various studies where multiple testing is used or in clinical trials where interim analyses are conducted, the *p*-value based on one hypothesis test is no longer appropriate. The simultaneous inference approach can be used and is covered well in Miller (1981b). Bounds on the *p*-value such as the Bonferroni inequality can be very conservative.

A bootstrap approach to p-value adjustment in the multiple testing framework was devised by Westfall (1985) in dealing with multiple binomial tests and the general approach to a variety of problems is given in the text by Westfall and Young (1993). They also provide a permutation approach to p-value adjustment. Software implementation is available in SAS Version 6.12.

8.5.1. Passive Plus DX Example

The Passive Plus DX premarket approval (PMA) clinical report provides an example where such p-value adjustment using the machinery of Westfall and Young could have been applied. The Passive Plus study is similar to the Tendril DX study described in Section 3.3.1. Passive Plus DX is a steroid eluting lead similar to Tendril DX. The only difference is the way the lead is placed in the heart and consequently the tips of the leads differ.

The Tendril leads are active fixation, which means they attach to the tissue with a screw in the tip. The Passive Plus leads are placed on the tissue (in the atrial appendage for atrial leads). Fibrous tissue grows around the lead to improve the placement over time. Dislodgement of the leads is always a possibility and is commonly detected through impedance measurements and/or threshold testing. In the event of a dislodgement a new procedure is required to reattach the lead. It is referred to as repositioning.

Just as with the Tendril lead, there is a nonsteroid version of the Passive Plus lead that serves as a concurrent control in a randomized control clinical trial with the same objective for showing improvement in capture threshold.

An unbalanced 3:1 ratio of steroid to nonsteroid leads was used so that most patients would get the investigational device, which is known from studies with other manufacturers to improve thresholds and reduce inflammation. Analyses were conducted in the same way as for the Tendril DX study.

Another aspect of these trials (both the Tendril DX study and the Passive Plus DX trial) is the use of interim reports. It is common to have meetings in the early phase of the study to make sure that there are no unusual adverse events that would raise safety issues that could lead to the termination of the study. Often there are data and safety monitoring boards and institutional review boards that want to review the data to get an idea of the risk–benefit trade-offs in the study.

Commonly, since these are multicenter studies, the investigators want the meetings because they are curious to see how the study is progressing and they want to share their experiences with the other investigators. These meetings are also used to ensure that all the centers are complying properly with the protocol. The FDA is also interested in these interim reports, which are generated at the time of these meetings.

The FDA is primarily interested in the safety aspects (report on complications). The study would not be terminated early for better than expected performance, based on interim report results, since the design was a prospective fixed sample size design.

For a group sequential design, the interim points could be chosen prospectively at patient enrollment times when sufficiently good performance would dictate successful termination of the study. However, the medical device center at the FDA usually doesn't allow such group sequential trials.

In the Passive Plus study, there were two or three interim reports and a final report. Because repeated comparisons were made on capture thresholds for the treatment and control groups, the FDA views the study as a repeated significance test and requires that the sponsor provide a *p*-value adjustment to account for the multiplicity of tests.

Although the Bonferroni inequality provides a legitimate conservative bound on the *p*-value, it would be advantageous to the sponsor (St. Jude Medical) to provide an acceptable but more accurate adjustment to the individual *p*-value. The bootstrap adjustment using Proc MULTTEST in SAS was planned as the method of adjustment.

The results of the test showed that the individual *p*-values for the main efficacy variables were less than 0.0001 and hence the question became moot. Even the Bonferroni bound based on three interim reports and one final report would only multiply the *p*-value by a factor of 4 and hence the adjusted *p*-value is less than 0.0004. The bootstrap adjusted *p*-value would be less than the Bonferroni bound.

8.5.2. Consulting Example

In this example a company conducted a clinical trial for a medical treatment in one country, but due to slow enrollment they chose to extend the trial into several other countries. In the first country, which we shall call country E, the new treatment appeared to be more effective than the control, but this was not the case in the other countries (which we shall call A, B, C, and D). Fisher's exact test was used to compare the failure rates in each of the other countries with country E. However, this required four pairwise tests, which were not part of the original plan and hence *p*-value adjustment is appropriate. In this case the Bonferroni bound is too conservative and so a bootstrap *p*-value adjustment was used. We provide a comparison of the *p*-value based on a single test (referred to as the raw *p*-value) with both the Bonferroni and the bootstrap *p*-value adjustment.

Table 8.1 provides the treatment data. For each pair of countries (A vs. E, B vs. E, C vs. E, and D vs. E), *p*-values and adjusted *p*-values were determined using Proc MULTTEST in SAS. The results comparing the three *p*-values are provided in Table 8.2. The bootstrap *p*-values are obtained using the default number of bootstrap replications (20,000).

Table 8.2 shows the conservativeness of the Bonferroni bound. Clearly in each case *p*-value adjustment makes a difference. In the case of country E versus country B the *p*-value is clearly significantly small by either method. In E versus A, however, the raw *p*-value suggested statistical significance at the 5% or 10% significance levels while Bonferroni does not. Note that the bootstrap *p*-value is statistically significant at the 10% level.

Table 8.1   Comparison of Treatment Failure Rates

Country	Treatment Failure Rate
A	40% (18/45)
B	41% (58/143)
C	29% (20/70)
D	29% (51/177)
E	22% (26/116)

Table 8.2   Comparison of *p*-Value Adjustments

Countries	Raw *p*-Value	Bonferroni *p*-Value	Bootstrap *p*-Value
E versus A	0.0307	0.1229	0.0855
E versus B	0.0021	0.0085	0.0062
E versus C	0.3826	1.0000	0.7654
E versus D	0.2776	1.0000	0.6193

These results show that country E is different from at least one and possibly two of the other countries. This could be used by the sponsor to argue the merits of the treatment compared to the control in country E (ignoring the results from the other countries).

8.6. PROCESS CAPABILITY INDICES

In many manufacturing companies, process capability indices, which measure how well the production process behaves relative to specification limits, are popular performance measures. These indices were used by the Japanese in the 1980s as part of their quality improvement movement. American automobile companies were the first to do this, but now it is being incorporated in product performance qualification that is part of the required process validation for manufactured devices that are regulated by the FDA.

In the late 1980s and the 1990s there has been a major movement in industry toward improved quality. This has had both good and bad aspects. By following the teachings of Deming, Juran, Taguchi, and some of the other quality guru's, many useful statistical methods including control charts, design of experiments, tolerance intervals, response surfaces methods, evolutionary operation, and process capability indices are being introduced in the manufacturing processes at many companies where they have never been seen before.

Unfortunately, however, there seems to be a bandwagon effect where programs are developed under the buzzwords such as Total Quality Management (TQM) or six sigma. Some companies have been successful with this approach, including General Motors, General Electric, Motorola, and AT&T. For each success story you will find a company with a strong statistics organization or at least an appropriate use of statistical expertise. There have been numerous failures where programs were implemented too quickly and without careful thought.

The problem is that a number of these companies would institute these programs without bringing in the necessary statistical expertise. Some procedures that require Gaussian assumptions are being blindly applied to processes where the Gaussian distribution is not applicable. This has certainly been the story with process capability indices and it has led to major problems.

As the senior statistician at Pacesetter, there were numerous occasions when engineers would come to me to complain that they could not get their leads to pass a qualification test either because the process capability index was too low or the statistical tolerance interval had limits that fell outside the specification limits. In most cases they had a legitimate complaint. They would test 20 to 30 leads and have none of the test measurements fall outside the specification limits.

Without exception, I would find that the problem was with the statistical method being employed. Gaussian assumptions were crucial to the method and the data did not fit the Gaussian model.

In the case of the tolerance intervals, the intervals were too wide in one direction because the data were either highly skewed or had a very short tail on one end. Application of nonparametric tolerance intervals usually solved the problem (when the increased sample size requirement was tolerable).

When they applied process capability indices to these data, they generally wanted to see numbers like 1.0 or 1.33, which were standards based on the Gaussian distribution. Consequently, their capability indices failed to meet these standards.

Here the bootstrap can play a very helpful role by removing the requirement for normal distributions. Kotz and Johnson (1993) were motivated to write their book on process capability indices to clear up much of the confusion that was caused by applying the Gaussian assumption where it didn't belong. They also wanted to show that these indices, along with some modifications, do have a useful place in quality assurance, as long as the right statistical methods are employed and we do not rely on just one to characterize the capability of a process.

A good historical account can be found in Ryan (1989, Chap. 7). Generally, capability indices are estimated from data and the distributional theory of these estimates is best covered in Kotz and Johnson (1993). One of the most popular indexes when both lower and upper specifications exist is C_{pk}.

Let m be the process mean and σ the process standard deviation (the process is assumed to be stable and under control); let LSL denote the lower specification limit and let USL denote the upper specification limit. Then C_{pk} is defined to be the minimum of $(\text{USL} - m)/(3\sigma)$ and $(m - \text{LSL})/(3\sigma)$. In practice, m and σ are estimated from data and hence the process capability index is estimated.

For Gaussian distributions, confidence intervals can be generated and hypothesis tests performed based on tabulated values for the estimated indices. In practice, the process may have a highly skewed distribution or have one or two short tails.

Non-Gaussian processes have been treated in a variety of ways. Many of these are discussed in Kotz and Johnson (1993, pp. 135–161). The seminal paper of Kane (1986) devotes only a short paragraph to this topic in which Kane concedes that nonnormality is a common occurrence in practice and that the resulting confidence intervals could be sensitive to departures from normality.

Kane states: "Alas, it is possible to estimate the percentage of parts outside the specification limits, either directly or with a fitted distribution. This percentage can be related to an equivalent capability for a process having a normal distribution."

The point Kane is making is that in some cases a lower capability index for a particular distribution (say, one with short tails) has significantly less probability outside the specification limits than for a process with a normal distribution with the same mean and variance. Either the associated probabilities outside the specification limits should be specified or the capability index should be reinterpreted. Kane is suggesting transforming the index for the actual distribution to a comparable one for a corresponding Gaussian distribution. This approach is suggested because managers are now used to thinking in terms of these standard values for Gaussian indices.

Another approach is taken by Gunter in a series of four papers in the journal *Quality Progress*. In two of Gunter's (1989a, b) papers he emphasizes the difference between "perfect" (precisely normal) and "occasionally erratic" processes (i.e., a mixture of two normal distributions with the mixture proportion p close to one for the basic normal distribution and $1 - p$ small, the mixture proportion for the erratic normal distribution). The erratic distribution has a different mean and a larger variance than the basic distribution.

Gunter considered three types of distribution—(1) a highly skewed distribution, central chi-square with 4.5 degrees of freedom, (2) a heavily tailed symmetric distribution, Student t with 8 degrees of freedom, and (3) a uniform distribution. Using these exact distributions, Gunter shows what the expected number of nonconforming parts outside the 3σ limits of the mean are out of one million parts chosen from these distributions (which are standardized by a shift and scale transformation so that they all have mean zero and variance 1).

The results are strikingly different, indicating how important the tails of the distribution are in relation to the interpretation of process capability indices. For the chi-square distribution 14,000 are outside the limits but all of them are above the 3σ limit, none below -3σ. For the t distribution 4000 are outside the limit (2000 below -3σ and 2000 above $+3\sigma$).

For the uniform case none are outside the limits! This is to be contrasted to the 2700 in the case of normality (1350 below -3σ and 1350 above plus 3σ). The t distribution had significantly more cases in each tail, the skewed chi-square distribution with 14,000 had many more in total but all on one side, and the uniform case had such short tails that none fall outside the 3σ limits.

To consider cases where the true parameters are unknown and must be estimated requires simulation. English and Taylor (1990) did extensive simulations using normal, triangular, and the short-tailed uniform distributions. We notice a major difference, as can be seen in Table 4.1 of Kotz and Johnson (1993, p. 139). The table in Kotz and Johnson (1993) is taken from the English and Taylor (1990) paper. It shows results similar to Gunter's, as we described above.

This points to one of the factors that caused problems in the use of capability indices. Companies accepted that their indices were characterizing Gaussian distributions without realizing that they were even making the assumption.

This is implicit in setting a value of 1.33 as a goal for an index. The real goal is to have a very small percentage of the product outside the specification limit. But instead of trying to directly measure that goal, a C_{pk} is estimated and the value is compared to a standard that is based on its interpretation for Gaussian data.

A second problem with the naive use of sample estimates of process capability indices is that the companies treat the estimate as though it were the true value of the index. They forget that there is always some statistical uncertainty in the estimate whose magnitude depends on the number of tests. This uncertainty must be taken into account in the inference process.

Taking a theoretical approach, Kocherlakota, Kocherlakota, and Kirmani (1992) derive the exact distribution for C_p (a slightly different process capability index) in two special cases (one of which is the mixture of two normal distributions where the two distributions have the same variance). Price and Price (1992) study the expected value of the estimated C_{pk} via simulation for a large number of distributions.

Another approach for dealing with indices from production processes with non-Gaussian distributions is to bootstrap. Bootstrapping is fairly straightforward. USL and LSL are fixed. Ordinary bootstrap samples of size n are generated and the mean and standard deviation are estimated for each bootstrap sample.

The index C_{pk}, for example, is then computed for each bootstrap sample based on the above definition with the sample estimates of mean and standard deviation replacing the population parameters. Bootstrap confidence intervals or hypothesis tests for C_{pk} can then be generated using the methods discussed in Chapter 3. This approach is very general and can be applied to any of the commonly used capability indices and not just for C_{pk}.

Various approaches are discussed in Kotz and Johnson (1993, pp. 161–164) and in the original papers by Franklin and Wasserman (1991) and Price and Price (1992). They involve the use of the confidence interval methods that were covered in Chapter 4, including Efron's percentile method and the bias-corrected percentile method.

In reviewing this work, Kotz and Johnson (1993) point to Schenker's (1985) paper as an indication of problems with the bias-corrected method. Apparently no one had taken the step to apply BC_a or bootstrap iteration although Kotz and Johnson (1993) point to Hall's work on bootstrap iteration as an approach that could possibly provide additional improvement.

Improvements in statistical software packages now make it easier to routinely apply process capability indices. For example, SAS in Release 6.12 has a procedure called Proc CAPABILITY, which can be used to calculate various capability indices, provide standard univariate analyses, and test normality assumptions.

The following example illustrates the use of bootstrap confidence intervals in capability studies based on the depth of lesions in catheter ablation tests. It is clear from the Shapiro–Wilk tests, the stem and leaf diagrams, the boxplots, and the normal probability plots that the data depart sufficiently from a Gaussian distribution to require a robust or nonparametric approach like the bootstrap.

See Fig. 8.2 (SAS output from Proc UNIVARIATE), which shows the results of the Shapiro–Wilk test, provides the stem and leaf graph, the boxplot and the normal probability plot for catheter number 12 using SAS. The p-value from the test is 0.0066. The stem and leaf plot and the boxplot show a positive skewness (the sample estimate of skewness is 0.7031).

The data are a summary of 30 samples of lesions generated with catheter number 12 on a beef heart. These catheters are used to create lesions generated from tissue heating caused by RF energy applied to particular locations in a live human heart. The purpose of the lesion is to destroy nerve tissue, which otherwise would stimulate the heart, causing improper fast beating in either the atrial or the ventricular chamber. If the generated lesion is too small the nerve tissue will not be destroyed. If it is too large there is a danger of causing damage to the heart.

To assure that the catheter along with the RF generator will produce safe lesions in humans, tests are conducted on live animals and/or on beef hearts taken from slain cattle. There is no way to measure the dimensions of the lesions inside a living heart. Consequently, a surrogate measure of effectiveness of the catheter and the RF generator is obtained through the measurement of the lesion dimensions in beef hearts.

In these tests the length, width, and depth of each lesion are measured. The performance of the catheter and the generator in these tests is a basis for predicting their performance for the catheter in humans. For catheter number 12, the average lesion depth was 6.650 mm (millimeters) with a standard deviation of 0.852 mm. The minimum value of the lesion depth was 5.000 mm

Univariate Procedure Summary for Navistar DS Lesion Dimensions

Univariate Procedure

Variable = DEPTH

Moments

N	30	Sum Wgts	30
Mean	6.65	Sum	199.5
Std Dev	0.852481	Variance	0.726724
Skewness	0.703112	Kurtosis	0.819517
USS	1347.75	CSS	21.075
CV	12.81927	Std Mean	0.155641
T:Mean=0	42.72652	Pr>|T|	0.0001
Num ^=0	30	Num>0	30
M(Sign)	15	Pr>=|M|	0.0001
Sgn Rank	232.5	Pr>=|S|	0.0001
W:Normal	0.895608	Pr<W	0.0066

Quantiles(Def=5)

100% Max	9	99%	9
75% Q3	8	95%	7
50% Med	8	90%	6.75
25% Q1	6	10%	6
0% Min	5.5	5%	5
		1%	5
Range	4		
Q3-Q1	1		
Mode	6		

Extremes

Lowest	Obs	Highest	Obs
5	1)	7	30)
5.5	5)	8	8)
6	26)	8	9)
6	24)	8	13)
6	23)	9	14)

Stem	Leaf	#	Boxplot
9	0	1	0
8			
8	000	3		
7			
7	00000000000	11	+-----+
6	55	2	*--+--*
6	00000000000	11	+-----+
5	5	1		
5	0	1	0

Normal Probability Plot

Figure 8.2 Summary statistics for Navistar DS lesion depth dimension from original SAS univariate procedure output.

139

and the maximum value was 9.000 mm. Similar results were found for other catheters.

In theory, there should be a maximum value beyond which lesion size would be so large as to present a danger to the patient while too small a value could indicate an ineffective catheter or catheter–generator system. Consequently, upper and lower specification limits could possibly be determined. In practice no such limits have been determined and the correlation with lesions for human subjects cannot be determined.

For illustration purposes, we shall assume an upper specification limit of 10 mm and a lower specification limit of 4 mm and that the target value is 7 mm. Figure 8.3 is SAS output from Proc CAPABILITY, which shows the estimate of the C_{pk} index and other capability indices based on the 30 sample lesions in a beef heart using catheter number 12. For C_{pk}, we observe an estimate of 1.036191. We shall use the bootstrap to estimate the variability of this estimate and to provide confidence intervals.

Next, we generate bootstrap samples of size 30 to determine the sampling distribution of the capability index. The bootstrap percentile method confidence interval is then generated to illustrate the technique.

The C_{pk} index is a natural index to use here since the specification limits are two-sided. We want a strong enough lesion to be sure that it is effective. Yet we do not want the lesion to be too deep as to make it cause a perforation or other complication for the patient.

Another problem in the medical device industry that could involve a two-sided specification is the seal strength of sterile packaging. The seal requires enough strength so that the sterility of the package will not be compromised in shipping. On the other hand, it must not be too difficult to open when needed at the time of a procedure.

We now return to the lesion data example. We apply bootstrap resampling and Efron's percentile method approach. A bootstrap histogram for C_{pk} is shown in Fig. 8.4. I used the statistical package Resampling Stats, a product of Resampling Stats, Inc., to generate the 95% bootstrap percentile method confidence limits for C_{pk}. The computation of the capability indices required generating some code in the Resampling Stats Programming Language. The actual code that I used is presented as Fig. 8.5 for the bootstrap percentile method.

The figures show both the code used in the execution of the program and the output of the run. In each case I used 10,000 bootstrap replications to produce the confidence limits and the program executed very rapidly (19.7 seconds total execution time for the percentile method). Resampling Stats provides a very readable user's manual (see Resampling Stats, 1997).

The results showed that the estimated C_{pk} was 1.0362. The percentile method confidence interval is [0.84084, 1.4608], showing a great deal of uncertainty in the estimate. Perhaps to get a good estimate of C_{pk} a sample size much larger than 30 should be used.

Capability Indices for Lesion Data

CAPABILITY

Variable = DEPTH

Moments

N	30	Sum Wgts	30
Mean	6.65	Sum	199.5
Std Dev	0.852481	Variance	0.726724
Skewness	0.703112	Kurtosis	0.819517
USS	1347.75	CSS	21.075
CV	12.81927	Std Mean	0.155641
T:Mean=0	42.72652	Pr>|T|	0.0000
Sgn Rank	232.5	Pr>=|S|	0.0000
Num ^=0	30		
W:Normal	0.895608	Pr<W	0.0066

Quantiles (Def=5)

100% Max	9	99%	9	
75% Q3	7	95%	8	
50% Med	6.75	90%	8	
25% Q1	6	10%	6	
0% Min	5	5%	5.5	
		1%	5	
Range	4			
Q3-Q1	1			
Mode	6			

Extremes

Lowest	Obs	Highest	Obs
5	1)	7	30)
5.5	5)	8	8)
6	26)	8	9)
6	24)	8	13)
6	23)	9	14)

Indices

CPL	1.036191	CPU	1.309902
CP	1.173046	CPK	1.036191
K	0.116667	CPM	1.085148

WARNING: Normality rejected at $\alpha = 0.05$

Specifications

LSL	4	% < LSL	0
Target	7		
USL	10	% > USL	0
		% Between	100

Figure 8.3 Capability statistics for Navistar DS lesion depth dimension from original SAS capability procedure output for catheter no. 12.

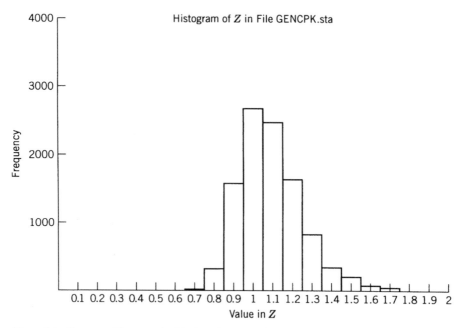

Figure 8.4 Bootstrap histogram for Navistar DS lesion depth dimension based on original sample of 30 for catheter number 12.

8.7. MISSING DATA

Many times in clinical studies complete data is not recorded on all patients. This is particularly true when quality of life surveys are conducted. In such studies many standard statistical analyses can be conducted only on complete data. Incomplete data can be analyzed directly using the likelihood approach and the EM algorithm.

Alternatively, imputation of all missing values can lead to an artificial complete data set that can then be analyzed by standard methods. However, it is well known that even if the imputation method and assumed missing data model are appropriate, single imputation tends to underestimate the variability of the result.

Multiple imputation is an accepted technique for reconstructing the appropriate variability in the result. In a real clinical study at Eli Lilly, Stacy David compared multiple imputation with other imputation methods using a Bayesian bootstrap. She reported her preliminary findings at an FDA sponsored workshop in Crystal City, Virginia, in September 1998.

Detailed results are not yet available. The Bayesian bootstrap was used in David's study to show the value of multiple imputation in certain missing data problems, particularly when compared to a single imputation method called LOCF, which is commonly used in drug development.

Start execution.

```
DATA (5.0 6.0 6.5 7.0 5.5 6.0 6.5 8.0 8.0 7.0 7.0 6.0 8.0 9.0 6.0 6.0 6.0 6.0 7.0 7.0 6.0 7.0 6.0
6.0 7.0 6.0 7.0 7.0 7.0 7.0) DEPTH
MAXSIZE Z 15000
MEAN DEPTH MD
STDEV DEPTH SIGD
SUBTRACT 10 MD USLNUM
SUBTRACT MD 4 LSLNUM
MULTIPLY SIGD 3 DENOM
DIVIDE USLNUM DENOM CPU
DIVIDE LSLNUM DENOM CPL
PRINT CPU CPL MD SIGD
CONCAT CPU CPL CP
MIN CP CPK
PRINT CPK
REPEAT 10000
   SAMPLE 30 DEPTH BOOTD
   MEAN BOOTD MDS
      STDEV BOOTD SIGD$
   SUBTRACT 10 MD$ USLNUM$
         SUBTRACT MD$ 4 LSLNUM$
         MULTIPLY SIGD$ 3 DENOM$
   DIVIDE USLNUM$ DENOM$ CUP$
         DIVIDE LSLNUM$ DENOM$ CPL$
         CONCAT CPU$ CPL$ CP$
   MIN CP$ CPK$
   SCORE CPK$ Z

END
```

Vector no. 1: z

```
HISTOGRAM Z
PERCENTILE Z (2.5 97.5) K
PRINT CPK K
CPU  =  1.3099
CPL  =  1.0362
MD   =  6.65
SIGD =  0.85248
CPK  = 1.0362
```

Bin Center	Freq	Pct	Cum Pct
0.7	10	0.1	0.1
0.6	306	3.1	3.2
0.9	1554	15.5	18.7
1	2668	26.7	45.4
1.1	2461	24.6	70.0
1.2	1603	16.0	86.0
1.3	804	8.0	94.1
1.4	327	3.3	97.3
1.5	184	1.8	99.2
1.6	52	0.5	99.7
1.7	23	0.2	99.9
1.8	5	0.1	100.0
1.9	3	0.0	100.0

Note: Each bin covers all values within 0.05 of its center.

CPK = 1.0362
K   = 0.84084   1.4608

Successful execution (19.7 seconds)

Figure 8.5 Resampling Stats code and output for the bootstrap percentile method 95% confidence interval for Navistar DS lesion depth dimension based on data from catheter number 12.

8.8. POINT PROCESSES

A point process is a collection of events or occurrences over a particular period of time. Examples include the time of failures for a medical device, the time to recurrence of an arrhythmia, the lifetime of a patient with a known disease, and the time to occurrence of an earthquake. There are a wealth of possible examples.

Point processes are studied either in terms of the time between events or by the number of events occurring in a given interval of time. The simplest point process is the Poisson process, which occurs under simple assumptions including (1) rarity of events and (2) that the expected number of events in an interval of length t equals λt, where λ is known as the rate or intensity of the process.

The simple Poisson process gets its name because it has a Poisson distribution with parameter λ for the number of events occurring in an arbitrarily chosen interval of unit length. For a Poisson process, the times between events are independent and identically distributed exponential random variables with rate parameter λ.

The mean time between events for a Poisson process is λ^{-1}. The Poisson process is a special case of larger classes of point processes such as stationary processes, renewal processes, and regenerative processes.

In addition to being able to define point processes on the time line, we can define them in higher dimensions where time is replaced by spatial coordinates. In general, point processes can be determined by their joint probability distributions over subsets (intervals) of the time line or regions in space.

There are also classes of nonstationary point processes. A point process that has its instantaneous rate vary with time but still has joint distributions that are Poisson and independent over disjoint regions or intervals is called a nonhomogeneous or inhomogeneous Poisson process. Inhomogeneous Poisson processes are nonstationary.

Detailed theoretical treatment of point processes can be found in Daley and Vere-Jones (1988) and Bremaud (1981). Murthy (1974) presents results for renewal processes along with medical and engineering applications. Bootstrap applications to homogeneous and inhomogeneous Poisson processes are presented in Davison and Hinkley (1997, pp. 415–426).

There are some very simplistic approaches to resampling a point process that Davison and Hinkley describe. The simplest is to randomly select n samples from the set of observed events. This relies on the independence assumption for Poisson processes and hence is somewhat restrictive.

A different approach, for an inhomogeneous Poisson process, would be to compute a smoothed estimate of the intensity (rate) function based on the original data and generate realizations of the inhomogeneous Poisson process with the estimated intensity function via Monte Carlo. This would be a form of parametric bootstrapping. Davison and Hinkley take such an approach with a point process of neurophysiological data.

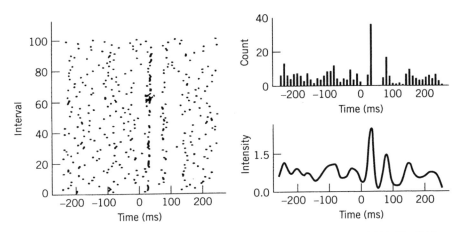

Figure 8.6 Neurophysiological point process. [From Davison and Hinkley (1997, Fig. 8.15, p. 418), with permission from Cambridge University Press.]

Figure 8.6 is taken from Davison and Hinkley (1997). The data were generated by Dr. S. J. Boniface of the Clinical Neurophysiology Unit at the Radcliff Infirmary in Oxford, England. It is a study of human subjects responding to a stimulus. The stimulus was applied 100 times and points in the figure indicate times in a 250-millisecond interval for which the response (firing of a motoneuron) was observed. Theoretical considerations indicate that the process obtained by superimposing the 100 intervals will be Poisson with a time varying intensity (an inhomogeneous Poisson process). The left panel in the figure shows the sample of this point process. The two right panels show a histogram of the superimposed occurrences (the upper right panel) and a rescaled kernel estimate of the intensity function $\lambda(t)$.

We would like to put confidence bounds on this estimated intensity function. Two-sided 95% bootstrap confidence bands are presented in Fig. 8.7, which is also from Davison and Hinkley (1997).

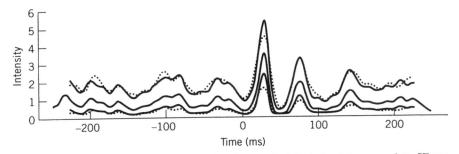

Figure 8.7 Confidence bands for the intensity of neurophysiological point process data. [From Davidson and Hinkley (1997, Fig. 8.15, p. 418), with permission from Cambridge University Press.]

8.9. HISTORICAL NOTES

The application of the bootstrap to kriging was done by Switzer and Eynon and documented in unpublished work at Stanford [later published as Eynon and Switzer (1983)]. The results were summarized in Diaconis and Efron (1983), who referred to the unpublished work. Cressie (1991) deals with kriging and bootstrapping in the case of spatial data and remains the best source for information on resampling methods applied to spatial data.

Hall (1985, 1988c) provides thoughtful accounts for applying bootstrap approaches with spatial data. There are many other excellent texts that deal with spatial data including Bartlett (1975), Diggle (1983), Ripley (1981, 1988), Cliff and Ord (1973, 1981), and Upton and Fingleton (1985, 1989). Matheron (1975) provides some fundamental probability for spatial data. Mardia, Kent, and Bibby (1979) deal in part with spatial data in the context of multivariate analysis.

The idea of applying the bootstrap to determine the variability of subset selection procedures was due to Gong. Her analysis of the Gregory data was an important part of her Stanford dissertation. Results on this theme are reported in Efron and Gong (1981, 1983) and Gong (1982, 1986).

Gong (1986) is essentially Gail Gong's Stanford Ph.D. dissertation. In emphasizing the variety of applications and power of computer-intensive techniques such as the bootstrap, Diaconis and Efron (1983) also discuss Gong's results.

Miller (1990) deals with model selection procedures in regression. A general approach to model selection based on information theory and called stochastic complexity is given in Rissanen (1989).

Applications of the bootstrap to determine the number of distributions in a mixture model can be found in McLachlan and Basford (1988). A simulation study of the power of the bootstrap test for a mixture of two normals versus a single normal distribution is given in McLachlan (1987).

Work on censored data applications of the bootstrap began with Turnbull and Mitchell (1978) who considered complicated censoring mechanisms. Efron (1981a) is a seminal paper on the subject.

The more advanced confidence interval procedures such as Efron (1987) and Hall (1988b) could probably provide improvement over the procedures discussed in Efron (1981a). Reid (1981) provides an approach to estimating the median of the Kaplan–Meier survival curve based on influence functions.

Akritas (1986) compared variance estimates for median survival using the Efron (1981a) approach with the Reid (1981) approach. Altman and Andersen (1989), Chen and George (1985), and Sauerbrei and Schumacher (1992) apply case resampling methods to survival data models such as the Cox proportional hazard model. There is no known theory to support this approach.

Applications of survival analysis methods and related reliability studies can be found in Miller (1981a) and Meeker and Escobar (1998), to name two out of many. Csorgo, Csorgo, and Horvath (1986) apply results for empirical processes to reliability problems.

The bootstrap application to p-value adjustment was first given in Westfall (1985) and was generalized in Westfall and Young (1993). Their approaches using both bootstrap and permutation methods have been implemented in statistical software. In particular, SAS Version 6.12 has a procedure called MULTTEST, which provides bootstrap and permutation resampling approaches to multiple testing p-value adjustment along with the various classical bounds (e.g., Sidak and Bonferroni) as illustrated in Section 8.5 for Bonferroni. Bootstrap application to multiple testing problems is also discussed in Noreen (1989).

Initial work on the distribution of process capability estimates is due to Kane (1986). The bootstrap work was described in Franklin and Wasserman (1991, 1992, 1994), Wasserman, Mohsen, and Franklin (1991), and Price and Price (1992). A review of this work and a general account of capability indices can be found in Kotz and Johnson (1993).

The Bayesian bootstrap was originated by Rubin (1981). Dempster, Laird, and Rubin (1977) is the seminal paper on the EM algorithm that popularized and solved a number of important missing data problems. The literature on the EM algorithm now encompasses many important problems. A detailed up-to-date account can be found in the text by McLachlan and Krishnan (1997).

Rubin (1987) is the treatise on the multiple imputation approach to missing data problems. Efron (1994) is a key reference to resampling approaches to missing data problems. Rubin and Schenker (1998) present a current account of imputation methods including the usage of the Bayesian bootstrap. Key papers include their original works (Rubin and Schenker, 1987, 1991). Application of bootstrap techniques to imputation problems in survey sampling is presented in Shao and Sitter (1996).

Press (1989) is a good reference for the Bayesian approach to inference and Gelman, Carlin, Stern, and Rubin (1995) give a modern treatment with the recent advances in computation using Markov chain Monte Carlo methods. Maritz and Lwin (1989) present empirical Bayes methods as does the more recent text by Carlin and Louis (1996).

For inhomogeneous Poisson processes, examples sketched in this chapter along with an outline of the related theory can be found in Cowling, Hall, and Phillips (1996). Ventura, Davison, and Boniface (1997) describe another method with application to neurophysiological data. Diggle, Lange, and Benes (1991) provide an application of the bootstrap to a point process problem in neuroanatomy. Point processes are also referred to as counting processes by some authors.

Reliability analysis and life data analysis involve the study of censored survival times. These can be studied in terms of time to event or in terms of the number of events in an interval.

The relationship $P\{N(t) \geq n\} = P[X_1 \leq t, X_2 \leq t, \ldots, X_n \leq t]$, where $N(t)$ is the number of events in the interval $[0, t]$ and $X_1, X_2, \ldots, X_n$ are the times for the first n events, relates the counting process $N(t)$ with the time to events X_i for $i = 1, 2, \ldots, n$. This means that the properties of the lifetimes can be mathematically determined from the counting process. So results from the theory of counting processes can be used to obtain results about the times to events. One book that takes this approach is Anderson, Borgan, Gill, and Keiding (1993).

CHAPTER 9

When Does Bootstrapping Fail?

The question we raise in the title of this chapter shall be addressed in the narrow context of the nonparametric version of the bootstrap (i.e., sampling with replacement from the empirical distribution function F_n). There has been some controversy over exactly what situations are appropriate for bootstrapping.

In complex survey sampling problems and in certain regression problems some researchers have argued against the use of the bootstrap. On the other hand, early articles such as Diaconis and Efron (1983) and Efron and Gong (1983) painted a very rosy picture for the general application of the bootstrap.

It is not my intention to resolve the controversy here. The reader is referred to LePage and Billard (1992) for current research, which addresses this issue, although it deals more with extending the limits then identifying them. Mammen (1992b) provides mathematical conditions that allow the bootstrap to work. Needless to say, much still remains unresolved and hopefully these limitations of the bootstrap will be better defined after more research has been completed.

There are, however, some results known today of which the practitioner should be aware. We shall briefly describe, in Sections 9.2 and 9.3, examples where the "bootstrap algorithm" fails (i.e., replacing population parameters with sample estimates and sample estimates with estimates from a bootstrap sample). These examples are intended to caution the practitioner against blind application of bootstrap methods.

Before discussing these examples, we will point out in Section 9.1 why it is unwise to apply the bootstrap in very small samples. In general, if the sample size is very small (e.g., less than 5 in the case of a single parameter estimate), it is unwise to draw inferences from the estimate or to rely on an estimate of standard error to describe variability of the estimate.

Nothing magically changes when one is willing to apply the bootstrap Monte Carlo approximation to the data. Unfortunately, the name "bootstrap" may suggest, to some, "getting something for nothing." Of course, the bootstrap cannot do that! The proper interpretation should be "getting the most from the little that is available."

9.1. TOO SMALL OF A SAMPLE SIZE

For the nonparametric bootstrap, the bootstrap samples are, of course, drawn from a discrete set. Many exact and approximate results about these "atoms" of the bootstrap distribution can be found in results from classical occupancy theory or the theory of multinomial distributions. See Chernick and Murthy (1985) and Hall (1992a, Appendix I).

In Chapter 2, we saw examples where bootstrap bias correction worked better than cross-validation in the estimation of the error rate of a linear discriminant function. Although we have good reasons not to trust the bootstrap in very small samples and theoretical justification is asymptotic, the results were surprisingly good even for sample sizes as small as 14 in the two-class problem.

A main concern in small samples is that with only a few values to select from, the bootstrap sample will underrepresent the true variability as observations are frequently repeated and the bootstrap samples themselves repeat.

The bootstrap distribution is defined to be the distribution of the possible samples of size n drawn with replacement from the sample size of n. The number of possible samples or atoms is shown in Hall (1992a, Appendix I) to be

$$\binom{2n-1}{n} = (2n-1)!/[n!(n-1)!].$$

Even for n as small as 20, this is a very large number and consequently, when generating resamples by Monte Carlo, the probability of repeating a particular resample is small. For $n = 10$, the number of possible atoms is 92,378.

As a concrete example, Hall (1992a, Appendix I) shows that when $n = 20$ and the number of bootstrap repetitions $B = 2000$, the probability is greater than 0.95 that none of the bootstrap samples will repeat. This suggests that, for many practical problems, the bootstrap distribution may be regarded as though it were continuous. It also may suggest why researchers have found that smoothing the empirical distribution helps less than they expected.

Because the bootstrap distribution grows so rapidly with increasing n, exact computations of the bootstrap estimate are usually not possible (there are, of course, some exceptions where theoretical tricks are applied, e.g., the sample median) and the Monte Carlo approximations are necessary. For very small n, say, $n \leqslant 8$, it is feasible to calculate the exact bootstrap estimate [see Fisher and Hall (1990) for a discussion of such algorithms or Diaconis and Holmes (1994) for complete enumeration using Gray codes].

The practitioner should be guided by common statistical practice here. Samples of size less than 10 are usually too small to rely on sample estimates even in "nice" parametric cases. So we should expect that such sample sizes are also too small for bootstrap estimates to be of much use.

In many practical contexts, the number 30 is used as a "minimum" sample size. Justification for this can be found by noting the closeness of the t distribution to the Gaussian when $n \geqslant 30$. This suggests that, when sampling from Gaussian populations, the effect of the estimated standard deviation has essentially disappeared.

Also, the normal approximation to distributions such as the binomial or the sum of independent uniform random variables is usually very accurate for $n \geqslant 30$. In the case of the binomial, we must exclude the highly skewed cases where the success probability p is close to either 0 or 1. Certainly for p in the interval $[0.40, 0.60]$ 30 samples is quite adequate.

It would seem that, for many practical cases, sample size considerations should not be altered when applying the bootstrap. In nonparametric problems, larger sample sizes are required to make up for the lack of information that is implicit in parametric assumptions. Although it is always dangerous to set "rules of thumb" for sample sizes, I would suggest that in most cases it would be wise to take $n \geqslant 50$, if possible.

The best rule of thumb advice regarding the Monte Carlo approximation is to take $B = 100$ at least for bias and standard error estimation and to take $B = 1000$ or more for confidence intervals. One can always cite exceptions to these rules in the literature and variance reduction techniques may help to reduce B as we have discussed in Chapter 7. Also keep in mind the discussion regarding results in Booth and Sarkar (1998) and Efron (1987). In light of the speed of computing today and the results from Booth and Sarkar (1998), I would boost my recommendation to $B = 1000$ for standard error estimation and $B = 10,000$ for confidence interval estimation.

9.2. DISTRIBUTIONS WITH INFINITE MOMENTS

Singh (1981) and Bickel and Freedman (1981) showed that if $X_1, X_2, \ldots, X_n$ are independent and identically distributed random variables with finite second moments and if $Y_1, Y_2, \ldots, Y_n$ are chosen by simple random sampling with replacement from the sample $X_1, X_2, \ldots, X_n$, then letting

$$H_n(x, \omega) = P\left(\frac{\bar{Y}_n - \bar{X}_n}{S_n} \leqslant x \mid X_1, X_2, \ldots, X_n\right)$$

where

$$\bar{Y}_n = n^{-1}\sum_{i=1}^{n} Y_i, \quad \bar{X}_n = n^{-1}\sum_{i=1}^{n} X_i, \quad S_n^2 = n^{-1}\sum_{i=1}^{n}(X_i - \bar{X}_n)^2$$

and $\Phi(x)$ is the cumulative standard normal distribution,

$$\sup_{-\infty < x < \infty} |H_n(x, \omega) - \Phi(x)| \to 0 \quad \text{with probability 1 as } n \to \infty$$

Here ω denotes the random outcome that leads to the values $X_1, X_2, \ldots,$ X_n and $H_n(x, \omega)$ is a random probability distribution. For fixed $X_1, X_2, \ldots, X_n$ (i.e., a particular ω), $H_n(x, \omega)$ is a cumulative probability distribution. It is well known (i.e., the central limit theorem) that

$$G_n(x) = P\left(\frac{\bar{X}_n - \mu}{\sigma} \leqslant x\right)$$

converges to $\Phi(x)$.

The bootstrap principle replaces μ and σ with $\bar{X}_n$ and S_n, and $\bar{X}_n$ with $\bar{Y}_n$. So we hope that $H_n(x, \omega)$ would also converge to $\Phi(x)$ for almost all ω. That is what the results of Bickel and Freedman (1981) and Singh (1981) tell us.

Atheya (1987b) asks whether the same kind of result holds in cases where the variance is not finite and so the central limit theorem does not apply. In the case when $X_1, X_2, \ldots, X_n$ have a distribution F satisfying

$$1 - F(x) \sim x^{-\alpha}L(x)$$

and

$$f(-x) \sim cx^{-\alpha}L(x)$$

as $x \to \infty$, where L is a slowly varying function as $x \to \infty$ and c is a nonnegative constant, $\bar{X}_n$ appropriately normalized converges to a stable law for $0 < \alpha \leqslant 2$. When $\alpha = 2$, we get the usual central limit theorem. For $\alpha < 2$ the variance does not exist, but the norming constants and the limiting distribution are well defined. See Feller (1971) for more details.

Theorem 1 of Athreya (1987b) proves the basic result for $1 < \alpha < 2$; the case $0 < \alpha \leqslant 1$ was given in an earlier unpublished report. The theorem tells us that when we apply the bootstrap algorithm to the appropriately normalized mean we get convergence to a random probability distribution rather than a fixed probability distribution and so if

$$H_n(x, \omega) = P[T_n \leqslant x | X_1, X_2, \ldots, X_n]$$

where

$$T_n = nX_{(n)}^{-1}(\bar{Y}_n - \bar{X}_n)$$

and

$$X_{(n)} = \max(X_1, X_2, \ldots, X_n)$$

and $G(x)$ is the limiting stable law, we would have hoped that

$$\sup_{-\infty < x < \infty} |H_n(x, \omega) - G(x)| \to 0$$

with probability one. Unfortunately, since $H_n(x, \omega)$ converges to a random probability distribution we do not get the hoped for result. No fixed distribution other than G can work either since the resulting asymptotic distribution is random.

A similar example for the maximum was given by Bickel and Freedman (1981) and Knight (1989). Angus (1993) has extended their counterexample to the minimum and maximum of independent identically distributed random variables, which we discuss in the next section.

Athreya (1987b) points out that difficulties with the bootstrap mean in heavy-tailed situations can be overcome by the use of trimmed means or a smaller number of observations in the resample (less than n but still of order n).

9.3. ESTIMATING EXTREME VALUES

Suppose we have a sample size of n from a probability distribution F, which is in the domain of attraction of one of the three extreme value types. This assumption requires that certain conditions on the tail behavior of F are satisfied.

See Galambos (1978) for details on the limiting distributions and proof of Gnedenko's theorem for the maximum or minimum of a sequence of independent identically distributed random variables. By Gnedenko's theorem, the maximum $X_{(n)}$ or the minimum $X_{(1)}$ of the sequence of observations X_1, X_2, $X_3, \ldots,$ X_n has an extreme value limiting distribution when appropriately normalized.

Angus (1993) derives the limiting random measure for the maximum and minimum in the case of each of the three extreme value types when applying the bootstrap to the sample extreme values (minimum or maximum). He does this by applying the bootstrap principle to the appropriately normalized extremes in a way analogous to what Athreya (1987b) did for the normalized sample mean in the case of stable laws with $0 < \alpha < 2$.

A key fact that sheds light on the result is that for a fixed integer $r \geqslant 1$, if we let $X_{(r)}$ denote the rth smallest observation from the sample $X_1, X_2, X_3, \ldots,$ X_n, then

$$P[Y_{(1)} = X_{(r)} | X_1, X_2, \ldots, X_n] = P[Y_{(n)} = X_{(n-r+1)} | X_1, X_2, \ldots, X_n]$$

$$= \left(1 - \frac{(r-1)}{n}\right)^n \left[1 - \left(1 - \frac{1}{n}\right)^n\right]$$

where $Y_{(1)}$ is the minimum of a bootstrap sample $Y_1, Y_2, \ldots, Y_n$ drawn by sampling with replacement from $X_1, X_2, \ldots, X_n$ and $Y_{(n)}$ is the corresponding maximum from that same bootstrap sample. Taking the limit as $n \to \infty$ yields the limiting value $e^{-r}(e - 1)$.

Suppose that a_n^* and b_n^* are the appropriate bootstrap analogs to the a_n and b_n for the maximum in Gnedenko's theorem. We would then expect that

$$P\left[\frac{Y_{(n)} - a_n^*}{b_n^*} \leqslant t\right] \xrightarrow{d} H(t, \omega)$$

as $n \to \infty$, where H places probability mass $e^{-r}(e - 1)$ at the random point Z_r for $r = 1, 2, \ldots$, where $\xrightarrow{d}$ denotes convergence in distribution and Z_r is determined by the sequence $X_1, X_2, X_3, \ldots$. Then unconditionally

$$(Y_{(n)} - a_n^*)/b_n^* \xrightarrow{d} Z$$

where

$$P[Z \leqslant t] = (e - 1) \sum_{r=1}^{\infty} e^{-r} P[Z_r \leqslant t]$$

Angus proceeds to determine these conditional random measures and the unconditional distributions in the various cases.

The key point is that $H(t, \omega)$ is again a random probability distribution rather than a fixed distribution. So again the bootstrap principle fails since $(X_{(n)} - a_n)/b_n$ converges in distribution to a fixed extreme value distribution, say, G, but $(Y_{(n)} - a_n^*/b_n^*)$ converges to a random probability distribution H, which necessarily differs from G.

Hall (1992a) developed a bootstrap theory for functionals that are sufficiently smooth to admit Edgeworth and Cornish–Fisher expansions. These counterexamples obviously fail the smoothness conditions. However, this does not tell the whole story since the smoothness conditions are far from necessary conditions for the bootstrap principle to work.

The best that can be said is that there are many practical situations where the bootstrap has been shown to work well empirically and additionally there is a theory which proves that under certain conditions it is guaranteed to work asymptotically. Unfortunately, the counterexamples provided in this chapter show that there are situations where the bootstrap principle fails.

This suggests that the practitioner must be very careful when applying the ordinary bootstrap procedure. In some of these cases there may be modifications that allow a form of bootstrapping to work. This theory needs further development to make the guidelines clear.

9.4. SURVEY SAMPLING

In the case of sample surveys, the target population is always finite. If the population size is N and the sample size is n and n/N is not very small, then the variance of estimates such as averages is smaller than the variance in the usual theory of infinite populations. Recall that for independent identically distributed observations X_i for $i = 1, 2, \ldots, n$ the sample mean X_b has expected value equal to μ, the population mean, and variance equal to σ^2/n, where σ^2 is the population variance.

For a finite population, a random sample of size n taken from a population of size N also has mean μ but the variance is only $(\sigma^2/n)(N - n)/N$ if we define

$$\mu = (X_1 + X_2 + \cdots + X_n + X_{n+1} + \cdots + X_N)/N$$

and

$$\sigma^2 = [(X_1 - \mu)^2 + (X_2 - \mu)^2 + \cdots + (X_N - \mu)^2]/(N - 1)$$

The factor $f = n/N$ is called the sampling fraction and $(N - n)/N = 1 - f$ is called the finite population correction. The reason for the smaller variance is that the observations have a slight positive correlation (they are not independent) due to the fact that the following equation holds:

$$\mu = (X_1 + X_2 + \cdots + X_n + X_{n+1} + \cdots + X_N)/N$$
$$= nX_b/N + (X_{n+1} + \cdots + X_N)/N$$

or

$$X_b = [\mu - (X_{n+1} + \cdots + X_N)/N]N/n$$

See Cochran (1977, pp. 23–24) for a derivation of the variance formula and page 22 of the same reference for the mean using slightly different notation.

Now if one applies the ordinary bootstrap to the sample mean by sampling with replacement from the n observations in the original sample, the bootstrap will mimic the sampling of independent observations and, consequently, the bootstrap estimate of variance will approach σ^2/n and not $(1 - f)\sigma^2/n$. However, for the finite population, we would want it to approach $(1 - f)\sigma^2/n$. In this sense the bootstrap fails.

Fortunately, in most practical problems, f is known. So simply multiplying the bootstrap estimate of variance by $1 - f$ should work.

Another approach that works is to take a smaller sample of size m out of the n original observations by sampling with replacement. For an appropriate choice of m, this resampling scheme will lead to a consistent estimate of the variance.

The same problem carries over to other estimators. Of course, if f is very small, say 1% to 5%, the effect on the variance is not terribly great. Davison and Hinkley (1997) point out that for many practical cases f could be between 10% and 50% and then it matters a great deal.

Stratification is common in survey sampling as estimates by group are commonly of interest. Also, stratification is useful in reducing variability of estimates.

Stratified bootstrapping is very simple. We just sample with replacement within each strata. Further discussion about bootstrapping for finite populations can be found in Davison and Hinkley (1997, pp. 92–100).

In survey sampling a randomization scheme that samples from groups proportional to a measure of size is sometimes done for convenience and other times for very sound statistical reasons. In either case, variance estimators are required that account for the sampling mechanism. In a recent paper, Kaufman (1998) provides a bootstrap variance estimator for a systematic sampling scheme that samples proportional to size.

9.5. BOOTSTRAP DIAGNOSTICS

In parametric problems, like regression analysis, there are many assumptions that are required for the models to work well (e.g., homogeneity of variance, normally distributed residuals, no trends in the residuals). These assumptions can be checked using various statistical tests or through numerical or graphical diagnostic tools. Similar diagnostics can be applied in the case of parametric bootstraps.

However, in the case of the nonparametric bootstrap, the number of assumptions are minimal and it is difficult to characterize conditions when the bootstrap can be expected to work or fail. Consequently, it was believed for a long time that no diagnostics could be developed for the nonparametric bootstrap.

Efron (1992) introduced the concept of a jackknife-after-bootstrap measure, which could be used as a tool for assessing the nonparametric bootstrap. The idea is to illustrate what the effect of leaving out individual observations has on bootstrap calculations. This idea addresses a basic question: Once a nonparametric bootstrap calculation has been performed, how different might the result have been if a single observation, say, y_j had been left out of the original data?

An intuitive approach might be to do the resampling over again with the original data replaced by the data set with y_j left out. We would then compare the two bootstrap estimates to determine the effect. However, we really would want to see what the effect is for each observation left out. This would be even more computationally intensive since we would be repeating the bootstrap resampling n times.

However, this brute force approach is not necessary, as computationally, we can do something equivalent without repeating the bootstrap sampling. What we do is take advantage of the information in the original bootstrap samples that have the observation y_j left out.

If we did enough bootstrap replications originally (say, 1000 to 5000), there should be many cases where each observation is left out. In fact, in any bootstrap sample there is approximately a 0.368 (e^{-1}) probability that any particular observation is not included. You may recall some discussion of this in the heuristic justification of the .632 estimator for error rates in discriminant analysis. So for each observation, there should be a subset consisting of approximately 36.8% of the bootstrap samples in which that observation does not appear.

To illustrate, suppose we are estimating the bias of an estimator t. Let B be the bootstrap estimate of the bias based on the entire bootstrap sample. Consider the subset N_j of bootstrap samples that do not contain the observation y_j. Let B_{-j} denote the bootstrap estimate of the bias for t based on the same bootstrap calculation using only the subset N_j of bootstrap samples. The jackknife estimate scales by the sample size n and provides an estimator of the effect as $n(B_{-j} - B)$.

This can be repeated for all j. Each value is analogous to the pseudovalue in ordinary jackknife estimation. Such an effect is very much akin to empirical influence function estimates. Because this jackknife is applied after bootstrapping it is called a jackknife-after-bootstrap measure.

In fact, one suggested diagnostic is obtained by plotting these jackknife-after-bootstrap measures with empirical influence values, possibly standardized. More details along with an example can be found in Davison and Hinkley (1997, pp. 114–116).

9.6. HISTORICAL NOTES

Bickel and Freedman (1981) and Knight (1989) consider the example of the bootstrap distribution for the maximum of a sample from the uniform distribution on the interval $(0, \theta)$. Since the parameter θ is the upper endpoint of the distribution, the maximum $X_{(n)}$ increases to θ (almost surely) as $n \to \infty$. This is a case where regularity conditions fail. They show that the bootstrap distribution for the normalized maximum converges to a random probability measure. This is also discussed in Schervish (1995, Example 5.80, p. 330).

Consequently, Bickel and Freedman (1981) provide an early counterexample to the correctness of the bootstrap principle. Work by Angus (1993) extends this result to general cases of the limiting distribution for maxima and minima of independent identically distributed sequences.

Angus's work was motivated by the results in Athreya (1987b), who worked out the counterexamples for the sample mean when second moments fail to exist but the limiting stable law is well defined. Knight (1989) provides an alternative proof of Athreya's result.

A good general account of extreme value theory can be found in Leadbetter, Lindgren, and Rootzen (1983), Galambos (1978, 1987), and Resnick (1987).

Reiss (1989) presents the application of the bootstrap to quantile estimation and gives the exact distribution of the bootstrap distribution for the sample quantiles (Reiss, 1989, pp. 220–226). Another good account is given by Reiss and Thomas (1997), who provide examples and software. They also discuss bootstrap confidence intervals (p. 82–84). The book edited by Adler, Feldman, and Taqqu (1998) contains two articles that show ways to bootstrap when distributions have heavy tails (see Pictet, Dacorogna, and Muller, 1998, pp. 283–310; LePage, Podgorski, Ryznar, and White, 1998, pp. 339–358).

A simple illustration of the inconsistency for the bootstrap estimate of the minimum is given by Reiss (1989, p. 221). Castillo (1988) presents some of the theory along with engineering applications.

Singh (1981) and Bickel and Freedman (1981) were the first to show that the bootstrap principle works for the sample mean when finite second moments exist. This was an important theoretical result, which gave Efron's bootstrap a big shot in the arm, since it provided stronger justification than simple heuristics and analogies to the jackknife and other similar methods.

Since then, efforts (Yang, 1988; Angus, 1989) have been made to make the proof of the asymptotic normality of the bootstrap mean in the finite variance case simpler and more self-contained.

The treatise by Mammen (1992b) attempts to show when the bootstrap can be relied on, based on asymptotic theory and simulation results. The bootstrap conference at Ann Arbor in 1990 attempted to look at new applications and limitations of the bootstrap. See LePage and Billard (1992) for a collection of published papers from that meeting.

Combinational results from classical occupancy theory can be found in Feller (1971) and Johnson and Kotz (1977). Chernick and Murthy (1985) apply these results to obtain the repetition frequencies for the bootstrap samples. Hall (1992a, Appendix I) discusses the atoms of the bootstrap distribution. In addition, general combinational results applicable to the bootstrap distribution can be found in Roberts (1984).

Regarding finite populations, we have already mentioned Cochran (1977) as the classic reference. Variance estimation by balanced subsampling methods goes back to McCarthy (1969). The first attempt to apply the bootstrap directly was made by Gross (1980). His method was called the "population bootstrap" and was restricted to cases where N/n is an integer. Bickel and Freedman (1984) and Chao and Lo (1994) apply the approach that Davison and Hinkley advocate. Booth, Butler, and Hall (1994) describe the construction of studentized bootstrap confidence limits in the finite population context.

Presnell and Booth (1994) give a critical discussion of earlier literature and based on the superpopulation model approach to survey sampling, they describe the superpopulation bootstrap. A modified sample size approach was presented by McCarthy and Snowden (1985) and the "mirror-matched" method by Sitter (1992a). A rescaling approach was introduced by Rao and Wu (1988).

A comprehensive account of both the jackknife and the bootstrap approaches to survey sampling can be found in Shao and Tu (1995, Chap. 6). Kovar (1985, 1987) and Kovar, Rao, and Wu (1988) performed simulation studies to compare various resampling variance estimators and confidence intervals in the case of stratified one-stage simple random sampling with replacement. Shao and Tu (1995) provide a summary of these studies and their findings. Shao and Sitter (1996) apply the bootstrap to imputation problems in the survey sampling context.

As mentioned earlier, the jackknife-after-bootstrap diagnostics were introduced in Efron (1992). Different graphical diagnostics for the reliability of the bootstrap have very recently been developed in an asymptotic framework in Beran (1997). Linear plots for diagnosis of problems with the bootstrap are presented in Davison and Hinkley (1997, p. 119). This approach is due to Cook and Weisberg (1994).

Bibliography

Aastveit, A. H. (1990). Use of bootstrapping for estimation of standard deviation and confidence intervals of genetic variance and covariance components. *Biom. J.* **32**, 515–527.   *(1)

Abel, U., and Berger, J. (1986). Comparison of resubstitution, data splitting, the bootstrap and the jackknife as methods for estimating validity indices of new marker tests: a Monte Carlo study. *Biom. J.* **28**, 899–908.   *(2)

Abramovitch, L., and Singh, K. (1985). Edgeworth corrected pivotal statistics and the bootstrap. *Ann. Statist.* **13**, 116–132.   *(3)

Acutis, M., and Lotito, S. (1997). Possible application of resampling methods to statistical analysis of agronomic data. *Riv. Agron.* **31**, 810–816.

Aczel, A. D., Josephy, N. H., and Kunsch, H. R. (1993). Bootstrap estimates of the sample bivariate autocorrelation and partial autocorrelation distributions. *J. Statist. Comput. Simul.* **46**, 235–249.

Adams, D. C., Gurevitch, J., and Rosenberg, M. S. (1997). Resampling tests for meta-analysis of ecological data. *Ecology* **78**, 1277–1283.   *(4)

Adkins, L. C. (1990). Small sample performance of jackknife confidence intervals for the James–Stein estimator. *Commun. Statist. Simul. Comput.* **19**, 401–418.

Adkins, L. C., and Hill, R. C. (1990). An improved confidence ellipsoid for the linear regression model. *J. Statist. Comput. Simul.* **36**, 9–18.

Adler, R. J., Feldman, R. E., and Taqqu, M. S. (editors) (1998). *A Practical Guide to Heavy Tails*. Birkhauser, Boston.   *(5)

Aebi, M., Embrechts, P., and Mikosch, T. (1994). Stochastic discounting, aggregate claims and the bootstrap. *Adv. Appl. Probab.* **26**, 183–206.

Aegerter, P., Muller, F., Nakache, J. P., and Boue, A. (1994). Evaluation of screening methods for Down's syndrome using bootstrap comparison of ROC curves. *Comput. Methods Prog. Biomed.* **43**, 151–157.   *(6)

Aerts, M., and Gijbels, I. (1993). A three-stage procedure based on bootstrap critical points. *Commun. Statist. Seq. Anal.* **12**, 93–113.

Aerts, M., Janssen, P., and Veraverbeke, N. (1994). Bootstrapping regression quantiles. *J. Nonparametric Statist.* **4**, 1–20.

Agresti, A. (1990). Categorical Data Analysis. Wiley, New York.   *(7)

Ahn, H., and Chen, J. J. (1997). Tree-structured logistic model for overdispersed binomial data with application to modeling development effects. *Biometrics* **53**, 435–455.

Aitkin, M., Anderson, D., and Hinde, J. (1981). Statistical modelling of data on teaching styles (with discussion). *J. R. Statist. Soc. A* **144**, 419–461.

Aitkin, M., and Tunnicliffe Wilson, G. (1980). Mixture models, outliers and the EM algorithm. *Technometrics* **22**, 325–331.   *(8)

Akritas, M. G. (1986). Bootstrapping the Kaplan–Meier estimator. *J. Am. Statist. Assoc.* **81**, 1032–1038.   *(9)

Alemayehu, D. (1987a). Approximating the distributions of sample eigenvalues. *Proc. Comput. Sci. Statist.* **19**, 535–539.

Alemayehu, D. (1987b). Bootstrap method for multivariate analysis. *Proc. Statist. Comp.* 321–324, American Statistical Association, Alexandria.

Alemayehu, D. (1988). Bootstrapping the latent roots of certain random matrices. *Commun. Statist. Simul. Comput.* **17**, 857–869.

Alemayehu, D., and Doksum, K. (1990). Using the bootstrap in correlation analysis with application to a longitudinal data set. *J. Appl. Statist.* 17, 357–368.

Alkuzweny, B. M. D., and Anderson, D. A. (1988). A simulation study of bias in estimation of variance by bootstrap linear regression model. *Commun. Statist. Simul. Comput.* **17**, 871–886.

Allen, D. L. (1997). Hypothesis testing using an L_1 - distance bootstrap design. *Amer. Statist.* **51**, 145–150.

Albanese, M. T., and Knott, M. (1994). Bootstrapping latent variable models for binary response. *Br. J. Math. Statist. Psychol.* **47**, 235–246.

Altman, D. G., and Andersen, P. K. (1989). Bootstrap investigation of the stability of a Cox regression model. *Statist. Med.* **8**, 771–783.   *(10)

Amari, S. (1985). *Differential Geometrical Methods in Statistics.* Springer-Verlag, Berlin. *(11)

Ames, G. A., and Muralidhar, K. (1991). Bootstrap confidence intervals for estimating audit value from skewed populations and small samples. *Simulation* **56**, 119–128. *(12)

Anderson. P. K., Borgan, O., Gill, R. D., and Keiding, N. (1993). *Statistical Models Based on Counting Processes.* Springer-Verlag, New York.   *(13)

Anderson, T. W. (1984). *An Introduction to Multivariate Statistical Analysis*, 2nd ed. Wiley, New York.   *(14)

Andrade, I., and Proenca, I. (1992). Search for a break in the Portuguese GDP 1833–1985 with bootstrap methods. In *Bootstrapping and Related Techniques. Proceedings, Trier, FRG. Lecture Notes in Economics and Mathematical Systems,* (K.-H. Jockel, G. Rothe, and W. Sendler, editors), **376**, 133–142. Springer-Verlag, Berlin.

Andrews, D. F., Bickel, P. J., Hampel, F. R., Huber, P. J., Rogers, W. H., and Tukey, J. W. (1972). *Robust Estimates of Location: Survey and Advances.* Princeton University Press, Princeton.   *(15)

Andrieu, G., Caraux, G., and Gascuel, O. (1997). Confidence intervals of evolutionary distances between sequences and comparison with usual approaches including the bootstrap method. *Mol. Biol. Evol.* **14**, 875–882.   *(16)

Angus, J. E. (1989). A note on the central limit theorem for the bootstrap mean. *Commun. Statist. Theory Methods* **18**, 1979–1982.   *(17)

Angus, J. E. (1993). Asymptotic theory for bootstrapping the extremes. *Commun. Statist. Theory Methods* **22**, 15–30.   *(18)

Angus, J. E. (1994). Bootstrap one-sided confidence intervals for the log-normal mean. *Statistician* **43**, 395–401.

Archer, G., and Chan, K. (1996). Bootstrapping uncertainty in image analysis. In *Proceedings in Computational Statistics. 12th Symposium*, 193–198.

Arcones, M. A., and Gine, E. (1989). The bootstrap of the mean with arbitrary bootstrap sample size. *Ann. Inst. Henri Poincare* **25**, 457–481.

Arcones, M. A., and Gine, E. (1991). Some bootstrap tests of symmetry for univariate continuous distributions. *Ann. Statist.* **19**, 1496–1511.

Arcones, M. A., and Gine, E. (1992). On the bootstrap of M-estimators and other statistical functionals. In *Exploring the Limits of Bootstrap* (R. LePage and L. Billard, editors), 13–47. Wiley, New York.   *(19)

Arcones, M. A., and Gine, E. (1994). U-processes indexed Vapnik–Cervonenkis classes of functions with applications to asymptotics and bootstrap of U-statistics with estimated parameters. *Stoch. Proc.* **52**, 17–38.

Armstrong, J. S., Brodie, R. J., and McIntyre, S. H. (1987). Forecasting methods for marketing: review of empirical research. *Int. J. For.* **3**, 355–376.

Arvesen, J. N. (1969). Jackknifing U-statistics. *Ann. Math. Statist.* **40**, 2076–2100.

Ashour, S. K., Jones, P. W., and El-Sayed, S. M. (1991). Bootstrap investigation on mixed exponential model using weighted least squares. In *26th Annual Conference on Statistics, Computer Science and Operations Research. Vol. 1; Mathematical Statistics*, **1**, 79–103. Cairo University.

Athreya, K. B. (1983). Strong law for the bootstrap. *Statist. Probab. Lett.* **1**, 147–150.   *(20)

Athreya, K. B. (1987a). Bootstrap of the mean in the infinite variance case. In *Proceedings of the First World Congress of the Bernoulli Society* (Y. Prohorov and V. Sazonov, editors), **2**, 95–98. VNU Science Press, The Netherlands.

Athreya, K. B. (1987b). Bootstrap estimation of the mean in the infinite variance case. *Ann. Statist.* **15**, 724–731.   *(21)

Athreya, K. B., and Fuh, C. D. (1992a). Bootstrapping Markov chains: countable case. *J. Statist. Plann. Inf.* **33**, 311–331.   *(22)

Athreya, K. B., and Fuh, C. D. (1992b). Bootstrapping Markov chains. In *Exploring the Limits of Bootstrap* (R. LePage and L. Billard, editors), 49–64, Wiley, New York.   *(23)

Athreya, K. B., Ghosh, M., Low, L. Y., and Sen, P. K. (1984). Laws of large numbers for bootstrapped U-statistics. *J. Statist. Plann. Inf.* **9**, 185–194.

Atwood, C. L. (1984). Approximate tolerance intervals based on maximum likelihood estimates. *J. Am. Statist. Assoc.* **79**, 459–465.

Atzinger, E., Brooks, W., Chernick, M. R., Elsner, B., and Foster, W. (1972). A Compendium on Risk Analysis Techniques, U. S. Army Materiel Systems Analysis Activity. Special Publication No. 4.   *(24)

Azzalini, A., Bowman, A. W., and Hardle, W. (1989). Nonparametric regression for model checking. *Biometrika* **76**, 1–11.

Babu, G. J. (1984). Bootstrapping statistics with linear combinations of chi-squares as weak limit. *Sankhya A* **46**, 85–93.

Babu, G. J. (1986). A note on bootstrapping the variance of the sample quantiles. *Ann. Inst. Statist. Math.* **38**, 439–443.

Babu, G. J. (1989). Applications of Edgeworth expansions to bootstrap: a review. In *Statistical Data Analysis and Inference* (Y. Dodge, editor), 223–237. Elsevier Science Publishers, Amsterdam.

Babu, G. J. (1992). Subsample and half-sample methods. *Ann. Inst. Statist. Math.* **44**, 703–720.   *(25)

Babu, G. J. (1995). Bootstrap for nonstandard cases. *J. Statist. Plann. Inf.* **43**, 197–203.

Babu, G. J. (1998). Breakdown theory for estimators based on bootstrap and other resampling schemes. Preprint, Department of Statistics, Pennsylvania State University.   *(26)

Babu, G. J., and Bai, Z. D. (1996). Mixtures of global and local Edgeworth expansions and their applications. *J. Multivar. Anal.* **59**, 282–307.

Babu, G. J., and Bose, A. (1989). Bootstrap confidence intervals. *Statist. Probab. Lett.* **7**, 151–160.   *(27)

Babu, G. J., and Feigelson, E. (1996). *Astrostatistics*. Chapman & Hall, New York. *(28)

Babu, G. J., Pathak, P. K., and Rao, C. R. (1992). A note of Edgeworth expansion for the ratio of sample means. *Sankhya A* **54**, 309–322.

Babu, G. J., Pathak, P. K., and Rao, C. R. (1998). Second order corrections of the sequential bootstrap. Preprint, Department of Statistics, Pennsylvania State University.   *(29)

Babu, G. J., and Rao, C. R. (1993). Bootstrap methodology. In *Handbook of Statistics* (C. R. Rao, editor), **9**, 627–659. North-Holland, Amsterdam.   *(30)

Babu, G. J., and Singh, K. (1983). Nonparametric inference on means using the bootstrap. *Ann. Statist.* **11**, 999–1003.   *(31)

Babu, G. J., and Singh, K. (1984a). On one term Edgeworth correction by Efron's bootstrap. *Sankhya A* **46**, 219–232.   *(32)

Babu, G. J., and Singh, K. (1984b). Asymptotic representations related to jackknifing and bootstrapping L-statistics. *Sankhya A* **46**, 195–206.   *(33)

Babu, G. J., and Singh, K. (1985). Edgeworth expansions for sampling without replacement from finite populations. *J. Multivar. Anal.* **17**, 261–278.   *(34)

Babu, G. J., and Singh, K. (1989). On Edgeworth expansions in the mixture cases. *Ann. Statist.* **17**, 443–447.   *(35)

Bahadur, R., and Savage, L. (1956). The nonexistence of certain statistical procedures in nonparametric problems. *Ann. Math. Statist.* **27**, 1115–1122.   *(36)

Bai, C. (1988). Asymptotic properties of some sample reuse methods for prediction and classification. Ph. D. dissertation. Department of Mathematics, University of California, San Diego.

Bai, C., Bickel, P. J., and Olshen, R. A. (1990). Hyperaccuracy of bootstrap-based prediction. In *Probability in Banach Spaces: Proceedings of the Seventh International Conference* (E. Eberlein, J. Kuelbs, and M. B. Marcus, editors), 31–42. Birkhauser, Boston.

Bai, Z., and Rao, C. R. (1991). Edgeworth expansion of a function of sample means. *Ann. Statist.* **19**, 1285–1315.   *(37)

Bai, Z., and Rao, C. R. (1992). A note on Edgeworth expansion for ratio of sample means. *Sankhya A* **54**, 309–322.   *(38)

Bai, Z., and Zhao, L. (1986). Edgeworth expansions of distribution function of independent random variables. *Sci. Sin. A* **29**, 1–22.

Bailer, A. J., and Oris, J. T. (1994). Assessing toxicity of pollutants in aquatic systems. In *Case Studies in Biometry* (N. Lange, L. Ryan, L. Billard, D. Brillinger, L. Conquest, and J. Greenhouse, editors), 25–40. Wiley, New York.   *(39)

Bailey, R. A., Harding, S. A., and Smith, G. L. (1989). Cross validation. *Encyclopedia of Statistical Sciences*, Supplement Volume, 39–44. Wiley, New York.

Bailey, W. A. (1992). Bootstrapping for order statistics sans random numbers (operational bootstrapping). In *Exploring the Limits of Bootstrap* (R. LePage and L. Billard, editors), 309–318. Wiley, New York.   *(40)

Bajgier, S. M. (1992). The use of bootstrapping to construct limits on control charts. *Proc. Dec. Sci. Inst.*, 1611–1613.   *(41)

Baker, S. G., and Chu, K. C. (1990). Evaluating screening for the early detection and treatment of cancer without using a randomized control group. *J. Am. Statist. Assoc.* **85**, 321–327.   *(42)

Banks, D. L. (1988). Histospline smoothing the Bayesian bootstrap. *Biometrika* **75**, 673–684.   *(43)

Banks, D. L. (1989). Bootstrapping II. *Encyclopedia of Statistical Sciences*, Supplement Volume, 17–22. Wiley, New York.   *(44)

Banks, D. L. (1993). Book Review of *Exploring the Limits of Bootstrap*. (R. LePage and L. Billard, editors), *J. Am. Statist. Assoc.* **88**, 708–710.

Barabas, B., Csorgo, M. J., Horvath, L., and Yandell, B. S. (1986). Bootstrapped confidence bands for percentile lifetime. *Ann. Inst. Statist. Math.* **38**, 429–438.

Barbe, P., and Bertail, P. (1995). *The Weighted Bootstrap*. Lecture Notes in Statistics. Springer-Verlag, New York.   *(45)

Barlow, W. E., and Sun, W. H. (1989). Bootstrapped confidence intervals for linear relative risk models. *Statist. Med.* **8**, 927–935.   *(46)

Barnard, G. (1963). Comment on "The spectral analysis of point processes" by M. S. Bartlett. *J. R. Statist. Soc. B* **25**, 294.   *(47)

Bar-Ness, Y., and Punt, J. B. (1996). Adaptive "bootstrap" CDMA multi-user detector. *Wireless Pers. Commun.* **3**, 55–71.   *(48)

Barndorff-Nielsen, O. E. (1986). Inference on full or partial parameters based on standardized signed log likelihood ratio. *Biometrika* **73**, 307–322.

Barndorff-Nielsen, O. E., and Cox, D. R. (1989). *Asymptotic Techniques for Use in Statistics*. Chapman & Hall, London.   *(49)

Barndorff-Nielsen, O. E., and Cox, D. R. (1994). *Inference and Asymptotics*. Chapman & Hall, London.   *(50)

Barndorff-Nielsen, O. E., and Hall, P. (1988). On the level-error after Barlett adjustment of the likelihood ratio statistic. *Biometrika* **75**, 374–378.

Barndorff-Nielsen, O. E., James, I. R., and Leigh, G. M. (1989). A note on a semiparametric estimation of mortality. *Biometrika* **76**, 803–805.

Barnett, V., and Lewis, T. (1995). *Outliers in Statistical Data*, 3rd ed. Wiley, Chichester. *(51)

Barraquand, J. (1995). Monte Carlo integration, quadratic resampling and asset pricing. *Math. Comput. Simul.* **38**, 173–182.

Bartlett, M. S. (1975). *The Statistical Analysis of Spatial Pattern*. Chapman & Hall, London.   *(52)

Basawa, I. V., Green, T. A., McCormick, W. P., and Taylor, R. L. (1990). Asymptotic bootstrap validity for finite Markov chains. *Commun. Statist. Theory Methods* **19**, 1493–1510.   *(53)

Basawa, I. V., Mallik, A. K., McCormick, W. P., and Taylor, R. L. (1989). Bootstrapping explosive autoregressive processes. *Ann. Statist.* **17**, 1479–1486.   *(54)

Basawa, I. V., Mallik, A. K., McCormick, W. P., Reeves, J. H., and Taylor, R. L. (1991a). Bootstrapping unstable first order autoregressive processes. *Ann. Statist.* **19**, 1098–1101.   *(55)

Basawa, I. V., Mallik, A. K., McCormick, W. P., Reeves, J. H., and Taylor, R. L. (1991b). Bootstrap test of significance and sequential bootstrap estimators for unstable first order autoregressive processes. *Commun. Statist. Theory Methods* **20**, 1015–1026. *(56)

Basford, K. E., and McLachlan, G. J. (1985a). Estimation of allocation rates in a cluster analysis context. *J. Am. Statist. Assoc.* **80**, 286–293.   *(57)

Basford, K. E., and McLachlan, G. J. (1985b). Cluster analysis in randomized complete block design. *Commun. Statist. Theory Methods* **14**, 451–463.   *(58)

Basford, K. E., and McLachlan, G. J. (1985c). The mixture method of clustering applied to three-way data. *J. Classification* **2**, 109–125.   *(59)

Basford, K. E., and McLachlan, G. J. (1985d). Likelihood estimation with normal mixture models. *Appl. Statist.* **34**, 282–289.   *(60)

Bates, D. M., and Watts, D. G. (1988). *Nonlinear Regression Analysis and Its Applications*. Wiley, New York.   *(61)

Bau, G. J. (1984). Bootstrapping statistics with linear combinations. *Sankhya A* **46**, 195–206.   *(62)

Bauer, P. (1994). Book Review of *Resampling-Based Multiple Testing: Examples and Methods for p-Value Adjustment* (by P. Westfall and S. S. Young). *Statist. Med.* **13**, 1084–1086.

Beadle, E. R., and Djuric, P. M. (1997). A fast weighted Bayesian bootstrap filter for nonlinear model state estimation. *IEEE Trans. Aerosp. Electron. Syst.* **33**, 338–343.

Bean, N. G. (1995). Dynamic effective bandwidths using network observation and the bootstrap. *Aust. Telecommun. Res.* **29**, 43–52.

Bedrick, E. J., and Hill, J. R. (1992). A generalized bootstrap. In *Exploring the Limits of Bootstrap* (R. LePage and L. Billard, editors), 319–326. Wiley, New York. *(63)

Belsley, D. A., Kuh, E., and Welsch, R. E. (1980). *Regression Diagnostics: Identifying Influential Data Sources of Collinearity.* Wiley, New York.   *(64)

Belyaev, Y. K. (1996). Resampling from realisations of random processes and fields, with applications to statistical inference. In *4th World Congress of the Bernoulli Society. Vienna.*

Benichou, J., Byrne, C., and Gail, M. (1997). An approach to estimating exposure-specific rate of breast cancer from a two-stage case–control study within a cohort. *Statist. Med.* **16**, 133–151.

Bensmail, H., and Celeux, G. (1996). Regularized Gaussian discriminant analysis through eigenvalue decomposition. *J. Am. Statist. Assoc.* **91**, 1743–1749.

Bentler, P. M., and Yung, Y. -F. (1994). Bootstrap-corrected ADF test statistics in covariance structure analysis. *Br. J. Math. Statist. Psychol.* **47**, 63–84.

Beran, R. J. (1982). Estimated sampling distributions: the bootstrap and competitors. *Ann. Statist.* **10**, 212–225.   *(65)

Beran, R. J. (1984a). Jackknife approximations to bootstrap estimates. *Ann. Statist.* **12**, 101–118.   *(66)

Beran, R. J. (1984b). Bootstrap methods in statistics. *Jahreshber. D. Dt. Math. Verein.* **86**, 14–30.   *(67)

Beran, R. J. (1985). Stochastic procedures: bootstrap and random search methods in statistics. *Proceedings of the 45th Session of the ISI,* **4**, 25. 1, Amsterdam.

Beran, R. J. (1986). Simulated power functions. *Ann. Statist.* **14**, 151–173.   *(68)

Beran, R. J. (1987). Prepivoting to reduce level error of confidence sets. *Biometrika* **74**, 457–468.   *(69)

Beran, R. J. (1988a). Weak convergence: statistical applications. *Encyclopedia of Statistical Sciences,* **9**, 537–539, Wiley, New York.

Beran, R. J. (1988b). Balanced simultaneous confidence sets. *J. Am. Statist. Assoc.* **83**, 679–686.

Beran, R. J. (1988c). Prepivoting test statistics: a bootstrap view of asymptotic refinements. *J. Am. Statist. Assoc.* **83**, 687–697.   *(70)

Beran, R. J. (1990a). Refining bootstrap simultaneous confidence sets. *J. Am. Statist. Assoc.* **85**, 417–426.   *(71)

Beran, R. J. (1990b). Calibrating prediction regions. *J. Am. Statist. Assoc.* **85**, 715–723. *(72)

Beran, R. J. (1992). Designing bootstrap prediction regions. In *Bootstrapping and Related Techniques. Proceedings, Trier, FRG* (K.-H. Jockel, G. Rothe, and W. Sendler, editors). *Lecture Notes in Economics and Mathematical Systems,* **376**, 23–30. Springer-Verlag, Berlin.   *(73)

Beran, R. J. (1993). Book Review of *The Bootstrap and Edgeworth Expansion* (by P. Hall). *J. Am. Statist. Assoc.* **88**, 375–376.

Beran, R. J. (1994). Seven stages of bootstrap. In *25th Conference on Statistical Computing: Computational Statistics* (P. Dirschedl and R. Ostermann, editors), 143–158. Physica-Verlag, Heidelberg.

Beran, R. J. (1996). Bootstrap variable-selection and confidence sets. In *Robust Statistics, Data Analysis and Computer Intensive Methods* (P. J. Huber and H. Rieder, editors). *Lecture Notes in Statistics,* **109**, 1–16. Springer-Verlag, New York.

Beran, R. J. (1997). Diagnosing bootstrap success. *Ann. Inst. Statist. Math.* **49**, 1–24.

Beran, R. J., and Ducharme, G. R. (1991). *Asymptotic Theory for Bootstrap Methods in Statistics.* Les Publications Centre de Recherches Mathematiques, Universite de Montreal, Montreal.   *(74)

Beran, R. J., and Fisher, N. I. (1998). Nonparametric comparison of mean directions or mean axes. *Ann. Statist.* **26**, 472–493.

Beran, R. J., LeCam, L., and Millar, P. W. (1987). Convergence of stochastic empirical measures. *J. Multivar. Anal.* **23**, 159–168.

Beran, R. J., and Millar, P. W. (1985). Asymptotic theory of confidence sets. In *Proceedings Berkeley Conference in Honor of J. Neyman and J. Kiefer* (L. M. LeCam and R. A, Olshen, editors), **2**, 865–887. Wadsworth, Monterey, CA.

Beran, R. J., and Millar, P. W. (1986). Confidence sets for a multivariate distribution. *Ann. Statist.* **14**, 431–443.

Beran, R. J., and Millar, P. W. (1987). Stochastic estimation and testing. *Ann. Statist.* **15**, 1131–1154.

Beran, R. J., and Millar, P. W. (1989). A stochastic minimum distance test for multivariate parametric models. *Ann. Statist.* **17**, 125–140.

Beran, R. J., and Srivastava, M. S. (1985). Bootstrap tests and confidence regions for functions of a covariance matrix. *Ann. Statist.* **13, 95**–115. Correction. *Ann. Statist.* **15**, 470–471.   *(75)

Bernard, V. L. (1987). Cross-sectional dependence and problems in inference in market-based accounting. *J. Acct. Res.* **25**, 1–48.

Bernard, J. T., and Veall, M. R. (1987). The probability distribution of future demand: the case of hydro Quebec. *J. Bus. Econ. Statist.* **5**, 417–424.

Bertail, P. (1992). La methode du bootstrap quelques applications et results theoretiques. Ph. D. dissertation. University of Paris IX.

Bertail, P. (1997). Second-order properties of an extrapolated bootstrap without replacement under weak assumptions. *Bernoulli* **3**, 149–179.

Besag, J. E., and Diggle, P. J. (1977). Simple Monte Carlo tests for spatial patterns. *Appl. Statist.* **26**, 327–333.   *(76)

Besag, J., and Clifford, P. (1989). Generalized Monte Carlo significance tests. *Biometrika* **76**, 633–642.

Besag, J., and Clifford, P. (1991). Sequential Monte Carlo *p*-values. *Biometrika* **78**, 301–304.

Besse, P., and de Falguerolles, A. (1993). Application of resampling methods to the choice of dimension in principal component analysis. In *Computer Intensive Methods in Statistics,* 167–176.

Bhatia, V. K., Jayasankar, J., and Wahi, S. D. (1994). Use of bootstrap technique for variance estimation of heritability estimators. *Ann. Agric. Res.* **15**, 476–480.

Bhattacharya, R. N. (1987). Some aspects of Edgeworth expansions in statistics and probability. In *New Perspectives in Theoretical and Applied Statistics* (M. Puri, J. P. Vilaplana, and W. Wertz, editors), 157–170, Wiley, New York.

Bhattacharya, R. N., and Ghosh, J. K. (1978). On the validity of the formal Edgeworth expansion. *Ann. Statist.* **6**, 435–451.   *(77)

Bhattacharya, R. N., and Qumsiyeh, M. (1989). Second-order and L^p-comparisons between the bootstrap and empirical Edgeworth expansion methodologies. *Ann. Statist.* **17**, 160–169.   *(78)

Bianchi, C., Calzolari, G., and Brillet, J,-L. (1987). Measuring forecast uncertainty: a review with evaluation based on a macro model of the French economy. *Int. J. For.* **3**, 211–227.

Bickel, P. J. (1992). Theoretical comparison of different bootstrap-t confidence bounds. In *Exploring the Limits of Bootstrap* (R. LePage and L. Billard, editors), 65–76, Wiley, New York.   *(79)

Bickel, P. J., and Freedman, D. A. (1980). On Edgeworth expansions for the bootstrap. Technical Report. Department of Statistics, University of California, Berkeley.

Bickel, P. J., and Freedman, D. A. (1981). Some asymptotic theory for the bootstrap. *Ann. Statist.* **9**, 1196–1217.   *(80)

Bickel, P. J., and Freedman, D. A. (1983). Bootstrapping regression models with many parameters. In *Festschrift for Erich Lehmann* (P. J. Bickel, K. Doksum, and J. L. Hodges, editors), 28–48. Wadsworth, Belmont, CA.   *(81)

Bickel, P. J., and Freedman, D. A. (1984). Asymptotic normality and the bootstrap in stratified sampling. *Ann. Statist.* **12**, 470–482.   *(82)

Bickel, P. J., and Ghosh, J. K. (1990). A decomposition for the likelihood ratio statistic and the Bartlett correction. A Bayesian argument. *Ann. Statist.* **18**, 1070–1090.

Bickel, P. J., Gotze, F., and van Zwet, W. R. (1997). Resampling fewer than n observations: gains, losses, and remedies for losses. *Statist. Sin.* **7**, 1–32.

Bickel, P. J., Klassen, C. A. J., Ritov, Y., and Wellner, J. A. (1993). *Efficient and Adaptive Estimation for Semiparametric Models.* Johns Hopkins University Press, Baltimore.

Bickel, P. J., and Krieger, A. M. (1989). Confidence bands for a distribution function using the bootstrap. *J. Am. Statist. Assoc.* **84**, 95–100.   *(83)

Bickel, P. J., and Ren, J. -J. (1996). The m out of n bootstrap and goodness of fit tests with double censored data. In *Robust Statistics, Data Analysis and Computer Intensive Methods* (P. J. Huber and H. Rieder, editors). Lecture Notes in Statistics, **109**, 35–48. Springer-Verlag, New York.   *(84)

Bickel, P. J., and Yahav, J. A. (1988). Richardson extrapolation and the bootstrap. *J. Am. Statist. Assoc.* **83**, 387–393.   *(85)

Biddle, G., Bruton, C., and Siegel, A. (1990). Computer-intensive methods in auditing: bootstrap difference and ratio estimation. *Auditing: J. Practice Theory* **9**, 92–114. *(86)

Biometrics Editors (1997). Book Review of *Modern Digital Simulation Methodology, II: Univariate and Bivariate Distribution Fitting, Bootstrap Methods & Applications*, (E. J. Dudewicz, editor). *Biometrics* **53**, 1564.

Bliese, P., Halverson, R., and Rothberg, J. (1994). Within-group agreement scores: using resampling procedures to estimate expected variance. In *Academy of Management Best Papers Proceedings*, 303–307.

Bloch, D. A. (1997). Comparing two diagnostic tests against the same "gold standard" in the same sample. *Biometrics* **53**, 73–85.

Bloch, D. A., and Silverman, B. W. (1997). Monotone discriminant functions and their application in rheumatology. *J. Am. Statist. Assoc.* **91**, 144–153.

Bloomfield, P. (1976). *Fourier Analysis of Time Series: An Introduction*. Wiley, New York.   *(87)

Bogdanov, Y. I., and Bogdanova, N. A. (1997). Bootstrap, data structures, and process control in microelectronics. *Russ. Microelectron.* **26**, 155–158.

Bollen, K. A., and Stine, R. A. (1993). Bootstrapping goodness-of-fit measures in structural equation models. In *Testing Structural Equation Models* (K. A. Bollen and J. S. Long, editors). 111–135. Sage Publications, Beverly Hills, CA.   *(88)

Bolviken, E., and Skovlund, E. (1996). Confidence intervals from Monte Carlo tests. *J. Am. Statist. Assoc.* **91**, 1071–1078.

Bonate, P. L. (1993). Approximate confidence intervals in calibration using the bootstrap. *Anal. Chem.* **65**, 1367–1372.

Bondesson, L., and Holm, S. (1985). Bootstrap-estimation of the mean square error of the ratio estimate for sampling without replacement. In *Contributions to Probability and Statistics in Honor of Gunnar Blom* (J. Lanke and G. Lindgren, editors), 85–96. Studentlitteratur, Lund.

Bone, P. F., Sharma, S., and Shimp, T. A. (1987). A bootstrap procedure for evaluating goodness-of-fit indices of structural equation and confirmatory factor models. *J. Marketing Res.* **26**, 105–111.

Boomsma, A. (1986). On the use of bootstrap and jackknife in covariance structure analysis. In *Compstat 1986*. Physica-Verlag, Heidelberg.

Boos, D. D. (1980). A new method for constructing approximate confidence intervals from *M*-estimates. *J. Am. Statist. Assoc.* **75**, 142–145.

Boos, D. D., and Brownie, C. (1989). Bootstrap methods for testing homogeneity of variances. *Technometrics* **31**, 69–82.

Boos, D. D., Janssen, P., and Veraverbeke, N. (1989). Resampling from centered data in the two sample problem. *J. Statist. Plann. Inf.* **21**, 327–345.

Boos, D. D., and Monahan, J. F. (1986). Bootstrap methods using prior information. *Biometrika* **73**, 77–83.

Booth, J. G. (1994). Book Review of *Resampling-Based Multiple Testing: Examples and Methods for p-Value Adjustment* (by P. Westfall and S. S. Young). *J. Am. Statist. Assoc.* **89**, 354–355.

Booth, J. G. (1996). Bootstrap methods for generalized linear mixed models with applications to small area estimation. In *Statistical Modelling* (G. U. H. Seeber, B. J. Francis, R. Hatzinger, and G. Steckel-Berger, editors). *Lecture Notes in Statistics*, **104**, 43–51. Springer-Verlag, New York.

Booth, J. G., and Butler, R. W. (1990). Randomization distributions and saddlepoint approximations in generalized linear models. *Biometrika* **77**, 787–796.

Booth, J. G., Butler, R. W., and Hall, P. (1994). Bootstrap methods for finite populations. *J. Am. Statist. Assoc.* **89**, 1282–1289.   *(89)

Booth, J. G., and Do, K.-A. (1994). Automatic importance resampling for the bootstrap. In *7th Conference on the Scientific Use of Statistical Software. SoftStat 93* (F. Faulbaum, editor), 519–526.

Booth, J. G., and Hall, P. (1993a). An improvement on the jackknife distribution function estimator. *Ann. Statist.* **21**, 1476–1485.

Booth, J. G., and Hall, P. (1993b). Bootstrap confidence regions for functional relationships in error-in-variables models. *Ann. Statist.* **21**, 1780–1791.

Booth, J. G., and Hall, P. (1994). Monte-Carlo approximation and the iterated bootstrap. *Biometrika* **81**, 331–340.

Booth, J. G., Hall, P., and Wood, A. T. A. (1992). Bootstrap estimation of conditional distributions. *Ann. Statist.* **20**, 1594–1610.

Booth, J. G., Hall, P., and Wood, A. T. A. (1993). Balanced importance sampling for the bootstrap. *Ann. Statist.* **21**, 286–298.   *(90)

Booth, J. G., and Sarkar, S. (1998). Monte Carlo approximation of bootstrap variances. *Am. Statist.* **52**, 354–357.   *(91)

Borowiak, D. (1983). A multiple model discrimination procedure. *Commun. Statist. Theory Methods* **12**, 2911–2921.

Borchers, D. L. (1996). Line transect abundance estimation with uncertain detection on the trackline. Ph. D. thesis. University of Capetown, South Africa.

Borchers, D. L., Buckland, S. T., Goedhart, P. W., Clarke, E. D., and Hedley, S. L. (1998). Horvitz–Thompson estimators for double-platform line transect surveys. *Biometrics* **54**, 1221–1237.

Borrello, G. M., and Thompson, B. (1989). A replication bootstrap analysis of the structure underlying perceptions of stereotypic love. *J. Gen. Psychol.* **116**, 317–327.

Bose, A. (1988). Edgeworth correction by bootstrap in autoregressions. *Ann. Statist.* **16**, 1709–1726.   *(92)

Bose, A. (1990). Bootstrap in moving average models. *Ann. Inst. Statist. Math.* **42**, 753–768.   *(93)

Bose, A., and Babu, G. J. (1991). Accuracy of the bootstrap approximation. *Probab. Theory Relat. Fields* **90**, 301–316.

Boukai, B. (1993). A nonparametric bootstrapped estimate of the changepoint. *J. Nonparametric Statist.* **3**, 123–134.

Box, G. E. P., and Jenkins, G. M. (1970). *Time Series Analysis: Forecasting and Control.* Holden Day, San Francisco.   *(94)

Box, G. E. P., and Jenkins, G. M. (1976). *Time Series Analysis: Forecasting and Control,* 2nd ed. Holden Day, San Francisco.   *(95)

Box, G. E. P., Jenkins, G. M., and Reinsel, G. C. (1994). *Time Series Analysis: Forecasting and Control,* 3rd ed. Prentice Hall, Englewood Cliffs, N.J.   *(96)

Bradley, D. W. (1985). The effects of visibility bias on time-budget estimates of niche breadth and overlap. *Auk.* **102**, 493–499.

Bratley, P., Fox, B. L., and Schrage, L. E. (1987). *A Guide to Simulation,* 2nd ed. Springer Verlag, New York.   *(97)

Braun, W. J., and Kulperger, P. J. (1997). Properties of a Fourier bootstrap method for time series. *Commun. Statist. Theory Methods* **26**, 1326–1336.   *(98)

Breidt, F. J., Davis, R. A., and Dunsmuir, W. T. M. (1995) Improved bootstrap prediction intervals for autoregressions. *J. Time Series Anal.* **16**, 177–200.

Breiman, L. (1988). Submodel selection and evaluation in regression: the conditional case and little bootstrap. Technical Report 169, Department of Statistics, University of California, Berkeley.

Breiman, L. (1992). The little bootstrap and other methods for dimensionality selection in regression: x-fixed prediction error. *J. Am. Statist. Assoc.* **87**, 738–754.   *(99)

Breiman, L. (1995). Better subset regression using the nonnegative garrote. *Technometrics* **38**, 170–177.

Breiman, L. (1996). Bagging predictors. *Mach. Learn.* **24**, 123–140.

Breiman, L., and Friedman, J. H. (1985). Estimating optimal transformations for multiple regression and correlation. *J. Am. Statist. Assoc.* **80**, 580–619.   *(100)

Breiman, L., Friedman, J. H., Olshen, R. A., and Stone, C. J. (1984). *Classification and Regression Trees*. Wadsworth, Belmont, CA.   *(101)

Breiman, L., and Spector, P. (1992). Submodel selection and evaluation in regression. X-random case. *Int. Statist. Rev* **60**, 291–319.

Breitung, J. (1992). Nonparametric bootstrap tests: some applications. In *Bootstrapping and Related Techniques. Proceedings, Trier, FRG.* (K.-H. Jockel, G. Rothe, and W. Sendler, editors). *Lecture Notes in Economics and Mathematical Systems*, **376**, 87–97. Springer-Verlag, Berlin.

Bremaud, P. (1981). *Point Processes and Queues: Martingale Dynamics*. Springer-Verlag, New York.   *(102)

Brennan, T. F., and Milenkovic, P. H. (1995). Fast minimum variance resampling. *Conference Proceedings of the 1995 International Conference on Acoustics, Speech and Signal Processing*, **2**, 905–908.

Bretagnolle, J. (1983). Lois limites du bootstrap de certaines fonctionelles. *Ann. Inst. Henri Poincare Sec. B* **19**, 281–296.

Brey, T. (1990). Confidence limits for secondary prediction estimates: application of the bootstrap to the increment summation method. *Mar. Biol.* **106**, 503–508.   *(103)

Brillinger, D. R. (1981). *Time Series Data Analysis and Theory: Expanded Edition*. Holt, Reinhart and Winston, New York.   *(104)

Brockwell, P. J., and Davis, R. A. (1991). *Time Series Methods*. 2nd ed. Springer-Verlag, New York.   *(105)

Brostrom, G. (1997). A martingale approach to the changepoint problem. *J. Am. Statist. Assoc.* **92**, 1177–1183.

Brown, J. K. M. (1994). Bootstrap hypothesis tests for evolutionary trees and other dendrograms. *Proc. Natl. Acad. Sci.* **91**, 12293–12297.

Brownstone, D. (1992). Bootstrapping admissible linear model selection procedures. In *Exploring the Limits of Bootstrap* (R. LePage and L. Billard, editors), 327–344, Wiley, New York.   *(106)

Brumback, B. A., and Rice, J. A. (1998). Smoothing spline models for the analysis of nested and crossed samples of curves (with discussion). *J. Am. Statist. Assoc.* **93**, 961–994.

Bryand, J., and Day, R. (1991). Empirical Bayes analysis for systems of mixed models with linked autocorrelated random effects. *J. Am. Statist. Assoc.* **86**, 1007–1012.

Buckland, S. T. (1980). A modified analysis of the Jolly–Seber capture recapture model. *Biometrics* **36**, 419–435.   *(107)

Buckland, S. T. (1983). Monte Carlo methods for confidence interval estimation using the bootstrap technique. *BIAS* **10**, 194–212.   *(108)

Buckland, S. T. (1984). Monte Carlo confidence intervals. *Biometrics* **40**, 811–817.   *(109)

Buckland, S. T. (1985). Calculation of Monte Carlo confidence intervals. *Appl. Statist.*

34, 296–301.   *(110)

Buckland, S. T. (1993). Book Review of *Exploring the Limits of Bootstrap* (R. LePage and L. Billard, editors). *J. Appl. Statist.* **20**, 332–333.

Buckland, S. T. (1994a). Book Review of *Computer Intensive Statistical Methods: Validation, Model Validation and Bootstrap* (by J. S. U. Hjorth). *Biometrics* **50**, 586–587.

Buckland, S. T. (1994b). Book Review of *An Introduction to the Bootstrap* (by B. Efron and R. J. Tibshirani). *Biometrics* **50**, 890–891.

Buckland, S. T., Burnham, K. P., and Augustin, N. H. (1997). Model selection: an integral part of inference. *Biometrics* **53**, 603–618.

Buckland, S. T., and Garthwaite, P. H. (1990). Algorithm AS 259: estimating confidence intervals by the Robbins–Monro search process. *Appl. Statist.* **39**, 413–424.

Buckland, S. T., and Garthwaite, P. H. (1991). Quantifying precision of mark-recapture estimates using the bootstrap and related methods. *Biometrics* **47**, 255–268.

Buhlmann, P. (1994). Blockwise bootstrap empirical process for stationary sequences. *Ann. Statist.* **22**, 995–1012.

Buhlmann, P. (1997). Sieve bootstrap for time series. *Bernoulli* **3**, 123 148.

Buhlmann, P., and Kunsch, H. R. (1995). The blockwise bootstrap for general parameters of a stationary time series. *Scand. J. Statist.* **22**, 35–54.   *(111)

Bull, S. B., and Greenwood, C. M. T. (1997). Jackknife bias reduction for polychotomous logistic regression. *Statist. Med.* **16**, 545–560.

Bunke, O., and Droge, B. (1984). Bootstrap and cross-validation estimates of the prediction error for linear regression models. *Ann. Statist.* **12**, 1400–1424.

Bunke, O., and Riemer, S. (1983). A note on bootstrap and other empirical procedures for testing linear hypotheses without normality. *Statistics* **14**, 517–526.

Bunt, M., Koch, I., and Pope, A. (1995). How many and where are they? A bootstrap-based segmentation strategy for estimating discontinuous functions. In *Conference Proceedings DICTA-95*, 110–115.

Burke, M. D., and Gombay, E. (1988). On goodness-of-fit and the bootstrap. *Statist. Probab. Lett.* **6**, 287–293.

Burke, M. D., and Horvath, L. (1986). Estimation of influence functions. *Statist. Probab. Lett.* **4**, 81–85.

Burke, M. D., and Yuen, K. C. (1995). Goodness-of-fit tests for the Cox model via bootstrap method. *J. Statist. Plann. Inf.* **47**, 237–256.

Burr, D. (1994). A comparison of certain bootstrap confidence intervals in the Cox model. *J. Am. Statist. Assoc.* **89**, 1290–1302.   *(112)

Burr, D., and Doss, H. (1993). Confidence bands for the median survival time as a function of the covariates in the Cox model. *J. Am. Statist. Assoc.* **88**, 1330–1340.

Butler, R., and Rothman, E. D. (1980). Prediction intervals based on reuse of the sample. *J. Am. Statist. Assoc.* **75**, 881–889.

Buzas, J. S. (1997). Fast estimators of the jackknife. *Am. Statist.* **51**, 235–240.

Caers, J., Beirlant, J., and Vynckier, P. (1997). Bootstrap confidence intervals for tail indices. *Comput. Statist. Data Anal.* **26**, 259–277.

Cammarano, P., Palm, P., Ceccarelli, E., and Creti, R. (1992). Bootstrap probability that archaea are monophyletic. In *Probability and Bayesian Statistics in Medicine and*

Biology (I. Barrai, G. Coletti, and M. Di Bacco, editors). Applied Mathematics Monographs No. 4, 12–35. Giardini, Pisa.

Campbell, G. (1994). Advances in statistical methodology for the evaluation of diagnostic and laboratory tests. *Statist. Med.* **13**, 499–508.

Cano, R. (1992). On the Bayesian bootstrap. In *Bootstrapping and Related Techniques. Proceedings, Trier, FRG* (K.-H. Jockel, G. Rothe, and W. Sendler, editors). *Lecture Notes in Economics and Mathematical Systems*, **376**, 159–161. Springer-Verlag, Berlin.

Canty, A. J., and Davison, A. C. (1997). Implementation of saddlepoint approximations to bootstrap distributions. In *The 28th Symposium on the Interface Between Computer Science and Statistics* (L. Billard and N. I. Fisher, editors), **28**, 248–253. Springer-Verlag, New York.

Cao-Abad, R. (1989). On wild bootstrap confidence intervals. Unpublished manuscript.

Cao-Abad, R. (1991). Rate of convergence for the wild bootstrap in nonparametric regression. *Ann. Statist.* **19**, 2226–2231.   *(113)

Cao-Abad, R., and Gonzalez-Manteiga, W. (1993). Bootstrap methods in regression smoothing. *J. Nonparametric Statist.* **2**, 379–388.

Carlin, B. P., and Louis, T. A. (1996). *Bayes and Empirical Bayes Methods for Data Analysis.* Chapman & Hall, London.   *(114)

Carlin, B. P., and Gelfand, A. E. (1990). Approaches for empirical Bayes confidence intervals. *J. Am. Statist. Assoc.* **85**, 105–114.

Carlin, B. P., and Gelfand, A. E. (1991). A sample reuse method for accurate parametric empirical Bayes confidence intervals. *J. R. Statist. Soc. B* **53**, 189–200.

Carlstein, E. (1986). The use of subseries values for estimating the variance of a general statistic from a stationary sequence. *Ann. Statist.* **14**, 1171–1194.   *(115)

Carlstein, E. (1988a). Typical values. *Encyclopedia of Statistical Sciences*, **9**, 375–377, Wiley, New York.

Carlstein, E. (1988b). Bootstrapping ARMA models: some simulations. *IEEE Trans. Sys.* **16**, 294–299.

Carlstein, E. (1992). Resampling techniques for stationary time series: some recent developments. In *IMA Volumes in Mathematics and Their Applications: New Directions in Time Series Analysis.* Springer-Verlag, New York.

Carpenter, J. R. (1996). Simulated confidence regions for parameters in epidemiological models. Ph. D. thesis, Department of Statistics, Oxford University.

Carroll, R. J. (1979). On estimating variances of robust estimates when the errors are asymmetric. *J. Am. Statist. Assoc.* **74**, 674–679.

Carroll, R. J., Kuchenhoff, H., Lombard, F., and Stefanski, L. A. (1996). Asymptotics for the SIMEX estimator in nonlinear measurement error models. *J. Am. Statist.* **91**, 242–250.

Carroll, R. J., and Ruppert, D. (1988). *Transformation and Weighting in Regression.* Chapman & Hall, New York.   *(116)

Carroll, R. J., Ruppert, D., and Stefanski, L. A. (1995). *Measurement Error in Nonlinear Models.* Chapman & Hall, New York.   *(117)

Carson, R. T. (1985). SAS macros for bootstrapping and cross-validating regression equations. *SAS SUGI* **10**, 1064–1069.

Carter, E. M., and Hubert, J. J. (1985). Analysis of parallel-line assays with multivariate responses. *Biometrics* **41**, 703–710.

Castillo, E. (1988). *Extreme Value Theory in Engineering*. Academic Press, New York. *(118)

Castillo, E., and Hadi, A. S. (1995). Modeling lifetime data with application to fatigue models. *J. Am. Statist. Assoc.* **90**, 1041–1054.

Chambers, J., and Hastie, T. J. (editors) (1991). *Statistical Models in S*. Wadsworth, Belmont, CA. *(119)

Chan, E. (1996). An application of a bootstrap to the Lisrel modeling of two-wave psychological data. In *1st Conference on Applied Statistical Science* (M. Ahsanullah and D. S. Bhoj, editors), **1**, 165–174. Nova Science, Commack, NY.

Chan, Y. M., and Srivastava, M. S. (1988). Comparison of powers for the sphericity tests using both the asymptotic distribution and the bootstrap method. *Commun. Statist. Theory Methods* **17**, 671–690.

Chang, M. N., and Rao, P. V. (1993). Improved estimation of survival functions in the new-better-than-used class. *Technometrics* **35**, 192–203.

Chang, S. I., and Eng, S. S. A. H. (1997). Classification of process variability using neural networks with a bootstrap sampling scheme. In *Proceedings of the 6th Industrial Engineering Research Conference* (G. L. Curry, B. Bidanda, and S. Jagdale, editors), 83–88.

Chao, A. (1984). Nonparametric estimation of the number of classes in a population. *Scand. J. Statist.* **11**, 265–270.

Chao, A., and Huwang, L.-C. (1987). A modified Monte-Carlo technique for confidence limits of system reliability using pass-fail data. *IEEE Trans. Reliab.* **R-36**, 109–112. *(120)

Chao, M.-T., and Lo, S. -H. (1985). A bootstrap method for finite populations. *Sankhya A* **47**, 399–405. *(121)

Chao, M.-T., and Lo, S. -H. (1994). Maximum likelihood summary and the bootstrap method in structured finite populations. *Statist. Sin.* **4**, 389–406. *(122)

Chapman, P., and Hinkley, D. V. (1986). The double bootstrap, pivots and confidence limits. Technical Report 34, Center for Statistical Sciences, University of Texas at Austin.

Chateau, F., and Lebart, L. (1996). Assessing sample variability in the visualization techniques related to principal component analysis: bootstrap and alternative simulation methods. In *COMPSTAT 96 12th Biannual Symposium* (A. Prat, editor), **12**, 205–210. Physica-Verlag, Heidelberg.

Chatfield, C. (1988). *Problem Solving: A Statistician's Guide*. Chapman & Hall, London.

Chatterjee, S. (1984). Variance estimation in factor analysis, an application of bootstrap. *Br. J. Math. Statist. Psychol.* **37**, 252–262.

Chatterjee, S., and Chatterjee, S. (1983). Estimation of misclassification probabilities by bootstrap methods. *Commun. Statist. Simul. Comput.* **12**, 645–656. *(123)

Chatterjee, S., and Hadi, A. S. (1988). *Sensitivity Analysis in Linear Regression*. Wiley, New York. *(124)

Chaubey, Y. P. (1993). Book Review of *Resampling-Based Multiple Testing: Examples and Methods for p-Value Adjustment* (by P. H. Westfall and S. S. Young). *Technometrics* **35**, 450–451.

Chaudhuri, P. (1996). On a geometric notion of quantiles for multivariate data. *J. Am. Statist. Assoc.* **91**, 862–872.

Chen, C., Davis, R. A., Brockwell, P. J., and Bai, Z. D. (1993). Order determination for autoregressive processes using resampling methods. *Statist. Sin.* **3**, 481–500. *(125)

Chen, C., and George, S. L. (1985). The bootstrap and identification of prognostic factors via Cox's proportional hazards regression model. *Statist. Med.* **4**, 39–46. *(126)

Chen, G., and Yang, G. L. (1993). Conditional bootstrap procedure for reconstruction of the incubation period of AIDS. *Math. Biosci.* **117**, 253–269.

Chen, H., and Liu, H. K. (1991). Checking the validity of the bootstrap. *Proceedings of the 23rd Symposium on the Interface Between Computer Science and Statistics,* 293–296. Springer-Verlag, New York.

Chen, H., and Loh, W. Y. (1991). Consistency of bootstrap for the transformed two sample *t* tests. *Commun. Statist. Theory Methods* **20**, 997–1014.

Chen, H., and Sitter, R. R. (1993). Edgeworth expansions and the bootstrap for stratified sampling without replacement from a finite population. *Can. J. Statist.* **21**, 347–357.

Chen, H., and Tu, D. (1987). Estimating the error rate in discriminant analysis: by the delta, jackknife and bootstrap methods. *Chin. J. Appl. Probab. Statist.* **3**, 203–210.

Chen, K., and Lo, S. H. (1996). On bootstrap accuracy with censored data. *Ann. Statist.* **24**, 569–595.

Chen, L. (1995). Testing the mean of skewed distributions. *J. Am. Statist. Assoc.* **90**, 767–772.

Chen, S. X. (1994). Comparing empirical likelihood and bootstrap hypothesis tests. *J. Multivar. Anal.* **51**, 277–293.

Chen, Z. (1990). A resampling approach for bootstrap hypothesis testing. Unpublished Manuscript.

Chen, Z., and Do, K. -A. (1992). Importance resampling for smoothed bootstrap. *J. Statist. Comput. Simul.* **40**, 107–124.

Chen, Z., and Do, K. -A. (1994). The bootstrap methods with saddlepoint approximations and importance resampling. *Statist. Sin.* **4**, 407–421.

Cheng, P. (1993). Weak convergence and bootstrap of bivariate product limit estimators under the bivariate competing risks case (in Chinese). *J. Math. Res. Exp.* **13**, 491–498.

Cheng, P., and Hong, S. -Y. (1993). Bootstrap approximation of parameter estimator in semiparametric regression models (in Chinese). *Sci. Sin.* **23**, 237–251.

Cheng, R. C. H. (1995). Bootstrap methods in computer simulation experiments. In *Proceedings of the 1995 Winter Simulation Conference WSC' 95*, 171–177.

Cheng, R. C. H., Holland, W., and Hughes, N. A. (1996). Selection of input models using bootstrap goodness-of-fit. In *Proceedings of the 1996 Winter Simulation Conference*, 199–206.

Chenier, T. C., and Vos, P. W. (1996). Beat Student's *t* and the bootstrap-using SAS software to generate conditional confidence intervals for means. *SUGI 21* **1**, 1311–1316.

Chernick, M. R. (1982). The influence function and its application to data validation. *Am. J. Math. Manage. Sci.* **2**, 263–288.   *(127)

Chernick, M. R., and Murthy, V. K. (1985). Properties of bootstrap samples. *Am. J. Math. Manage. Sci.* **5**, 161–170.   *(128)

Chernick, M. R., Murthy, V. K., and Nealy, C. D. (1985). Application of bootstrap and other resampling techniques: evaluation of classifier performance. *Pattern Recogn. Lett.* **3**, 167–178.   *(129)

Chernick, M. R., Murthy, V. K., and Nealy, C. D. (1986). Correction note to application of bootstrap and other resampling techniques: Evaluation of classifier performance. *Pattern Recogn. Lett.* **4**, 133–142.   *(130)

Chernick, M. R., Murthy, V. K., and Nealy, C. D. (1988a). Estimation of error rate for linear discriminant functions by resampling: non-Gaussian populations. *Comput. Math. Applic.* **15**, 29–37.   *(131)

Chernick, M. R., Murthy, V. K., and Nealy, C. D. (1988b). Resampling-type error rate estimation for linear discriminant functions: Pearson VII distributions. *Comput. Math. Applic.* **15**, 897–902.   *(132)

Chernick, M. R., Daley, D. J., and Littlejohn, R. P. (1988). A time-reversibility relationship between two Markov chains with exponential stationary distributions. *J. Appl. Probab.* **25**, 418–422.   *(133)

Chernick, M. R., Downing, D. J., and Pike, D. H. (1982). Detecting outliers in time series data. *J. Am. Statist. Assoc.* **77**, 743–747.   *(134)

Chin, L., Haughton, D., and Aczel, A. (1996). Analysis of student evaluation of teaching scores using bootstrap and permutation methods. *J. Comput. Higher Educ.* **8**, 69–84.

Cho, K., Meer, P., and Cabrera, J. (1997). Performance assessment through bootstrap. *IEEE Trans. Pattern Anal. Mach. Intell.* **19**, 1185–1198.

Choi, K. C., Nam, K. H., and Park, D. H. (1996). Estimation of capability index based on bootstrap method. *Microelectron. Reliab.* **36**, 256–259.   *(135)

Choi, S. C. (1986). Discrimination and classification: overview. *Comput. Math. Applic.* **12A**, 173–177.   *(136)

Christoffersson, J. (1997). Frequency domain resampling of time series data. Ph. D. thesis. Acta Universitatis Agriculturae Sueciae, Silvestria.

Chu, P.-S., and Wang, J. (1998). Interaval variability of tropical cyclone incidences in the vicinity of Hawaii using statistical resampling techniques. *Conference on Probability and Statistics in the Atmospheric Sciences*, **14**, 96–97.

Chung, K. L. (1974). *A Course in Probability Theory.* 2nd ed. Academic Press, New York.   *(137)

Ciarlini, P. (1997). Bootstrap algorithms and applications. In *Advanced Mathematical Tools in Metrology III* (P. Ciarlini, M. G. Cox, F. Pavese, and D. Richter, editors), 171–177. World Scientific, Singapore.

Ciarlini, P., Gigli, A., Regoliosi, G., Moiraghi, L., and Montefusco, A. (1996). Monitoring an industrial process: bootstrap estimation of accuracy of quality parameters. In *Advanced Mathematical Tools in Metrology II* (P. Ciarlini, M. G. Cox, F. Pavese, and D. Richter, editors), 123–129. World Scientific, Singapore.

Cirincione, C., and Gurrieri, G. A. (1997). Research methodology. Computer-intensive method in the social sciences. *Soc. Sci. Comput. Rev.* **15**, 83–97.

Clayton, H. R. (1994). Bootstrap approaches: some potential problems. *1994 Proceedings Decision Science Institute,* **2**, 1394–1396.

Clayton, H. R. (1996). Developing a robust parametric bootstrap bound for evaluating audit samples. *Proceedings of the Annual Meeting of the Decision Sciences Institute,* **2**, 1107–1109. Decision Sciences Institute, Atlanta.

Cleveland, W. S., and McGill, R. (1983). A color-caused optical illusion on a statistical graph. *Am. Statist.* **37**, 101–105.

Cliff, A. D., and Ord, J. K. (1973). *Spatial Autocorrelation.* Pion, London.   *(138)

Cliff, A. D., and Ord, J. K. (1981). *Spatial Processes: Models and Applications.* Pion, London.   *(139)

Coakley, K. J. (1991). Bootstrap analysis of asymmetry statistics for polarized beam studies. *Proceedings of the 23rd Symposium on the Interface Between Computer Science and Statistics,* 301–304, Springer-Verlag, New York.

Coakley, K. J. (1994). Area sampling scheme for improving maximum likelihood reconstructions for positive emission tomography images. *Proceedings of Medical Imaging 1994, The International Society for Optical Engineering (SPIE),* **2167**, 271–280.

Coakley, K. J. (1996). Bootstrap method for nonlinear filtering of EM-ML reconstructions of PET images. *Int. J. Imaging Syst. Technol.* **7**, 54–61.   *(140)

Cochran, W. (1977). *Sampling Techniques,* 3rd. ed. Wiley, New York.   *(141)

Cohen, A. (1986). Comparing variances of correlated variables. *Psychometrika* **51**, 379–391.

Cohen, A., Lo, S.-H., and Singh, K. (1985). Estimating a quantile of a symmetric distribution. *Ann. Statist.* **13**, 1114–1128.

Cole, M. J., and McDonald, J. W. (1989). Bootstrap goodness-of-link testing in generalized linear models. In *Statistical Modelling: Proceedings of GLIM 89 and the 4th International Workshop on Statistical Modelling* (A. Decarli, editor). *Lecture Notes in Statistics* **57**, Springer-Verlag, Berlin.

Collings, B. J., and Hamilton, M. A. (1988). Estimating the power of the two-sample Wilcoxon test for location shift. *Biometrics* **44**, 847–860.

Constantine, K., Karson, M. J., and Tse, S. -K. (1990). Confidence interval estimation of $P(Y < X)$ in the gamma case. *Commun. Statist. Simul. Comput.* **19**, 225–244.

Conti, P. L. (1993). Book Review of *Exploring the Limits of Bootstrap* (R. Lepage and L. Billard, editors). *Metron* **51**, 213–217.

Cook, R. D., and Weisberg, S. (1990). Confidence curves in nonlinear regression. *J. Am. Statist. Assoc.* **85**, 544–551.

Cook, R. D., and Weisberg, S. (1994). Transforming a response variable for linearity. *Biometrika* **81**, 731–737.   *(142)

Cowling, A., Hall, P., and Phillips, M. J. (1996). Bootstrap confidence regions for the intensity of a Poisson point process. *J. Am. Statist. Assoc.* **91**, 1516–1524.   *(143)

Cover, K. A., and Unny, T. E. (1986). Application of computer intensive statistics to parameter uncertainty in streamflow. *Water Res. B.* **22**, 495–507.

Cox, D. R. (1972). Regression models and life tables. *J. R. Statist. Soc. B* **34**, 187–202.   *(144)

Cox, D. R., and Reid, N. (1987a). Parameter orthogonality and approximate conditional inference (with discussion). *J. R. Statist. Soc. B* **49**, 1–39.

Cox, D. R., and Reid, N. (1987b). Approximations to noncentral distributions. *Can. J. Statist.* **15**, 105–114.

Crawford, S. (1989). Extensions to the CART algorithm. *Int. J. Mach. Studies* **31**, 197–217.

Cressie, N. (1991). *Statistics for Spatial Data.* Wiley, New York.   *(145)

Crivelli, A., Firinguetti, L., Montano, R., and Munoz, M. (1995). Confidence intervals in ridge regression by bootstrapping the dependent variable: a simulation study. *Commun. Statist. Simul. Comput.* **24**, 631–652.

Croci, M. (1994). Some semi-parametric bootstrap applications in unfavourable circumstances (in Italian). *Proc. Ital. Statist. Soc.* **2**, 449–456.

Crone, L. J., and Crosby, D. S. (1995). Statistical applications of a metric on subspaces to satellite meteorology. Technometrics **37**, 324–328.

Crosilla, F., and Pillirone, G. (1994). Testing the data quality of G. I. S. data base by bootstrap methods and other nonparametric statistics. *ISPRS Commission III Symposium on Spatial Information from Photogrammetry and Computer Vision, Munich Germany, Sept. 5–9, 1994. SPIE Proceedings* **2357**, 158–164.

Crosilla, F., and Pillirone, G. (1995). Non-parametric statistics and bootstrap methods for testing data quality of a geographic information system. In *Geodetic Theory Today: Symposium of the International Association of Geodesy* (F. Sansao, editor), **114**, 214–223.

Crowder, M. J., and Hand, D. J. (1990). *Analysis of Repeated Measures.* Chapman & Hall, London.

Crowley, P. H. (1992). Resampling methods for computer-intensive data analysis in ecology and evolution. *Annu. Rev. Ecol. Syst.* **23**, 405–407.

Csorgo, M. (1983). *Quantile Processes with Statistical Applications.* SIAM, Philadelphia.   *(146)

Csorgo, M., Csorgo, S., and Horvath, L. (1986). *An Asymptotic Theory for Empirical Reliability and Concentration Processes.* Springer-Verlag, New York.   *(147)

Csorgo, S., and Mason, D. M. (1989). Bootstrapping empirical functions. *Ann. Statist.* **17**, 1447–1471.   *(148)

Csorgo, M., and Zitikis, R. (1996). Mean residual life processes. *Ann. Statist.* **24**, 1717–1739.

Cuesta-Albertos, J. A., Gordaliza, A., and Matran, C. (1997). Trimmed k-means: an attempt to robustify quantizers. *Ann. Statist.* **25**, 553–576.

Cuevas, A., and Romo, J. (1994). Continuity and differentiability of statistical operators: some applications to the bootstrap. In *Proceedings of the 3rd World Conference of the Bernoulli Society and the 57th Annual Meeting of the Institute of Mathematical Statistics, Chapel Hill.*

Daggett, R. S., and Freedman, D. A. (1985). Econometrics and the law: a case study in the proof of antitrust damages. *Proc. Berk. Symp.* **VI**, 123–172.   *(149)

Dahlhaus, R., and Janas, D. (1996). A frequency domain bootstrap for ratio statistics in time series analysis. *Ann. Statist.* **24**, 1914–1933.   *(150)

Dalal, S. R., Fowlkes, E. B., and Hoadley, B. (1989). Risk analysis of the space shuttle: pre-*Challenger* prediction of failure. *J. Am. Statist. Assoc.* **84**, 945–957.

Daley, D. J., and Vere-Jones, D. (1988). *An Introduction to the Theory of Point Processes.* Springer-Verlag, New York.   *(151)

Dalgleish, L. I. (1995). Software review: bootstrapping and jackknifing with BOJA. *Statist. Comput.* **5**, 165–174.

Daniels, H. E. (1954). Saddlepoint approximations in statistics. *Ann. Math. Statist.* **25**, 631–650.   *(152)

Daniels, H. E., and Young, G. A. (1991). Saddlepoint approximation for the studentized mean, with an application to the bootstrap. *Biometrika* **78**, 169–179.   *(153)

Das, S., and Sen, P. K. (1995). Simultaneous spike-trains and stochastic dependence. *Sankhya B* **57**, 32–47.

Das Peddada, S., and Chang, T. (1992). Bootstrap confidence region estimation of motion of rigid bodies. *J. Am. Statist. Assoc.* **91**, 231–241.   *(154)

Datta, S. (1992). A note on the continuous Edgeworth expansions and the bootstrap. *Sankhya A* **54**, 171–182.

Datta, S. (1994). On a modified bootstrap for certain asymptotically nonnormal statistics. *Statist. Probab. Lett.* **24**, 91.

Datta, S., and McCormick, W. P. (1992). Bootstrap for finite state Markov chain based on i. i. d. resampling. In *Exploring the Limits of Bootstrap* (R. LePage and L. Billard, editors), 77–97. Wiley, New York.   *(155)

Datta, S., and McCormick, W. P. (1995). Bootstrap inference for a first-order autoregression with positive innovations. *J. Am. Statist. Assoc.* **90**, 1289–1300.   *(156)

Daudin, J. J., Duby, C., and Trecourt, P. (1988). Stability of principal component analysis studied by the bootstrap method. *Statistics* **19**, 241–258.

David, H. A. (1981). *Order Statistics.* Wiley, New York.   *(157)

Davis, C. E., and Steinberg, S. M. (1986). Quantile estimation. *Encyclopedia of Statistical Sciences*, **7**, 408–412. Wiley, New York.

Davis, R. B. (1995). Resampling: a good tool, no panacea. *MD Comput.* **12**, 89–91.

Davison, A. C. (1993). Book Review of *Exploring the Limits of Bootstrap* (R. LePage and L. Billard, editors). *J. R. Statist. Soc. A* **156**, 133.

Davison, A. C., and Hall, P. (1992). On the bias and variability of bootstrap and cross-validation estimates of error rate in discriminant analysis. *Biometrika* **79**, 279–284.   *(158)

Davison, A. C., and Hall, P. (1993). On studentizing and blocking methods for implementing the bootstrap with dependent data. *Aust. J. Statist.* **35**, 215–224.   *(159)

Davison, A. C., and Hinkley, D. V. (1988). Saddlepoint approximations in resampling methods. *Biometrika* **75**, 417–431.   *(160)

Davison, A. C., and Hinkley, D. V. (1992). Computer-intensive statistical methods. *Proceedings of 10th Symposium on Computational Statistics* **2**, 51–62.

Davison, A. C., and Hinkley, D. V. (1997). *Bootstrap Methods and Their Application.* Cambridge University Press, Cambridge.   *(161)

Davison, A. C., Hinkley, D. V., and Schechtman, E. (1986). Efficient bootstrap simulation. *Biometrika* **73**, 555–566.   *(162)

Davison, A. C., Hinkley, D. V., and Worton, B. J. (1992). Bootstrap likelihoods. *Biometrika* **79**, 113–130.   *(163)

Davison, A. C., Hinkley, D. V., and Worton, B. J. (1995). Accurate and efficient construction of bootstrap likelihoods. *Statist. Comput.* **5**, 257–264.

Day, N. E. (1969). Estimating the components of a mixture of two normal distributions. *Biometrika* **56**, 463–474.   *(164)

DeAngelis, D., and Young, G. A. (1992). Smoothing the bootstrap. *Int. Statist. Rev.* **60**, 45–56.   *(165)

DeAngelis, D., and Young, G. A. (1998). Bootstrap method. In *Encyclopedia of Biostatistics* (P. Armitage and T. Colton, editors), **1**, 426–433. Wiley, New York. *(166)

DeAngelis, D., Hall, P., and Young, G. A. (1993a). Analytical and bootstrap approximations to estimator distributions in L^1 regression. *J. Am. Statist. Assoc.* **88**, 1310–1316.   *(167)

De Angelis, D., Hall, P., and Young, G. A. (1993b). A note on coverage error of bootstrap confidence intervals for quantiles. *Math. Proc. Cambridge Philos. Soc.* **114**, 517–531.

Deaton, M. L. (1988). Simulation models (validation of). *Encyclopedia of Statistical Sciences*, **8**, 481–484. Wiley, New York.

DeBeer, C. F., and Swanepoel, J. W. H. (1989). A modified Durbin–Watson test for serial correlation in multiple regression under non-normality using bootstrap. *J. Statist. Comput. Simul.* **33**, 75–82.

DeBeer, C. F., and Swanepoel, J. W. H. (1993). A modified bootstrap estimator for the mean of an asymmetric distribution. *Can. J. Statist.* **21**, 79–87.

Deheuvels, P., Mason, D. M., and Shorack, G. R. (1993). Some results on the influence of extremes on the bootstrap. *Ann. Inst. Henri Poincare* **29**, 83–103.

de Jongh, P. J., and de Wet, T. (1985). A Monte Carlo comparison of regression trimmed means. *Commun. Statist. Simul. Comput.* **14**, 2457–2472.

de Jongh, P. J., and de Wet, T. (1986). Confidence intervals for regression parameters based on trimmed means. *S. Afr. Statist. J.* **20**, 137–164.

Delaney, N. J., and Chatterjee, S. (1986). Use of bootstrap and cross-validation in ridge regression. *J. Bus. Econ. Statist.* **4**, 255–262.

de la Pena, V., Gini, E., and Alemayehu. D. (1993). Bootstrap goodness-of-fit tests based on the empirical characteristic function. *Proceedings of the 25th Symposium on the Interface of Computer Science and Statistics*, **25**, 228–233.

Delicado, P., and del Rio, M. (1994). Bootstrapping the general linear hypothesis test. *Comput. Statist. Data Anal.* **18**, 305–316.

Dempster, A. P., Laird, N. M., and Rubin, D. B. (1977). Maximum likelihood from incomplete data via the EM algorithm (with discussion). *J. R. Statist. Soc. B* **39**, 1–38.   *(168)

Dennis, B., and Taper, M. L. (1994). Density dependence in time series observations of natural populations: estimation and testing. *Ecol. Monogr.* **64**, 205–224.

Depuy, K. M., Hobbs, J. R., Moore, A. H., and Johnston, J. W., Jr. (1982). Accuracy of univariate, bivariate and a "modified double Monte Carlo" technique for finding lower confidence limits of system reliability. *IEEE Trans. Reliab.* **R-31**, 474–477.

Desgagne, A., Castilloux, A.-M., Angers, J.-F., and Lelorier, J. (1998). The use of the bootstrap statistical method for the pharmacoeconomic cost analysis of skewed data. *Pharmacoeconomics* **13**, 487–497.

Dette, H., and Munk, A. (1998). Validation of linear regression models. *Ann. Statist.* **26**, 778–800.

Devorye, L. (1986). *Non-Uniform Random Variate Generation.* Springer-Verlag, New York.   *(169)

Devorye, L., and Gyorfi L. (1985). *Nonparametric Density Estimation.* Wiley, New York.   *(170)

de Wet, T., and Van Wyk, J. W. J. (1986). Bootstrap confidence intervals for regression coefficients when the residuals are dependent. *J. Statist. Comput. Simul.* **23**, 317–327.

Diaconis, P. (1985). Theories of data analysis: from magical thinking through classical statistics. In *Exploring Data Tables, Trends, and Shapes* (D. C. Hoaglin, F. Mosteller, and J. W. Tukey, editors), 1–36. Wiley, New York.

Diaconis, P., and Efron, B. (1983). Computer-intensive methods in statistics. *Sci. Amer.* **248**, 116–130.   *(171)

Diaconis, P., and Holmes, S. (1994). Gray codes for randomization procedures. *Statist. Comput.* **4**, 287–302.   *(172)

DiCiccio, T. J., and Efron, B. (1990). Better approximate confidence intervals in exponential families. Technical Report 345, Department of Statistics, Stanford University.   *(173)

DiCiccio, T. J., and Efron, B. (1992). More accurate confidence intervals in exponential families. *Biometrika* **79**, 231–245.   *(174)

DiCiccio, T. J., and Efron, B. (1996). Bootstrap confidence intervals (with discussion). *Statist. Sci.* **11**, 189–228.   *(175)

DiCiccio, T. J., Hall, P., and Romano, J. P. (1989). Comparison of parametric and empirical likelihood functions. *Biometrika* **76**, 465–476.

DiCiccio, T. J., Hall, P., and Romano, J. P. (1991). Empirical likelihood is Bartlett correctable. *Ann. Statist.* **19**, 1053–1061.

DiCiccio, T. J., Martin, M. A., and Young, G. A. (1992a). Fast and accurate approximate double bootstrap approximations. *Biometrika* **79,** 285–295.   *(176)

DiCiccio, T. J., Martin, M. A., and Young, G. A. (1992b). Analytic approximation for iterated bootstrap confidence intervals. *Statist. Comput.* **2**, 161–171.

DiCiccio, T. J., Martin, M. A., and Young, G. A. (1994). Analytic approximations to bootstrap distribution functions using saddlepoint methods. *Statist. Sin.* **4**, 281–295.

DiCiccio, T. J., and Romano, J. P. (1988). A review of bootstrap confidence intervals (with discussion). *J. R. Statist. Soc. B* **50**, 338–370. Correction *J. R. Statist. Soc. B* **51**, 470.   *(177)

DiCiccio, T. J., and Romano, J. P. (1989a). The automatic percentile method: accurate confidence limits in parametric models. *Can. J. Statist.* **17**, 155–169.

DiCiccio, T. J., and Romano, J. P. (1989b). On adjustments based on the signed root of the empirical likelihood ratio statistic. *Biometrika* **76**, 447–456.

DiCiccio, T. J., and Romano, J. P. (1990). Nonparametric confidence limits by resampling methods and least favorable families. *Int. Statist. Rev.* **58**, 59–76. *(178)

DiCiccio, T. J., and Tibshirani, R. (1987). Bootstrap confidence intervals and bootstrap approximations. *J. Am. Statist. Assoc.* **82**, 163–170.

Diebolt, J., and Ip, E. H. S. (1995). Stochastic EM: method and application. In *Markov Chain Monte Carlo in Practice* (W. R. Gilks, S. Richardson, and D. J. Spiegelhalter, editors), 259–273. Chapman & Hall, London.

Dielman, T. E., and Pfaffenberger, R. C. (1988). Bootstrapping in least absolute value regression: an application to hypothesis testing. *Commun. Statist. Simul. Comput.* **17**, 843–856.   *(179)

Diggle, P. J. (1983). *Statistical Analysis of Spatial Point Patterns*. Academic Press, New York.   *(180)

Diggle, P. J., Lange, N., and Benes, F. M. (1991). Analysis of variance for replicated spatial point patterns in clinical neuroanatomy. *J. Am. Statist. Assoc.* **78**, 618–625. *(181)

Dijkstra, D. A., and Veldkamp, J. H. (1988). Data-driven selection of regressors and the bootstrap. In *On Model Uncertainty and Its Statistical Implications.* (T. K. Dijkstra, editor). *Lecture Notes in Economics and Mathematical Systems*, **307**. Springer-Verlag, Berlin.

Dikta, G. (1990). Bootstrap approximation of nearest neighbor regression function estimates. *J. Multivar. Anal.* **32**, 213–229.   *(182)

Dikta, G., and Ghorai, J. K. (1990). Bootstrap approximation with censored data under the proportional hazard model. *Commun. Statist. Theory Methods* **19**, 573–581.

Dirschedl, P., and Grohmann, R. (1992). Exploring heterogeneous risk structure: comparison of a bootstrapped model selection and nonparametric classification technique. In *Bootstrapping and Related Techniques. Proceedings, Trier, FRG* (K.-H. Jockel, G. Rothe, and W. Sendler, editors). *Lecture Notes in Economics and Mathematical Systems*, **376**, 189–195. Springer-Verlag, Berlin.

Dixon, P. M. (1993). The bootstrap and the jackknife: describing the precision of ecological studies. In *Design and Analysis of Ecological Experiments* (S. M. Scheiner and J. Gurevitch, editors), 290–318. Chapman & Hall, New York.

Dixon, P. M., Weiner, J., Mitchell-Olds, T., and Woodley, R. (1987). Bootstrapping the Gini coefficient of inequality. *Ecology* **68**, 1548–1551.

Do, K.-A. (1992). A simulation study of balanced and antithetic bootstrap resampling methods. *J. Statist. Comput. Simul.* **40**, 153–156.   *(183)

Do, K.-A. (1996). A numerical study of the matched-block bootstrap for dependent data. In *27th Symposium on the Interface Between Computer Science and Statistics* (M. M. Meyer and J. L. Rosenberger, editors), **27**, 425–429.

Do, K.-A., and Hall, P. (1991a). On importance resampling for the bootstrap. *Biometrika* **78**, 161–167.   *(184)

Do, K.-A., and Hall, P. (1991b). Quasi-random resampling for the bootstrap. *Statist. Comput.* **1**, 13–22.   *(185)

Do, K.-A., and Hall, P. (1992). Distribution estimation using concomitants of order statistics with applications to Monte Carlo simulation for the bootstrap. *J. R. Statist. Soc. B* **54**, 595–607.

Dohman, B. (1990). Confidence intervals for small sample sizes: bootstrap vs. standard methods (in German). Diplom. thesis. University of Siegen.

Donegani, M. (1992). A bootstrap adaptive test for two-way analysis of variance. *Biom. J.* **34**, 141–146.

Donegani, M., and Unternahrer, M. (1991). Estimation of the variance of an adaptive estimator by bootstrapping. *Commun. Statist. Theory Methods* **20**, 3925–3931.

Dopazo, J. (1994). Estimating errors and confidence intervals for branch lengths in phylogenetic trees by a bootstrap approach. *J. Mol. Evol.* **38**, 300–304.

Dorfman, D. D., Berbaum, K. S., and Lenth, R. V. (1995). Multireader, multicase receiver operating characteristic methodology: a bootstrap analysis. *Acad. Radiol.* **2**, 626–633.

Doss, H., and Chiang, Y. C. (1994). Choosing the resampling scheme when bootstrapping: a case study in reliability. *J. Am. Statist. Assoc.* **89**, 298–308.

Doss, H., and Gill, R. D. (1992). An elementary approach to weak convergence for quantile processes, with applications to censored survival data. *J. Am. Statist. Assoc.* **87**, 869–877.

Douglas, S. M. (1987). Improving the estimation of a switching regressions model: an analysis of problems and improvements using the bootstrap. Ph. D. dissertation. University of North Carolina.

Draper, N. R., and Smith, H. (1981). *Applied Regression Analysis*, 2nd ed. Wiley, New York.   *(186)

Draper, N. R., and Smith, H. (1998). *Applied Regression Analysis*, 3rd ed. Wiley, New York.   *(187)

Droge, B. (1987). A note on estimating the MSEP in nonlinear regression. *Statistics* **18**, 499–520.

Duan, N. (1983). Smearing estimate: a nonparametric retransformation method. *J. Am. Statist. Assoc.* **78**, 605–610.   *(188)

Ducharme, G. R., and Jhun, M. (1986). A note on the bootstrap procedure for testing linear hypotheses. *Statistics* **17**, 527–531.

Ducharme, G. R., Jhun, M., Romano, J., and Truong, K. N. (1985). Bootstrap confidence cones for directional data. *Biometrika* **72**, 637–645.

Duda, R. O., and Hart, P. E. (1973). *Pattern Recognition and Scene Analysis*. Wiley, New York.   *(189)

Dudewicz, E. J. (1992). The generalized bootstrap. In *Bootstrapping and Related Techniques. Proceedings, Trier, FRG.* (K.-H. Jockel, G. Rothe, and W. Sendler, editors), *Lecture Notes in Economics and Mathematical Systems*, **376**, 31–37. Springer-Verlag, Berlin.   *(190)

Dudewicz, E. J. (editor) (1996). *Modern Digital Simulation Methodology, II: Univariate and Bivariate Distribution Fitting, Bootstrap Methods & Application*. American Sciences Press, Columbus.

Dudewicz, E. J., and Mishra, S. N. (1988). *Modern Mathematical Statistics*. Wiley, New York.   *(191)

Dudley, R. M. (1978). Central limit theorem for empirical measures. *Ann. Probab.* **6**, 899–929.

DuMouchel, W., and Waternaux, C. (1983). Some new dichotomous regression methods. In *Recent Advances in Statistics* (M. H. Rizvi, J. Rustagi, and D. Siegmund, editors), 529–555. Academic Press, New York.

Dumbgen, L. (1991). The asymptotic behavior of some nonparametric change-point estimators. *Ann. Statist.* **19**, 1471–1495.

Dutendas, D., Moreau, L., Ghorbel, F., and Allioux, P. M. (1995). Unsupervised Bayesian segmentation with bootstrap sampling application to eye fundus image coding. In *IEEE Nuclear Science Symposium and Medical Imaging Conference,* **4**, 1794–1796.

Eakin, B. K., McMillen, D. P., and Buono, M. J. (1990). Constructing confidence intervals using the bootstrap: an application to multi-product cost function. *Rev. Econ. Statist.* **72**, 339–344.

Eaton, M. L., and Tyler, D. E. (1991). On Wielandt's inequality and its application to the asymptotic distribution of the eigenvalues of a random symmetric matrix. *Ann. Statist.* **19**, 260–271.

Ecker, M. D., and Heltsche, J. F. (1994). Geostatistical estimates of scallop abundance. In *Case Studies in Biometry* (N. Lange, L. Ryan, L. Billard, D. Brillinger, L. Conquest, and J. Greenhouse, editors), 107–124. Wiley, New York.

Eckert, R. S., Carroll, R. J., and Wang, W. (1997). Transformations to additivity in measurement error models. *Biometrics* **53**, 262–272.

Eddy, W. F., and Gentle, J. E. (1985). Statistical computing: what's past is prologue. In *A Celebration of Statistics. The ISI Centenary Volume* (A. C. Atkinson and S. E. Fienberg, editors), 233–249. Springer-Verlag, New York.

Edgington, E. S. (1986). Randomization tests. *Encyclopedia of Statistical Sciences,* **7**, 530–538, Wiley, New York.

Edgington, E. S. (1980). *Randomization Tests,* Marcel Dekker, New York.   *(192)

Edgington, E. S. (1987). *Randomization Tests.* 2nd ed. Marcel Dekker, New York. *(193)

Edgington, E. S. (1995). *Randomization Tests.* 3rd ed. Marcel Dekker, New York. *(194)

Edler, L., Groger, P., and Thielmann, H. W. (1985). Computational statistics for cell survival curves II: evaluation of colony-forming ability of a group of cell strains by the APL function CFAGROUP. *Comp. Biomed.* **21**, 47–54.

Efron, B. (1978). Controversies in the foundations of statistics. *Am. Math. Month.* **85**, 231–246.   *(195)

Efron, B. (1979a). Bootstrap methods: another look at the jackknife. *Ann. Statist.* **7**, 1–26.   *(196)

Efron, B. (1979b). Computers and the theory of statistics: thinking the unthinkable. *SIAM Rev.* **21**, 460–480.   *(197)

Efron, B. (1981a). Censored data and the bootstrap. *J. Am. Statist. Assoc.* **76**, 312–319. *(198)

Efron, B. (1981b). Nonparametric estimates of standard error: the jackknife, the bootstrap and other methods. *Biometrika* **68**, 589–599.

Efron, B. (1981c). Nonparametric standard errors and confidence intervals (with discussion). *Can. J. Statist.* **9**, 139–172.

Efron, B. (1982a). *The Jackknife, the Bootstrap, and Other Resampling Plans*. SIAM, Philadelphia.   *(199)

Efron, B. (1982b). Computer-intensive methods in statistics. In *Some Recent Advances in Statistics* (J. T. de Oliveira and B. Epstein, editors), 173–181. Academic Press, London.   *(200)

Efron, B. (1983). Estimating the error rate of a prediction rule: improvements on cross-validation. *J. Am. Statist. Assoc.* **78**, 316–331.   *(201)

Efron, B. (1984). Comparing non-nested linear models. *J. Am. Statist. Assoc.* **79**, 791–803.

Efron, B. (1985). Bootstrap confidence intervals for a class of parametric problems. *Biometrika* **72**, 45–58.

Efron, B. (1986). How biased is the apparent error rate of a prediction rule? *J. Am. Statist. Assoc.* **81**, 461–470.

Efron, B. (1987). Better bootstrap confidence intervals (with discussion). *J. Am. Statist. Assoc.* **82**, 171–200.   *(202)

Efron, B. (1988a). Three examples of computer-intensive statistical inference. *Sankhya A* **50**, 338–362.   *(203)

Efron, B. (1988b). Computer-intensive methods in statistical regression. *SIAM Rev.* **30**, 421–449.   *(204)

Efron, B. (1988c). Bootstrap confidence intervals: Good or bad? *Psychol. Bull.* **104**, 293–296.   *(205)

Efron, B. (1990). More efficient bootstrap computations. *J. Am. Statist. Assoc.* **85**, 79–89.   *(206)

Efron, B. (1992a). Regression percentiles using asymmetric squared error loss. *Statist. Sin.* **1**, 93–125.   *(207)

Efron, B. (1992b). Six questions raised by the bootstrap. In *Exploring the Limits of Bootstrap* (R. LePage and L. Billard, editors), 99–126. Wiley, New York.   *(208)

Efron, B. (1992c). Jackknife-after-bootstrap standard errors and influence functions (with discussions). *J. R. Statist. Soc. B* **54**, 83–127.   *(209)

Efron, B. (1993). Statistics in the 21st Century. *Statist. Comput.* **3**, 188–190.

Efron, B. (1994). Missing data, imputation, and the bootstrap (with discussion). *J. Am. Statist. Assoc.* **89**, 463–479.   *(210)

Efron, B. (1996). Empirical Bayes methods for combining likelihoods (with discussion). *J. Am. Statist. Assoc.* **91**, 538–565.

Efron, B., and Feldman, D. (1991). Compliance as an explanatory variable in clinical trials. *J. Am. Statist. Assoc.* **86**, 9–26.

Efron, B., and Gong, G. (1981). Statistical thinking and the computer. *Proceedings Computer Science and Statistics* **13**, (W. F. Eddy, editor), **13**, 3–7. Springer-Verlag, New York.

Efron, B., and Gong, G. (1983). A leisurely look at the bootstrap, jackknife and cross-validation. *Am. Statist.* **37**, 36–48.   *(211)

Efron, B., Halloran, M. E., and Holmes, S. (1996). Bootstrap confidence levels for phylogenetic trees. *Proc. Natl. Acad. Sci.* **93**, 13429–13434.

Efron, B., and LePage, R. (1992). Introduction to bootstrap. In *Exploring the Limits of Bootstrap* (R. LePage and L. Billard, editors), 3–10, Wiley, New York.   *(212)

Efron, B., and Stein, C. (1981). The jackknife estimate of variance. *Ann. Statist.* **9**, 586–596.

Efron, B., and Tibshirani, R. (1985). The bootstrap method for assessing statistical accuracy. *Behaviormetrika* **17**, 1–35.   *(213)

Efron, B., and Tibshirani, R. (1986). Bootstrap methods for standard errors: confidence intervals and other measures of statistical accuracy. *Statist. Sci.* **1**, 54–77.   *(214)

Efron, B., and Tibshirani, R. (1993). *An Introduction to the Bootstrap*. Chapman & Hall, New York.   *(215)

Efron, B., and Tibshirani, R. (1996a). Computer-intensive statistical methods. In *Advances in Biometry* (P. Armitage and H. A. David, editors), 131–147. Wiley, New York.   *(216)

Efron, B., and Tibshirani, R. (1996b). The problem of regions. Stanford University Technical Report No. 192.   *(217)

Efron, B., and Tibshirani, R. (1997a). Improvements on cross-validation: The . 632+ bootstrap methods. *J. Am. Statist. Assoc.* **92**, 548–560.   *(218)

Efron, B., and Tibshirani, R. (1997b). Computer-intensive statistical methods. In *Encyclopedia of Statistical Sciences Update Volume 1* (S. Kotz, C. B. Read, and D. L. Banks, editors), 139–148. Wiley, New York.   *(219)

El-Sayed, S. M., Jones, P. W., and Ashour, S. K., (1991). Bootstrap censored data analysis of fractured femur patients under the mixed exponential model using maximum likelihood. In *26th Annual Conference on Statistics, Computer Science and Operations Research. Vol. 1; Mathematical Statistics,* **1**, 39–54. Cairo University.

English, J. R., and Taylor, G. D. (1990). Process capability analysis — a robustness study. Master of Science. Department of Industrial Engineering, University of Arkansas at Fayetteville.   *(220)

Eriksson, B. (1983). On the construction of confidence limits for the regression coefficients when the residuals are dependent. *J. Statist. Comput. Simul.* **17**, 297–309.

Eubank, R. L. (1986). Quantiles. *Encyclopedia of Statistical Sciences*, **7**, 424–432, Wiley, New York.

Eynon, B., and Switzer, P. (1983). The variability of rainfall acidity. *Can. J. Statist.* **11**, 11–24.   *(221)

Fabiani, M., Gratton, G., Corballis, P. M., Cheng, J., and Friedman, D. (1998). Bootstrap assessment of the reliability of maxima in surface maps of brain activity of individual subjects derived with electrophysiological and optical methods. *Behav. Res. Methods Instrum. Comput.* **30**, 78–86.

Falck, W., Bjornstad, O. N., and Stenseth, N. C. (1995). Bootstrap estimated uncertainty of the dominant Lyapunov exponent for Holarctic mocrotine rodents. *Proc. R. Soc. London B* **261**, 159–165.

Falk, M. (1986a). On the estimation of the quantile density function. *Statist. Probab. Lett.* **4**, 69–73.

Falk, M. (1986b). On the accuracy of the bootstrap approximation of the joint distribution of sample quantiles. *Commun. Statist. Theory Methods* **15**, 2867–2876.

Falk, M. (1988). Weak convergence of the bootstrap process for large quantiles. *Statist. Dec.* **6**, 385–396.

Falk, M. (1990). Weak convergence of the maximum error of the bootstrap quantile estimate. *Statist. Probab. Lett.* **10**, 301–305.

Falk, M. (1991). A note on the inverse bootstrap process for large quantiles. *Stoch. Proc. Their Appl.* **38**, 359–363.

Falk, M. (1992a). Bootstrapping the sample quantile: a survey. In *Bootstrapping and Related Techniques. Proceedings, Trier, FRG.* (K.-H. Jockel, G. Rothe, and W. Sendler, editors). *Lecture Notes in Economics and Mathematical Systems*, **376**, 165–172. Springer-Verlag, Berlin.   *(222)

Falk, M. (1992b). Bootstrap optimal bandwidth selection for kernel density estimates. *J. Statist. Plann. Inf.* **30**, 13–32.   *(223)

Falk, M., and Kaufmann, E. (1991). Coverage probabilities of bootstrap confidence intervals for quantile estimates. *Ann. Statist.* **19**, 485–495.   *(224)

Falk, M., and Reiss, R.-D. (1989a). Weak convergence of smoothed and nonsmoothed bootstrap quantile estimates. *Ann. Probab.* **17**, 362–371.

Falk, M., and Reiss, R.-D. (1989b). Bootstrapping the distance between smooth bootstrap and sample quantile distribution. *Probab. Theory Relat. Fields* **82**, 177–186.

Falk, M., and Reiss, R.-D. (1989c). Statistical inference of conditional curves: Poisson process approach. Preprint **231**, University of Seigen.

Falk, M., and Reiss, R.-D. (1992). Bootstrapping conditional curves. In *Bootstrapping and Related Techniques. Proceedings, Trier, FRG.* (K.-H. Jockel, G. Rothe, and W. Sendler, editors). *Lecture Notes in Economics and Mathematical Systems*, **376**, 173–180. Springer-Verlag, Berlin.

Fan, J., and Lin, J.-T. (1998). Test of significance when data are curves. *J. Am. Statist. Assoc.* **93**, 1007–1021

Fan, T.-H., and Hung, W.-L. (1997). Balanced resampling for bootstrapping finite Markov chains. *Commun. Statist. Simul. Comput.* **26**, 1465–1475.   *(225)

Fang, K. T., and Wang, Y. (1994). *Number-Theoretic Methods in Statistics.* Chapman & Hall, London.   *(226)

Faraway, J. J. (1990). Bootstrap selection of bandwidth and confidence bands for nonparametric regression. *J. Statist. Comput. Simul.* **37**, 37–44.

Faraway, J. J. (1992). On the cost of data analysis. *J. Comput. Graphical Statist.* **1**, 213–229.

Faraway, J. J., and Jhun, M. (1990). Bootstrap choice of bandwidth for density estimation. *J. Am. Statist. Assoc.* **85**, 1119–1122.   *(227)

Farewell, V. (1985). Nonparametric estimation of standard errors. *Encyclopedia of Statistical Sciences*, **6**, 328–331. Wiley, New York.

Farrell, P. J., MacGibbon, B., and Tomberlin, T. J. (1997). Empirical Bayes small-area estimation using logistic regression models and summary statistics. *J. Bus. Econ. Statist.* **15**, 101–110.

Feller, W. (1971). *An Introduction to Probability Theory and Its Applications*, **2**, 2nd ed. Wiley, New York.   *(228)

Felsenstein, J. (1985). Confidence limits on phylogenies: an approach using the bootstrap. *Evolution* **39**, 783–791.   *(229)

Felsenstein, J. (1992). Estimating effective population size from samples of sequences: a bootstrap Monte Carlo integration method. *Genet. Res. Cambridge* **60**, 209–220. *(230)

Ferguson, T. S. (1967). *Mathematical Statistics: A Decision Theoretic Approach.* Academic Press, New York. *(231)

Fernholtz, L. T. (1983). *von Mises Calculus for Statistical Functionals.* Lecture Notes in Statistics, **19**, Springer-Verlag, New York. *(232)

Ferreira, F. P., Stangenhaus, G., and Narula, S. C. (1993) Bootstrap confidence intervals for the minimum sum of absolute errors regression. *J. Statist. Comput. Simul.* **48**, 127–133.

Ferretti, N., and Romo, J. (1996). Unit root bootstrap tests for AR(1) models. *Biometrika* **83**, 849–860.

Field, C., and Ronchetti, E. (1990). *Small Sample Asymptotics. Institute of Mathematical Statistics Lecture Notes — Monograph Series,* **13**. Institute of Mathematical Statistics, Hayward. *(233)

Fiellin, D. A., and Feinstein, A. R. (1998). Bootstraps and jackknives: new computer-intensive statistical tools that require no mathematical theories. *J. Invest. Med.* **46**, 22–26. *(234)

Findley, D. F. (1986). On bootstrap estimates of forecast mean square errors for autoregressive processes. *Proc. Comput. Sci. Statist.* **17**, 11–17. *(235)

Firth, D., Glosup, J., and Hinkley, D. V. (1991). Model checking with nonparametric curves. *Biometrika* **78**, 245–252.

Fisher, G., and Sim, A. B. (1995). Some finite sample theory for bootstrap regression estimates. *J. Statist. Plann. Inf.* **43**, 289–300.

Fisher, N. I., and Hall, P. (1989). Bootstrap confidence regions for directional data. *J. Am. Statist. Assoc.* **84**, 996–1002. *(236)

Fisher, N. I., and Hall, P. (1990). On bootstrap hypothesis testing. *Aust. J. Statist.* **32**, 177–190. *(237)

Fisher, N. I., and Hall, P. (1991). Bootstrap algorithms for small samples. *J. Statist. Plann. Inf.* **27**, 157–169.

Fisher, N. I., and Hall, P. (1992). Bootstrap methods for directional data. In *The Art of Statistical Science* (K. V. Mardia, editor), 47–63, Wiley, New York.

Fisher, N. I., Hall, P., Jing, B.-Y., and Wood, A. T. A. (1996). Improved pivotal methods for constructing confidence regions with directional data. *J. Am. Statist. Assoc.* **91**, 1062–1070.

Fisher, N. I., Lewis, T., and Embleton, B. J. (1987). *Statistical Analysis of Spherical Data.* Cambridge University Press, Cambridge. *(238)

Fitzmaurice, G. M., Laird, N. M., and Zahner, G. E. P. (1996). Multivariate logistic models for incomplete binary responses. *J. Am. Statist. Assoc.* **91**, 99–108.

Flehinger, B. J., Reiser, B., and Yashchin, E. (1996). Inference about defects in the presence of masking. *Technometrics* **38**, 247–255.

Flury, B. D. (1988). *Common Principal Components and Related Multivariate Models.* Wiley, New York. *(239)

Flury, B. D. (1997). *A First Course in Multivariate Statistics.* Springer-Verlag, New York. *(240)

Flury, B. D., Nel, D. G., and Pienaar, I. (1995). Simultaneous detection of shift in means and variances. *J. Am. Statist. Assoc.* **90**, 1474–1481.

Fong, D. K. H., and Bolton, G. E. (1997). Analyzing ultimatum bargaining: a Bayesian approach to the comparison of two potency curves under shape constraints. *J. Bus. Econ. Statist.* **15**, 335–344.

Forster, J. J., McDonald, J. W., and Smith, P. W. F. (1996). Monte Carlo exact conditional tests for log-linear and logistic models. *J. R. Statist. Soc. B* **58**, 445–453.

Fortin, V., Bernier, J., and Bobee, B. (1997). Simulation, Bayes, and bootstrap in statistical hydrology. *Water Resour. Res.* **33**, 439–448.

Foster, D. H., and Bischof, W. F. (1997). Bootstrap estimates of the statistical accuracy of thresholds obtained from psychometric functions. *Spatial Vision* **11**, 135–139.

Foutz, R. V. (1980). A method for constructing exact tests from test statistics that have unknown null distributions. *J. Statist. Comput. Simul.* **10**, 187–193.

Frangos, C. C., and Schucany, W. R. (1990). Jackknife estimation of the bootstrap acceleration constant. *Comp. Statist. Data Anal.* **9**, 271–282.   *(241)

Frangos, C. C., and Schucany, W. R. (1995). Improved bootstrap confidence intervals in certain toxicological experiments. *Commun. Statist. Theory Methods* **24**, 829–844.

Frangos, C. C., and Stone, M. (1984). On jackknife cross-validatory and classical methods of estimating a proportion with batches of different sizes. *Biometrika* **71**, 361–366.

Frangos, C. C., and Swanepoel, C. J. (1994). Bootstrap confidence intervals for the slope parameter of a logistic model. *Commun. Statist. Simul. Comput.* **23**, 1115–1126.

Franke, J., and Hardle, W. (1992). On bootstrapping kernel spectral estimates. *Ann. Statist.* **20**, 121–145.

Franke, J., and Wendel, M. (1992). A bootstrap approach to nonlinear autoregression: some preliminary results. In *Bootstrapping and Related Techniques. Proceedings, Trier, FRG* (K. H. Jockel, G. Rothe, and W. Sendler, editors), *Lecture Notes in Economics and Mathematical Systems*, **376**, 101–105. Springer-Verlag, Berlin.

Franklin, L. A., and Wasserman, G. S. (1991). Bootstrap confidence interval estimates of C_{p_k}: an introduction. *Commun. Statist. Simult. Comput.* **20**, 231–242.   *(242)

Franklin, L. A., and Wasserman, G. S. (1992). Bootstrap lower confidence limits for capability indices. *J. Qual. Technol.* **24**, 196–210.   *(243)

Franklin, L. A., and Wasserman, G. S. (1994). Bootstrap lower confidence limits estimates for C_{jkp} (the new flexible process capability index). *Pakistan J. Statist. A* **10**, 33–45.   *(244)

Freedman, D. A. (1981). Bootstrapping regression models. *Ann. Statist.* **9**, 1218–1228.   *(245)

Freedman, D. A. (1984). On bootstrapping two-stage least squares estimates in stationary linear models. *Ann. Statist.* **12**, 827–842.   *(246)

Freedman, D. A., Navidi, W., and Peters, S. C. (1988). On the impact of variable selection in fitting regression equations. In *On Model Uncertainty and Its Statistical Implications.* (T. K. Dijkstra, editor). *Lecture Notes in Economics and Mathematical Systems,* **306**. Springer-Verlag, Berlin.

Freedman, D. A., and Peters, S. C. (1984a). Bootstrapping a regression equation: some empirical results. *J. Am. Statist. Assoc.* **79**, 97–106.   *(247)

Freedman, D. A., and Peters, S. C. (1984b). Bootstrapping an econometric model: some empirical results. *J. Bus. Econ. Statist.* **2**, 150–158.   *(248)

Fresen, J. L., and Fresen, J. W. (1986). Estimating the parameter in the Pauling equation. *J. Appl. Statist.* **13**, 27–37.

Friedman, J. H. (1991). Multivariate adaptive regression splines (with discussion). *Ann. Statist.* **19**, 1–141.

Friedman, J. H., and Stuetzle, W. (1981). Projection pursuit regression. *J. Am. Statist. Assoc.* **76**, 817–823.   *(249)

Friedman, L. W., and Friedman, H. H. (1995). Analyzing simulation output using the bootstrap method. *Simulation* **64**, 95–100.

Fuchs, C. (1978). On test sizes in linear models for transformed variables. *Technometrics* **20**, 291–299.   *(250)

Fujikoshi, Y. (1994). On the bootstrap approximations for Hotelling's T^2 statistic. In *5th Japan–China Symposium on Statistics* (M. Ichimura, S. Mao, and G. Fan, editors), **5**, 69–71.

Fukunaga, K. (1990). *Introduction to Statistical Pattern Recognition*, 2nd ed. Academic Press, San Diego.   *(251)

Fukunaga, K., and Hayes R. R. (1989). Estimation of classifier performance. *IEEE Trans. Pattern Anal. Mach. Intell.* **PAMI-11**, 1087–1101.

Fuller, W. A. (1976). *Introduction to Statistical Time Series*. Wiley, New York.   *(252)

Furlanello, C., Merler, S., Chemini, C., and Rizzoli, A. (1998). An application of the bootstrap 632 + rule to ecological data. In *Proceedings of the 9th Italian Workshop on Neural Nets* (M. Marinaro and R. Tagliaferri, editors), 227–232   *(253)

Gabriel, K. R., and Hsu, C. F. (1983). Evaluation of the power of rerandomization tests, with application to weather modification experiments. *J. Am. Statist. Assoc.* **78**, 766–775.

Gaenssler, P. (1987). Bootstrapping empirical measures indexed by Vapnik–Chervonenkis classes of sets. In *Probability Theory and Mathematical Statistics*, 467–481, VNU Science Press, Utrecht.

Gaenssler, P. (1992). Confidence bands for probability distributions on Vapnik–Chervonenkis classes of sets in arbitrary sample spaces using the bootstrap. In *Bootstrapping and Related Techniques. Proceedings, Trier, FRG* (K.-H. Jockel, G. Rothe, and W. Sendler, editors), *Lecture Notes in Economics and Mathematical Systems,* **376**, 57–61. Springer-Verlag, Berlin.

Galambos, J. (1978). *The Asymptotic Theory of Extreme Order Statistics*, Wiley, New York.   *(254)

Galambos, J. (1987). *The Asymptotic Theory of Extreme Order Statistics*, 2nd ed. Krieger, Malabar.   *(255)

Gallant, A. R. (1987). *Nonlinear Statistical Models*. Wiley, New York.   *(256)

Ganeshanandam, S., and Krzanowski, W. J. (1989). On selecting variables and assessing their performance in linear discriminant analysis. *Aust. J. Statist.* **31**, 433–448.

Ganeshanandam, S., and Krzanowski, W. J. (1990). Error-rate estimation in two-group discriminant analysis using the linear discriminant function. *J. Statist. Comput. Simul.* **36**, 157–176.

Gangopadhyay, A. K., and Sen, P. K. (1990). Bootstrap confidence intervals for conditional quantile functions. *Sankhya A* **52**, 346–363.

Ganoe, F. J. (1989). Statistical bootstrap with computer simulation: methodology for fuzzy-logic-based expert system validation. In *Proceedings of the 18th Annual Western Regional Meeting of the Decision Sciences Institute,* (V. V. Bellur and J. C. Rogers, editors), **18**, 211–213. Decision Sciences Institute.

Garcia-Cortes, L. A., Moreno, C., Varona, L., and Altarriba, J. (1995). Estimation of prediction-error variances by resampling. *J. Anim. Breeding Genet.* **112**, 176–182.

Garcia-Jurado, I., Gonzalez-Manteiga, W., Prada-Sanchez, J. M., Febrero-Bande, M., and Cao, R. (1995). Predicting using Box–Jenkins, nonparametric, and bootstrap techniques. *Technometrics* **37**, 303–310.

Garcia-Soidan, P. H., and Hall, P. (1997). On sample reuse methods for spatial data. *Biometrics* **53**, 273–281.

Garthwaite, P. H., and Buckland, S. T. (1992). Generating Monte Carlo confidence intervals by the Robbins–Monro process. *Appl. Statist.* **41**, 159–171.

Gatto, R., and Ronchetti, E. (1996). General saddlepoint approximations of marginal densities and tail probabilities. *J. Am. Statist. Assoc.* **91**, 666–673.

Geisser, S. (1975). The predictive sample reuse method with applications. *J. Am. Statist. Assoc.* **70**, 320–328.

Geisser, S. (1993). *Predictive Inference: An Introduction.* Chapman & Hall, London. *(257)

Geissler, P. H. (1987). Bootstrapping the lognormal distribution. *Proc. Comput. Sci. Statist.* **19**, 543–545.

Gelman, A., Carlin, J. B., Stern, H. S., and Rubin, D. B. (1995). *Bayesian Data Analysis.* Chapman & Hall, London.   *(258)

Gentle, J. E. (1985). Monte Carlo methods. *Encyclopedia of Statistical Sciences,* **5**, 612–617, Wiley, New York.

George, P. J., Oksanen, E. H., and Veall, M. R. (1995). Analytic and bootstrap approaches to testing a market saturation hypothesis. *Math. Comput. Simul.* **39**, 311–315.

George, S. L. (1985). The bootstrap and identification of prognostic factors via Cox's proportional hazards regression model. *Statist. Med.* **4**, 39–46.

Geweke, J. (1993). Inference and forecasting for chaotic non-linear time series. In *Non-linear Dynamics and Evolutionary Economics* (P. Chen and R. Day, editors). Oxford University Press, Oxford.

Geyer, C. J. (1991). Markov chain Monte Carlo maximum likelihood. In *Computing Science and Statistics: Proceedings of 23rd Symposium on the Interface* (E. M. Keramidas, editor), 156–163. Interface Foundation, Fairfax.

Geyer, C. J. (1995a). Likelihood ratio tests and inequality constraints. Technical Report 610, University of Minnesota, School of Statistics.

Geyer, C. J. (1995b). Estimation and optimization of functions. In *Markov Chain Monte Carlo in Practice* (W. R. Gilks, S. Richardson, and D. J. Spiegelhalter, editors), 241–258. Chapman & Hall, London.

Geyer, C. J., and Moller, J. (1994). Simulation procedures and likelihood inference for spatial point processes. *Scand. J. Statist.* **21**, 359–373.

Ghorbel, F., and Banga, C. (1994). Bootstrap sampling applied to image analysis. In *Proceedings of the 1994 IEEE International Conference on Acoustics, Speech and Signal Processing*, **6**, 81–84.

Ghosh, M. (1985). Berry–Esseen bounds for functions of *U*-statistics. *Sankhya A* **47**, 255–270.

Ghosh, M., and Meeden, G. (1997). *Bayesian Methods for Finite Population Sampling*. Chapman & Hall, London.   *(259)

Ghosh, M., Parr, W. C., Singh, K., and Babu, G. J. (1984). A note on bootstrapping the sample median. *Ann. Statist.* **12**, 1130–1135.

Giachin, E., Baggia, P., and Micca, G. (1994). Language models for spontaneous speech recognition: a bootstrap method for learning phrase bigrams. In *Proceedings of the 1994 International Conference on Spoken Language Processing*, **2**, 843–846.

Gifi, A. (name for a group of Dutch statisticians) (1990). *Nonlinear Multivariate Analysis*. Wiley, Chichester.   *(260)

Gigli, A. (1994a). Contributions to importance sampling and resampling. Ph. D. Thesis, Department of Mathematics, Imperial College, London.   *(261)

Gigli, A. (1994b). Efficient bootstrap methods: a review. Consiglio Nazionale delle Ricerche, Rome. Technical Report Number Quaderno-7.   *(262)

Gine, E. (1997). Lectures on some aspects of the bootstrap theory. In *Summer School of Probability* (E. Gine, G. R. Grimmett, and L. Saloff-Coste, editors). *Lecture Notes in Mathematics*, **1665**, 37–152. Springer-Verlag, New York.

Gine, E., and Zinn, J. (1989). Necessary conditions for bootstrap of the mean. *Ann. Statist.* **17**, 684–691.   *(263)

Gine, E., and Zinn, J. (1990). Bootstrap general empirical measures. *Ann. Probab.* **18**, 851–869.

Gine, E., and Zinn, J. (1991). Gaussian characterization of uniform Donsker classes of functions. *Ann. Probab.* **19**, 758–782.

Glasbey, C. A. (1987). Tolerance-distribution-free analysis of quantal dose–response data. *Appl. Statist.* **36**, 252–259.

Gleason, J. R. (1988). Algorithms for balanced bootstrap simulations. *Am. Statist.* **42**, 263–266.

Glick, N. (1978). Additive estimators for probabilities of correct classification. *Pattern Recogn.* **10**, 211–222.   *(264)

Gnanadesikan, R. (1977). *Methods for Statistical Data Analysis of Multivariate Observations*. Wiley, New York.   *(265)

Gnanadesikan, R. (1997). *Methods for Statistical Data Analysis of Multivariate Observations*. 2nd ed., Wiley, New York.   *(266)

Gnanadesikan, R., and Kettenring, J. R. (1982). Data-based metrics for cluster analysis. *Util. Mat.* **21A**, 75–99.

Golbeck, A. L. (1992). Bootstrapping current life table estimators. In *Bootstrapping and Related Techniques. Proceedings, Trier, FRG.* (K.-H. Jockel, G. Rothe, and W. Sendler, editors). *Lecture Notes in Economics and Mathematical Systems*, **376**, 197–201. Springer-Verlag, Berlin.

Goldstein, H. (1995). *Kendall's Library of Statistics 3: Multilevel Statistical Models*, 2nd ed. Edward Arnold, London.   *(267)

Gong, G. (1982). Some ideas in using the bootstrap in assessing model variability in regression. *Proc. Comput. Sci., Statist.* **14**, 169–173.   *(268)

Gong, G. (1986). Cross-validation, the jackknife, and the bootstrap: Excess error in forward logistic regression. *J. Am. Statist. Assoc.* **81**, 108–113.   *(269)

Gonzalez, L., and Manly, B. F. J. (1993). Bootstrapping for sample design with quality surveys. *ASA Proceedings of Quality and Productivity,* 262–265.

Gonzalez Manteiga, W., Prada Sanchez, J. M., and Romo, J. (1993). The bootstrap—a review. *Comput. Statist. Q.* **9**, 165–205.   *(270)

Good, P. (1989). Almost most powerful tests for composite alternatives. *Commun. Statist. Theory Methods* **18**, 1913–1925.

Good, P. (1994). *Permutation Tests.* Springer-Verlag, New York.   *(271)

Good, P. (1998). *Resampling Methods: A Practical Guide to Data Analysis.* Birkhauser, Boston.   *(272)

Good, P., and Chernick, M. R. (1993). Testing the equality of variances of two populations. Unpublished manuscript.   *(273)

Goodnight, C. J., and Schwartz, J. M. (1997). A bootstrap comparison of genetic covariance matrices. *Biometrics* **53**, 1026–1039.

Gotze, F., and Kunsch, H. R. (1996). Second-order correctness of the blockwise bootstrap for stationary observations. *Ann. Statist.* **24**, 1914–1933.   *(274)

Gould, W. R., and Pollock, K. H. (1997). Catch-effort estimation of population parameters under the robust design. *Biometrics* **53**, 207–216.

Graham, R. L., Hinkley, D. V., John, P. W. M., and Shi, S. (1990). Balanced design of bootstrap simulations. *J. R. Statist. Soc. B* **52**, 185–202.   *(275)

Graubard, B. I., and Korn, E. L. (1993). Hypothesis testing with complex survey data: the use of classical quadratic test statistics with particular reference to regression problems. *J. Am. Statist. Assoc.* **88**, 629–641.

Gray, H. L., and Schucany, W. R. (1972). *The Generalized Jackknife Statistic.* Marcel Dekker, New York.   *(276)

Green, P. J., and Silverman, B. W. (1994). *Nonparametric Regression and Generalized Linear Models: a Roughness Penalty Approach.* Chapman & Hall, London.

Green, R., Hahn, W., and Rocke, D. (1987). Standard errors for elasticities: a comparison of bootstrap and asymptotic standard errors. *J. Bus. Econ. Statist.* **5**, 145–149.   *(277)

Greenacre, M. J. (1984). *Theory and Application of Correspondence Analysis.* Academic Press, London.   *(278)

Gross, S. (1980) Median estimation in sample surveys. In *Proceedings of the Section on Survey Research Methods.* American Statistical Association, Alexandria.   *(279)

Gross, S. T., and Lai, T. L. (1996a). Bootstrap methods for truncated and censored data. *Statist. Sin.* **6**, 509–530.   *(280)

Gross, S. T., and Lai, T. L. (1996b). Nonparametric estimators and regression analysis with left-truncated and right-censored data. *J. Am. Statist. Assoc.* **91**, 1166–1180.

Gruet, M. A., Huet, S., and Jolivet, E. (1993). Practical use of bootstrap in regression. In *Computer Intensive Methods in Statistics,* **150–66**, *Statist. Comp. Physica,* Heidelberg.

Gu, C. (1987). What happens when bootstrapping the smoothing spline? *Commun. Statist. Theory Methods* **16**, 3275–3284.

Gu, C. (1992). On the Edgeworth expansion and bootstrap approximation for the Cox regression model under random censorship. *Can. J. Statist.* **20**, 399–414.

Guan, Z. (1993). The bootstrap of estimators of symmetric distribution functions. *Chin. J. Appl. Probab. Statist.* **9**, 402–408.

Guerra, R., Polansky, A. M., and Schucany, W. R. (1997). Smoothed bootstrap confidence intervals with discrete data. *Comput. Statist. Data Anal.* **26**, 163–176.

Guillou, A. (1995). Weighted bootstraps for studentized statistics. *C. R. Acad. Sci. Paris Ser. I Math* **320**, 1379–1384

Gunter, B. H. (1989a). The use and abuse of C_{pk}, *Qual. Prog.* **22 (3)**, 108–109.   *(281)

Gunter, B. H. (1989b). The use and abuse of C_{pk}, *Qual. Prog.* **22 (5)**, 79–80.   *(282)

Haeusler, E., Mason, D. M., and Newton, M. A. (1992). Weighted bootstrapping of means. *Cent. Wisk. Inf. Q.* **5**, 213–228.

Hagedorn, R. (1995). The long way to the statistical bootstrap model. In *NATO ASI Series B Physics. Hot Hadronic Matter: Theory and Experiment*, **346**, 13–46.

Hahn, G. J., and Meeker, W. Q. (1991). *Statistical Intervals: A Guide for Practitioners*. Wiley, New York.   *(283)

Hall, P. (1983). Inverting an Edgeworth expansion. *Ann. Statist.* **11**, 569–576.

Hall, P. (1985). Resampling a coverage pattern. *Stoch. Proc.* **20**, 231–246.   *(284)

Hall, P. (1986a). On the bootstrap and confidence intervals. *Ann. Statist.* **14**, 1431–1452.   *(285)

Hall, P. (1986b). On the number of bootstrap simulations required to construct a confidence interval. *Ann. Statist.* **14**, 1453–1462.   *(286)

Hall, P. (1987a). On bootstrap and likelihood-based confidence regions. *Biometrika* **74**, 481–493.

Hall, P. (1987b). On the bootstrap and continuity correction. *J. R. Statist. Soc. B* **49**, 82–89.

Hall, P. (1987c). Edgeworth expansion for Student's *t* statistic under minimal moment conditions. *Ann. Probab.* **15**, 920–931.

Hall, P. (1988a). On the bootstrap and symmetric confidence intervals. *J. R. Statist. Soc. B* **50**, 35–45.

Hall, P. (1988b). Theoretical comparison of bootstrap confidence intervals (with discussion). *Ann. Statist.* **16**, 927–985.   *(287)

Hall, P. (1988c). *Introduction to the Theory of Coverage Processes*. Wiley, New York. *(288)

Hall, P. (1988d). Rate of convergence in bootstrap approximations. *Ann. Probab.* **16**, 1665–1684.   *(289)

Hall, P. (1989a). On efficient bootstrap simulation. *Biometrika* **76**, 613–617.   *(290)

Hall, P. (1989b). Antithetic resampling for the bootstrap. *Biometrika* **76**, 713–724. *(291)

Hall, P. (1989c). Unusual properties of bootstrap confidence intervals in regression problems. *Probab. Theory Relat. Fields* **81**, 247–274.   *(292)

Hall, P. (1989d). On convergence rates in nonparametric problems. *Int. Statist. Rev.* **57**, 45–58.

Hall, P. (1990a). Pseudo-likelihood theory for empirical likelihood. *Ann. Statist.* **18**, 121–140.

Hall, P. (1990b). Using the bootstrap to estimate mean squared error and select smoothing parameters in nonparametric problems. *J. Multivar. Anal.* **32**, 177–203.

Hall, P. (1990c). Performance of bootstrap balanced resampling in distribution function and quantile problems. *Probab. Theory Relat. Fields* **85**, 239–260.

Hall, P. (1990d). Asymptotic properties of the bootstrap for heavy-tailed distributions. *Ann. Probab.* **18**, 1342–1360.

Hall, P. (1991a). Bahadur representations for uniform resampling and importance resampling, with applications to asymptotic relative efficiency. *Ann. Statist.* **19**, 1062–1072.   *(293)

Hall, P. (1991b). On relative performance of bootstrap and Edgeworth approximations of a distribution function. *J. Multivar. Anal.* **35**, 108–129.

Hall, P. (1991c). Balanced importance resampling for the bootstrap. Unpublished Manuscript.

Hall, P. (1991d). On bootstrap confidence intervals in nonparametric regression. Unpublished Manuscript.

Hall, P. (1991e). On Edgeworth expansions and bootstrap confidence bands in nonparametric curve estimation. Unpublished Manuscript.

Hall, P. (1991f). Edgeworth expansions for nonparametric density estimators, with applications to asymptotic relative efficiency. *Ann. Statist.* **19**, 1062–1072.

Hall, P. (1991g). Edgeworth expansions for nonparametric density estimators with applications. *Statistics* **22**, 215–232.

Hall, P. (1992a). *The Bootstrap and Edgeworth Expansion.* Springer-Verlag, New York.   *(294)

Hall, P. (1992b). On the removal of skewness by transformation. *J. R. Statist. Soc. B* **54**, 221–228.

Hall, P. (1992c). Efficient bootstrap simulations. In *Exploring the Limits of Bootstrap* (R. LePage and L. Billard, editors), 127–143. Wiley, New York.   *(295)

Hall, P. (1992d). On bootstrap confidence intervals in nonparametric regression. *Ann. Statist.* **20**, 695–711.

Hall, P. (1992e). Effect of bias estimation on coverage accuracy of bootstrap confidence intervals for a probability density. *Ann. Statist.* **20**, 675–694.

Hall, P. (1994). A short history of the bootstrap. In *Proceedings of the 1994 International Conference on Acoustics, Speech and Signal Processing*, **6**, 65–68.   *(296)

Hall, P. (1995). On the biases of error estimators in prediction problems. *Statist. Probab. Lett.* **24**, 257–262.

Hall, P. (1998). Block bootstrap. In *Encyclopedia of Statistical Sciences, Update Volume 2* (S. Kotz, C. B. Read, and D. L. Banks, editors), 83–84. Wiley, New York.   *(297)

Hall, P., DiCiccio, T. J., and Romano, J. P. (1989). On smoothing and the bootstrap. *Ann. Statist.* **17**, 692–704.

Hall, P., Hardle, W., and Simar, L. (1993). On the inconsistency of bootstrap distribution estimators. *Comput. Statist. Data Anal.* **16**, 11–18.   *(298)

Hall, P., and Hart, J. D. (1990). Bootstrap test for difference between means in nonparametric regression. *J. Am. Statist. Assoc.* **85**, 1039–1049.

Hall, P., and Horowitz, J. L. (1993). Corrections and blocking rules for the block bootstrap with dependent data. Technical Report SR11–93, Centre for Mathematics and Its Applications, Australian National University.

Hall, P., Horowitz, J. L., and Jing, B.-Y. (1995). On blocking rules for the bootstrap with dependent data. *Biometrika* **82**, 561–574.   *(299)

Hall, P., Huber, C., and Speckman, P. L. (1997). Covariate-matched one-sided tests for the difference between functional means. *J. Am. Statist. Assoc.* **92**, 1074–1083.

Hall, P., and Jing, B.-Y. (1996). On sample reuse methods for dependent data. *J. R. Statist. Soc. B* **58**, 727–737.   *(300)

Hall, P., and Keenan, D. M. (1989). Bootstrap methods for constructing confidence regions for hands. *Commun. Statist. Stochastic Models* **5**, 555–562.

Hall, P., and LaScala, B. (1990). Methodology and algorithms of empirical likelihood. *Int. Statist. Rev.* **58**, 109–127.

Hall, P., and Martin, M. A. (1988a). On bootstrap resampling and iteration. *Biometrika* **75**, 661–671.   *(301)

Hall, P., and Martin, M. A. (1988b). On the bootstrap and two-sample problems. *Aust. J. Statist.* **30A**, 179–192.

Hall, P., and Martin, M. A. (1989a). Exact convergence rate of the bootstrap quantile variance estimator. *Probab. Theory Relat. Fields* **80**, 261–268.

Hall, P., and Martin, M. A. (1989b). A note on the accuracy of bootstrap percentile method confidence intervals for a quantile. *Statist. Probab. Lett.* **8**, 197–200.

Hall, P., and Martin, M. A. (1991). On the error incurred using the bootstrap variance estimate when constructing confidence intervals for quantiles. *J. Multivar. Anal.* **38**, 70–81.

Hall, P., Martin, M. A., and Schucany, W. R. (1989). Better nonparametric bootstrap confidence intervals for the correlation coefficient. *J. Statist. Comput. Simul.* **33**, 161–172.   *(302)

Hall, P., and Mammen, E. (1994). On general resampling algorithms and their performance in distribution estimation. *Ann. Statist.* **22**, 2011–2030.

Hall, P., and Owen, A. B. (1989). Empirical likelihood confidence bands in curve estimation. Unpublished Manuscript.

Hall, P., and Owen, A. B. (1993). Empirical likelihood confidence bands in density estimation. *J. Comput. Graphical Statist.* **2**, 273–289.

Hall, P., and Padmanabhan, A. R. (1997). Adaptive inference for the two sample scale problem. *Technometrics* **39**, 412–422.   *(303)

Hall, P., and Pittelkow, Y. E. (1990). Simultaneous bootstrap confidence bands in regression. *J. Statist. Comput. Simul.* **37**, 99–113.

Hall, P., and Sheather, S. J. (1988). On the distribution of a studentized quantile. *J. R. Statist. Soc. B* **50**, 381–391.

Hall, P., and Titterington, D. M. (1989). The effect of simulation order on level accuracy and power of Monte Carlo tests. *J. R. Statist. Soc. B* **51**, 459–467.

Hall, P., and Weissman, I. (1997). On the estimation of extreme tail probabilities. *Ann. Statist.* **25**, 1311–1326.

Hall, P., and Wilson, S. R. (1991). Two guidelines for bootstrap hypothesis testing. *Biometrics* **47**, 757–762.   *(304)

Hall, P., and Wolff, R. C. L. (1995). Properties of invariant distributions and Lyapunov exponents for chaotic logistic maps. *J. R. Statist. Soc. B* **57**, 439–452.

Hamilton, J. D. (1994). *Time Series Analysis*. Princeton University Press, Princeton. *(305)

Hamilton, M. A., and Collings, B. J. (1991). Determining the appropriate sample size for nonparametric tests for location shift. *Technometrics* **33**, 327–337.

Hammersley, J. M., and Handscomb, D. C. (1964). *Monte Carlo Methods*. Methuen, London.   *(306)

Hammersley, J. M., and Morton, K. W. (1956). A new Monte Carlo technique: antithetic variates. *Proc. Cambridge Philos. Soc.* **52**, 449–475.   *(307)

Hampel, F. R. (1973). Some small-sample asymptotics. In *Proceedings of the Prague Symposium on Asymptotic Statistics* (J. Hajek, editor). Charles University, Prague. *(308)

Hampel, F. R. (1974). The influence curve and its role in robust estimation. *J. Am. Statist. Assoc.* **69**, 383–393.   *(309)

Hampel, F. R., Ronchetti, E. M., Rousseeuw, P. J., and Stahel, W. A. (1986). *Robust Statistics: The Approach Based on Influence Functions*. Wiley, New York.   *(310)

Hand, D. J. (1981). *Discrimination and Classification*. Wiley, Chichester.   *(311)

Hand, D. J. (1982). *Kernel Discriminant Analysis*. Wiley, Chichester.   *(312)

Hand, D. J. (1986). Recent advances in error rate estimation. *Pattern Recogn. Lett.* **4**, 335–340.   *(313)

Hardle W. (1989). Resampling for inference from curves. *Proceedings of the 47th Session ISI Paris*, **3**, 53–54.

Hardle, W. (1990a). *Applied Nonparametric Regression*. Cambridge University Press, Cambridge.   *(314)

Hardle, W. (1990b). *Smoothing Techniques with Implementation in S*. Springer-Verlag, New York.   *(315)

Hardle, W., and Bowman, A. W. (1988). Bootstrapping in nonparametric regression: local adaptive smoothing and confidence bands. *J. Am. Statist. Assoc.* **83**, 102–110. *(316)

Hardle, W., Hall, P., and Marron, S. (1988). How far are automatically chosen smoothing parameters from their optimum? (with discussion) *J. Am. Statist. Assoc.* **83**, 86–101.

Hardle, W., Huet, S., and Jolivet, E. (1990). Better bootstrap confidence intervals for regression curve estimation. Unpublished manuscript.

Hardle, W., and Kelly, G. (1987). Nonparametric kernel regression estimation — optimal choice of bandwidth. *Statistics* **18**, 21–35.

Hardle, W., and Mammen, E. (1991). Bootstrap methods in nonparametric regression. In *Nonparametric Functional Estimation and Related Topics. Proceedings of the NATO Advanced Study Institute,* Spetses, Greece (G. Roussas, editor).

Hardle, W., and Mammen, E. (1993). Comparing nonparametric versus parametric regression fits. *Ann. Statist.* **21**, 1926–1947.

Hardle, W., Mammen, E., and Muller, M. (1998). Testing parametric versus semiparametric modeling in generalized linear models. *J. Am. Statist. Assoc.* **93**, 1461–1474.

Hardle, W., and Marron, J. S. (1985). Optimal bandwidth selection in nonparametric regression function estimation. *Ann. Statist.* **13**, 1465–1481.

Hardle, W., and Marron, J. S. (1991). Bootstrap simultaneous error bars for non-parametric regression. *Ann. Statist.* **19**, 778–796.   *(317)

Hardle, W., Marron, J. S., and Wand, M. P. (1990). Bandwidth choice for densiity derivatives. *J. R. Statist. Soc. B* **52**, 223–232.

Hardle, W., and Nussbaum, M. (1992). Bootstrap confidence bands. In *Bootstrapping and Related Techniques. Proceedings, Trier, FRG* (K.-H. Jockel, G. Rothe, and W. Sendler, editors). *Lecture Notes in Economics and Mathematical Systems*, **376**, 63–70. Springer-Verlag, Berlin.

Harrell, F. E., and Davis, C. E. (1982). A new distribution-free quantile estimator. *Biometrika* **69**, 635–640.

Harshman, J. (1994). The effects of irrelevant characters on bootstrap values. *Syst. Biol.* **43**, 419–424.

Hart, J. D. (1997). *Nonparametric Smoothing and Lack-of-Fit Tests.* Springer-Verlag, New York.   *(318)

Hartigan, J. A. (1969). Using subsample values as typical values. *J. Am. Statist. Assoc.* **64**, 1303–1317.   *(319)

Hartigan, J. A. (1971). Error analysis by replaced samples. *J. R. Statist. Soc. B* **33**, 98–1130.   *(320)

Hartigan, J. A. (1975). Necessary and sufficient conditions for the asymptotic joint normality of a statistic and its subsample values. *Ann. Statist.* **3**, 573–580.   *(321)

Hartigan, J. A. (1990). Perturbed periodogram estimates of variance. *Int. Statist. Rev.* **58**, 1–7.

Hartigan, J. A., and Forsythe, A. (1970). Efficiency and confidence intervals generated by repeated subsample calculations. *Biometrika* **57**, 629–640.

Hasegawa, M., and Kishino, H. (1994). Accuracies of the simple methods for estimating the bootstrap probability of a maximum-likelihood tree. *Mol. Biol. Evol.* **11**, 142–145.

Hasselblad, V. (1966). Estimation of parameters for a mixture of normal distributions. *Technometrics* **8**, 431–444.   *(322)

Hasselblad, V. (1969). Estimation of finite mixtures of distributions from the exponential family. *J. Am. Statist. Assoc.* **64**, 1459–1471.   *(323)

Hastie. T. J., and Tibshirani, R. J. (1990). *Generalized Additive Models.* Chapman & Hall, London.   *(324)

Hauck, W. W., McKee, L. J., and Turner, B. J. (1997). Two-part survival models applied to administrative data for determining rate of and predictors for maternal–child transmission of HIV. *Statist. Med.* **16**, 1683–1694.

Haukka, J. K. (1995). Correction for covariate measurement error in generalized linear models—a bootstrap approach. *Biometrics* **51**, 1127–1132.

Hawkins, D. M. editor (1982). *Topics in Applied Multivariate Analysis.* Cambridge University Press, Cambridge.   *(325)

Hawkins, D. M., Simonoff, J. S., and Stromberg, A. J. (1994). Distributing a computationally intensive estimator: the case of exact LMS regression. *Comput. Statist.* **9**, 83–95.

Hayes, K. G., Perl, M. L., and Efron, B. (1989). Applications of the bootstrap statistical method to the tau-decay-mode problem. *Phys. Rev. D* **39**, 274–279.   *(326)

He, K. (1987). Bootstrapping linear *M*-regression models. *Acta Math. Sin.* **29**, 613–617.

Heitjan, D. F., and Landis, J. R. (1994). Assessing secular trends in blood pressure: a multiple-imputation approach. *J. Am. Statist. Assoc.* **89**, 750–759.

Heller, G., and Venkatraman, E. S. (1996). Resampling procedures to compare two survival distributions in the presence of right-censored data. *Biometrics* **52**, 1204–1213.

Helmers, R. (1991a). On the Edgeworth expansion and the bootstrap approximation for a studentized *U*-statistic. *Ann. Statist.* **19**, 470–484.

Helmers, R. (1991b). Bootstrap methods. Unpublished manuscript.   *(327)

Helmers, R., and Huskova, M. (1994). Bootstrapping multivariate U quantiles and related statistics. *J. Multivar. Anal.* **49**, 97–109.

Helmers, R., Janssen, P., and Serfling, R. (1988). Glivenko–Cantelli properties of some generalized empirical DF's and strong convergence of generalized *L*-statistics. *Probab. Theory Relat. Fields* **79**, 75–93.

Helmers, R., Janssen, P., and Serfling, R. (1990). Berry–Esseen and bootstrap results for generalized L-statistics. *Scand. J. Statist.* **17**, 65–78.

Helmers, R., Janssen, P., and Veraverbeke, N. (1992). Bootstrapping *U*-quantiles. In *Exploring the Limits of Bootstrap* (R. LePage and L. Billard, editors), 145–155. Wiley, New York.   *(328)

Henderson, R., Temple, A., and McKnespiey, P. (1997). Computer-intensive inference for calibration curves: experience at BNF. In *ESARDA 19th Annual Symposium on Safeguards and Nuclear Materials Management.* (C. Foggi and F. Genoni, editors), 753–758.

Herman, R. (1996) Using the bootstrap method for evaluating variations due to sampling in voltage regulation calculations. In *Proceedings of the Sixth Conference SAUPEC-96*, 235–238.

Hermida, R. C., Ayala, D. E., Fernandez, J. R., and Halberg, F. (1997). Nonparametric resampling and modelling procedure for testing circannual markers of depressive disorders. *Int. J. Model. Simul.* **17**, 223–230.

Hernandez, J. A., Ramirez, G., and Sanchez, A. (1997). A high-level language program to obtain the bootstrap corrected ADF test statistic. *Behav. Res. Methods Instrum. Comput.* **29**, 296–301.

Hesterberg, T. (1988). Variance reduction techniques for bootstrap and other Monte Carlo simulations. Ph. D. dissertation, Department of Statistics, Stanford University.   *(329)

Hesterberg, T. (1992). Efficient bootstrap simulations I: importance sampling and control variates. Unpublished Manuscript.

Hesterberg, T. (1994). Saddlepoint quantiles and distribution curves with bootstrap applications. *Comput. Stat.* **9**, 207–211.

Hesterberg, T. (1995a). Tail-specific linear approximations for efficient bootstrap simulations. *J. Comput. Graphical Statist.* **4**, 113–133.   *(330)

Hesterberg, T. (1995b). Weighted average importance sampling and defensive mixture distributions. *Technometrics* **37**, 185–194.   *(331)

Hesterberg, T. (1996). Control variates and importance sampling for efficient bootstrap simulations. *Statist. Comput.* **6**, 147–157.   *(332)

Hesterberg, T. (1997). Fast bootstrapping by combining importance sampling and concomitants. *Proc. Comput. Sci. Statist.* **29**, 72–78.   *(333)

Hewer, G., Kuo, W., and Peterson, L. (1996a). Multiresolution detection of small objects using bootstrap methods and wavelets. *Proceedings of the Conference on Signal and Data Processing of Small Targets, Orlando, Fl, April 9–11, 1996. SPIE Proceedings*, **2759**, 2–10.

Hewer, G., Kuo, W., and Peterson, L. (1996b). Adaptive wavelet detection of transients using the bootstrap. *Proc. SPIE* **2762**, 105–114.

Higgins, K. M., Davidian, M., and Giltinan, D. M. (1997). A two-step approach to measurement error in time-dependent covariates in nonlinear mixed effects models, with application to IGF-I pharmacokinetics. *J. Am. Statist. Assoc.* **92**, 436–448.

Hill, J. R. (1990). A general framework for model-based statistics. *Biometrika* **77**, 115–126.

Hillis, D. M., and Bull, J. J. (1993). An empirical test of bootstrapping as a method for assessing confidence in phylogenetic analysis. *Syst. Biol.* **42**, 182–192.

Hills, M. (1966). Allocation rules and their error rates. *J. R. Statist. Soc. B* **28**, 1–31.   *(334)

Hinkley, D. V. (1977). Jackknifing in unbalanced situations. *Technometrics* **19**, 285–292.

Hinkley, D. V. (1983). Jackknife methods. *Encyclopedia of Statistical Sciences*, **4**, 280–287. Wiley, New York.

Hinkley, D. V. (1984). A hitchhiker's guide to the galaxy of theoretical statistics. In *Statistics: An Appraisal* (H. A. David and H. T. David, editors), 437–453. Iowa State University Press, Ames.   *(335)

Hinkley, D. V. (1988). Bootstrap methods (with discussion). *J. R. Statist. Soc. B* **50**, 321–337.   *(336)

Hinkley, D. V. (1989). Bootstrap significance tests. *Proceedings of the 47th Session of International Statistics Institute, Paris*, 65–74.

Hinkley, D. V., and Schechtman, E. (1987). Conditional bootstrap methods in the mean-shift model. *Biometrika* **74**, 85–94.

Hinkley, D. V., and Shi, S. (1989). Importance sampling and the bootstrap. *Biometrika* **76**, 435–446.   *(337)

Hinkley, D. V., and Wei, B. C. (1984). Improvements of jackknife confidence limit methods. *Biometrika* **71**, 331–339.

Hirst, D. (1996). Error-rate estimation in multiple-group linear discriminant analysis. *Technometrics* **38**, 389–399.   *(338)

Hjort, N. L. (1985). Bootstrapping Cox's regression model. Technical Report NSF-241, Department of Statistics, Stanford University.

Hjort, N. L. (1992). On inference in parametric data models. *Int. Statist. Rev.* **60**, 355–387.

Hjorth, J. S. U. (1994). *Computer Intensive Statistical Methods: Validation, Model Selection and Bootstrap*. Chapman & Hall, London.   *(339)

Holbert, D., and Son, M.-S. (1986). Bootstrapping a time series model: some empirical results. *Commun. Statist. Theory Methods* **15**, 3669–3691.   *(340)

Holm, S. (1993). Abstract bootstrap confidence intervals in linear models. *Scand. J. Statist.* **20**, 157–170.

Hoffman, W. P., and Leurgans, S. E. (1990). Large sample properties of two tests for independent joint action of two drugs. *Ann. Statist.* **18**, 1634–1650.

Hope, A. C. A. (1968). A simplified Monte Carlo significance test procedure. *J. Roy. Statist. Soc. B*, 582–598.   *(341)

Horowitz, J. L. (1994). Bootstrap-based critical values for the information matrix test. *J. Econometrics* **61**, 395–411.

Horvath, L., and Yandell, B. S. (1987). Convergence rates for the bootstrapped product limit process. *Ann. Statist.* **15**, 1155–1173.

Hsieh, D. A., and Manski, C. F. (1987). Monte Carlo evidence on adaptive maximum likelihood estimation of a regression. *Ann. Statist.* **15**, 541–551.

Hsieh, J. J. (1992). A hazard process for survival analysis. In *Exploring the Limits of Bootstrap* (R. LePage and L. Billard, editors), 345–362. Wiley, New York. *(342)

Hsu, Y.-S., Walker, J. J., and Ogren, D. E. (1986). A stepwise method for determining the number of components in a mixture. *Math. Geog.* **18**, 153–160.

Hu, F. (1997). Estimating equations and the bootstrap. In *Estimating Functions*, (I. V. Basawa, V. P. Godambe, and R. L. Taylor, editors). *Institute of Mathematical Statistics Monograph Series*, **32**, 405–416.

Hu, F., and Zidek, J. V. (1995). A bootstrap based on the estimating equations of the linear model. *Biometrika* **82**, 263–275.

Huang, J. S. (1991). Efficient computation of the performance of bootstrap and jackknife estimators of the variance of L-statistics. *J. Statist. Comput. Simul.* **38**, 45–66.

Huang, J. S., Sen, P. K., and Shao, J. (1995). Bootstrapping a sample quantile when the density has a jump. *Statist. Sin.* **6**, 299–309.

Huang, X., Chen, S., and Soong, S.-J. (1998). Piecewise exponential survival trees with time-dependent covariates. *Biometrics* **54**, 1420–1433.

Hubbard, A. E., and Gilinsky, N. L. (1992). Mass extinctions as statistical phenomena: an examination of the evidence using χ^2 tests and bootstrapping. *Paleobiology* **18**, 148–160.

Huber, P. J. (1981). *Robust Statistics*. Wiley, New York.   *(343)

Huet, S., and Jolivet, E. (1989). Exactitude au second order des intervalles de cofiance bootstrap pour les parametres d'un modele de regression non lineare. *C. R. Acad. Sci. Paris Ser. I. Math.* **308**, 429–432.

Huet, S., Jolivet, E., and Messean, A. (1990). Some simulation results about confidence intervals and bootstrap methods in nonlinear regression. *Statistics* **21**, 369–432.

Hur, K., Oprian, C. A., Henderson, W. G., Thakkar, B., and Urbanski, S. (1996). A SAS/sup (R/) macro for validating a logistics model with split sample and bootstrap methods. *SUGI 21*, **1**, 953–956.

Hurvich, C. M., Simonoff, J. S., and Zeger, S. L. (1991). Variance estimation for sample autocovariances. Direct and resampling approaches. *Aust. J. Statist.* **33**, 23–42.

Hurvich, C. M., and Tsai, C.-L. (1990). The impact of model selection on inference in linear regression. *Am. Statist.* **44**, 214–217.

Huskova, M., and Janssen, P. (1993a). Generalized bootstrap for studentized U-statistics: a rank statistic approach. *Statist. Probab. Lett.* **16**, 225 233.

Huskova, M., and Janssen, P. (1993b). Consistency of the generalized bootstrap for degenerate U-statistics. *Ann. Statist.* **21**, 1811–1823.

Hwa-Tung, O., and Zoubir, A. M. (1997). Non-Gaussian signal detection from multiple sensors using the bootstrap. In *Proceedings of ICICS, 1997 International Conference on Information, Communications and Signal Processing: Trends in Information Systems Engineering and Wireless Multimedia Communications,* **1**, 340–344.

Hyde, J. (1980). Testing survival with incomplete observations. In *Biostatistics Casebook.* Wiley, New York.   *(344)

Iglewicz, B., and Shen, C. F. (1994). Robust and bootstrap testing procedures for bioequivalence. *J. Biopharmacol. Statist.* **4**, 65–90.

Izenman, A. J. (1985). Bootstrapping Kolmogorov–Smirnov statistics. *Proc. ASA Sec. Statist. Comp.,* 97–101.

Izenman, A. J. (1986). Bootstrapping Kolmogorov–Smirnov statistics II. *Proc. Comput. Sci. Statist.* **18**, 363–366.

Izenman, A. J., and Sommer, C. J. (1988). Philatelic mixtures and multimodal densities. *J. Am. Statist. Assoc.* **83**, 941–953.

Jacoby, W. G. (1992). PROC IML statements for creating a bootstrap distribution of OLS regression coefficients (assuming random regressors). Technical Report. University of South Carolina.

Jagoe, R. H., and Newman, M. C. (1997). Bootstrap estimation of community NOEC values. *Ecotoxicology* **6**, 293–306.

Jain, A. K., Dubes, R. C., and Chen, C. (1987). Bootstrap techniques for error estimation. *IEEE Trans. Pattern Anal. Mach. Intell.* **PAMI-9**, 628–633.   *(345)

James, G. S. (1951). The comparison of several groups of observations when the ratios of the population variances are unknown. *Biometrika* **38**, 324–329.   *(346)

James, G. S. (1954). Tests of linear hypotheses in univariate and multivariate analysis when the ratios of the population variances are unknown. *Biometrika* **41**, 19–43.   *(347)

James, G. S. (1955). Cumulants of a transformed variate. *Biometrika* **42**, 529–531.   *(348)

James, G. S. (1958). On moments and cumulants of systems of statistics. *Sankhya* **20**, 1–30.   *(349)

James, L. F. (1993). The bootstrap, Bayesian bootstrap and random weighted methods for censored data models. Ph. D. dissertation. State University of New York at Buffalo.

James, L. F. (1997). A study of a class of weighted bootstraps for censored data. *Ann. Statist.* **25**, 1595–1621.

Janas, D. (1991). A smoothed bootstrap estimator for a studentized sample quantile. Preprint, Sonderforschungsbereich 123, Universitat Heidelberg.

Janas, D. (1993). *Bootstrap Procedures for Time Series.* Verlag Shaker, Aachen.   *(350)

Janas, D., and Dahlhaus, R. (1994). A frequency domain bootstrap for time series. In *26th Conference on the Interface Between Computer Science and Statistics* (J. Sall and A. Lehman, editors), **26**, 423–425.

Janssen, P. (1994). Weighted bootstrapping of U-statistics. *J. Statist. Plann. Inf.* **38**, 31–42.

Jayasuriya, B. R. (1996). Testing for polynomial regression using nonparametric regression technique. *J. Am. Statist. Assoc.* **91**, 1626–1631.

Jennison, C. (1992). Bootstrap tests and confidence intervals for a hazard ratio when the number of observed failures is small, with applications to group sequential survival studies. In *Computer Science and Statistics: Proceedings of the 22nd Symposium on the Interface* (C. Page and R. LePage, editors), 89–97, Springer-Verlag, New York.

Jensen, J. L. (1992). A modified signed likelihood statistic and saddlepoint approximation. *Biometrika* **79**, 693–703.

Jensen, J. L. (1995). *Saddlepoint Approximations.* Clarendon Press, Oxford.   *(351)

Jensen, R. L., and Kline, G. M. (1994). The resampling cross-validation technique in exercise science: modelling rowing power. *Med. Sci. Sports Exerc.* **26**, 929–933.

Jeong, J., and Maddala, G. S. (1993). A perspective on application of bootstrap methods in econometrics. In *Handbook of Statistics, Volume 11: Econometrics* (G. S. Maddala, C. R. Rao, and H. D. Vinod, editors), 573–610. North-Holland, Amsterdam.   *(352)

Jeske, D. R., and Marlow, N. A. (1997). Alternative prediction intervals for Pareto proportions. *J. Qual. Technol.* **29**, 317–326.

Jhun, M. (1988). Bootstrapping density estimates. *Commun. Statist. Theory Methods* **17**, 61–78.

Jhun, M. (1990). Bootstrapping k-means clustering. *J. Jpn. Soc. Comput. Statist.* **3**, 1–14.

Jing, B.-Y., and Wood, A. T. A. (1996). Exponential empirical likelihood is not Bartlett correctable. *Ann. Statist.* **24**, 365–369.

Jockel, K.-H. (1986). Finite sample properties and asymptotic efficiency of Monte Carlo tests. *Ann. Statist.* **14**, 336–347.

Jockel, K.-H. (1990). Monte Carlo techniques and hypothesis testing. *In 1st International Conference Statistics and Computers 1987* (E. Dudewicz, editor), 910

Jockel, K.-H., Rothe, G., and Sendler, W. (editors) (1992). *Bootstrapping and Related Techniques. Lecture Notes in Economics and Mathematical Systems,* **376**, Springer-Verlag, Berlin.   *(353)

Johns, M. V. Jr. (1988). Importance sampling for bootstrap confidence intervals. *J. Am. Statist. Assoc.* **83**, 709–714.   *(354)

Johnson, M. E. (1987). *Multivariate Statistical Simulation.* Wiley, New York.   *(355)

Johnson, M. E. (1992). Some modelling and simulation issues related to bootstrapping. In *Bootstrapping and Related Techniques. Proceedings, Trier, FRG* (K.-H. Jockel, G. Rothe, and W. Sendler, editors). *Lecture Notes in Economics and Mathematical Systems,* **376**, 227–232. Springer-Verlag, Berlin.

Johnson, N. L., and Kotz, S. (1977). *Urn Models.* Wiley, New York.   *(356)

Johnsson, T. (1988). Bootstrap multiple test procedures. *J. Appl. Statist.* **15**, 335–339.

Johnstone, I. M., and Velleman, P. F. (1985). Efficient scores, variance decompositions and Monte Carlo swindles. *J. Am. Statist. Assoc.* **80**, 851–862.

Jolliffe, I. T. (1986). *Principal Component Analysis.* Springer-Verlag, New York.   *(357)

Jones, G. K. (1988). Sampling errors (computation of), *Encyclopedia of Statistical Sciences*, **8**, 241–246. Wiley, New York.

Jones, G., Wortberg, M., Kreissig, S. B., Hammock, B. D., and Rocke, D. M. (1996). Application of the bootstrap to calibration experiments. *Anal. Chem.* **68**, 763–770. *(358)

Jones, M. C., Marron, J. S., and Sheather, S. J. (1996). A brief survey of bandwidth selection for density estimation. *J. Am. Statist. Assoc.* **91**, 401–407.

Jones, P. W., Ashour, S. K., and El-Sayed, S. M. (1991). Bootstrap investigation on mixed exponential model using maximum likelihood. In *26th Annual Conference on Statistics, Computer Science and Operations Research. Vol. 1; Mathematical Statistics*, **1**, 55–78. Cairo University.

Jones, P., Lipson, K., and Phillips, B. (1994). A role for computer-intensive methods in introducing statistical inference. In *Proceedings of the 1st Scientific Meeting — International Association for Statistical Education* (L. Brunelli and G. Cicchitelli, editors), **1**, 255–264. Universita di Perugia, Perugia.

Joseph, L., and Wolfson, D. B. (1992). Estimation in multi-path change point problems. *Commun. Statist. Theory Methods* **21**, 897–913.

Joseph, L., Wolfson, D. B., du Berger, R., and Lyle, R. M. (1996). Changepoint analysis of a randomized trial on the effects of calcium supplementation on blood pressure. In *Bayesian Biostatistics* (D. A. Berry and D. K. Stangl, editors), 617–649. Marcel Dekker, New York.

Journel, A. G. (1994). Resampling from stochastic simulations (with discussion). *Environ. Ecol, Statist.* **1**, 63–91.

Junghard, O. (1990). Linear shrinkage in traffic accident models and their estimation by cross validation and bootstrap methods. *Linkoping Studies in Science and Technology, Thesis No.* **205**, Linkoping.

Jupp, P. E., and Mardia, K. V. (1989). Theory of directional statistics, 1975–1988. *Int. Statist. Rev.* **57**, 261–294.

Kabaila, P. (1993a). On bootstrap predictive inference for autoregressive process. *J. Time Ser. Anal.* **14**, 473–484.   *(359)

Kabaila, P. (1993b). Some properties of profile bootstrap confidence intervals. *Aust. J. Statist.* **35**, 205–214.

Kadiyala, K. R., and Qberhelman, D. (1990). Estimation of standard errors of empirical Bayes estimators in CAPM-type models. *Commun. Statist. Simul. Comput.* **19**, 189–206.

Kafadar, K. (1994). An application of nonlinear regression in R & D: a case study from the electronics industry. *Technometrics* **36**, 237–248.

Kaigh, W. D., and Cheng, C. (1991). Subsampling quantile estimator standard errors with applications. *Commun. Statist. Theory Methods* **20**, 977.

Kaigh, W. D., and Lachenbruch, P. (1982). A generalized quantile estimator. *Commun. Statist. Theory Methods* **11**, 2217–2238.

Kalbfleish, J. D., and Prentice, R. L. (1980). *The Statistical Analysis of Failure Time Data.* Wiley, New York.

Kanal, L. (1974). Patterns in pattern recognition: 1968–1974. *IEEE Trans. Inform. Theory* **2**, 472–479.   *(360)

Kane, V. E. (1986). Process capability indices. *J. Qual. Technol.* **24**, 41–52.   *(361)

Kang, S.-B. and Cho, Y.-S. (1997). Estimation of the parameters on a Pareto distribution by jackknife and bootstrap. *J. Inform. Optim. Sci.* **18**, 289–300.

Kaplan, E. L., and Meier, P. (1958). Nonparametric estimation from incomplete samples. *J. Am. Statist. Assoc.* **53**, 457–481.   *(362)

Kapoyannis, A. S., Ktorides, C. N., and Panagiotou, A. D. (1997). An extension of the statistical bootstrap model to include strangeness. *J. Phys. G Nucl. Particle Phys. Strangeness in Quark Matter 1997,* **23**, 1921–1932.

Kapoyannis, A. S., Ktorides, C. N., and Panagiotou, A. D. (1998). An extension of the statistical bootstrap model to include strangeness. Implications on particle ratios. *Phys. Rev. D* **58**, 1–17.

Karrison, T. (1990). Bootstrapping censored data with covariates. *J. Statist. Comput. Simul.* 36, 195–207.

Katz, A. S., Katz, S., and Lowe, N. (1994). Fundamentals of the bootstrap based analysis of neural network's accuracy. In *Proceedings of the World Congress on Neural Networks,* **3**, 673–678.

Kaufman, E. (1988). Asymptotic expansion of the distribution function of the sample quantile prepivoted by the bootstrap (in German). Diplom. thesis. University of Siegen, Siegen.

Kaufman, L., and Rousseeuw, P. J. (1990). *Finding Groups in Data: An Introduction to Cluster Analysis.* Wiley, New York.

Kaufman, S. (1993). A bootstrap variance estimator for the Schools and Staffing Survey. *ASA Proceedings of the Survey Research Section,* 675–680.

Kaufman, S. (1998). A bootstrap variance estimator for systematic PPS sampling. Unpublished paper.   *(363)

Kawano, H., and Higuchi, T. (1995). The bootstrap method in space physics: error estimation for the minimum variance analysis. *Geophys. Res. Lett.* **22**, 307–310.

Kay, J., and Chan, K. (1992). Bootstrapping blurred and noisy data. *Proceedings of the 10th Symposium Computer Statistics* (Y. Dodge and J. Whittaker, editors), **2**, 287–291. Physica, Vienna.

Keating, J. P., and Tripathi, R. C. (1985). Estimation of percentiles. *Encyclopedia of Statistical Sciences,* **6**, 668–674. Wiley, New York.

Kemp, A. W. (1997). Book Review of *Randomization, Bootstrap and Monte Carlo Methods in Biology,* 2nd ed, (by B. F. J. Manly). *Biometrics* **53**, 1560–1561.

Kendall, D. G., and Kendall, W. S. (1980). Alignments in two-dimensional random sets and points. *Adv. Appl. Probab.* **12**, 380–424.

Kim, D. (1993). Nonparametric kernel regression function estimation with bootstrap method. *J. Korean Statist. Soc.* **22**, 361–368.

Kim, H. T., and Truong, Y. K. (1998). Nonparametric regression estimates with censored data: local linear smoothers and their applications. *Biometrics* **54**, 1434–1444.

Kim, J.-H. (1990). Conditional bootstrap methods for censored data. Ph. D. thesis, Department of Statistics, Florida State University.

Kim, Y. B., Haddock, J., and Willemain, T. R. (1993). The binary bootstrap: inference with autocorrelated binary data. *Commun. Statist. Simul. Comput.* **22**, 205–216.

Kimber, A. (1994). Book Review of *An Introduction to the Bootstrap* (by B. Efron and R. Tibshirani). *Statistician* **43**, 600.

Kinateder, J. G. (1992). An invariance principle applicable to the bootstrap. In *Exploring the Limits of Bootstrap* (R. LePage and L. Billard, editors), 157–181. Wiley, New York.

Kindermann, J., Paass, G., and Weber, F. (1995). Query construction for neural networks using the bootstrap. In *ICANN '95* (F. Fogelman-Soulie and P. Gallinari, editors), **2**, 135–140.

Kinsella, A. (1987). The "exact" bootstrap approach to confidence intervals for the relative difference statistic. *Statistician* **36**, 345–347.

Kipnis, V. (1992). Bootstrap assessment of prediction in exploratory regression analysis. In *Exploring the Limits of Bootstrap* (R. LePage and L. Billard, editors), 363–387. Wiley, New York.

Kish, L., and Frankel, M. R. (1970). Balanced repeated replication for standard errors. *J. Am. Statist. Assoc.* **65**, 1071–1094.

Kish, L., and Frankel, M. R. (1974). Inference from complex samples (with discussion) *J. R. Statist. Soc. B* **36**, 1–37.

Kitamura, Y. (1997). Empirical likelihood methods with weakly dependent processes. *Ann. Statist.* **25**, 2084–2102.

Klenitsky, D. V., and Kuvshinov, V. I. (1996). Local fluctuations of multiplicity in the statistical-bootstrap model. *Phys. At. Nucl.* **59**, 129–134.

Klenk, A., and Stute, W. (1987). Bootstrapping of *L*-estimates. *Statist. Dec.* **5**, 77–87.

Klugman, S. A., Panjer, H. H., and Willmot, G. E. (1998). *Loss Models: From Data to Decisions.* Wiley, New York.

Knight, K. (1989). On the bootstrap of the sample mean in the infinite variance case. *Ann. Statist.* **17**, 1168–1175.   *(364)

Knight, K. (1997). Bootstrapping sample quantiles in non-regular cases. *Statist. Probab. Lett.* **37**, 259–267.

Knoke, J. D. (1986). The robust estimation of classification error rates. *Comput. Math. Applic.* **12A**, 253–260.

Knox, R. G., and Peet, R. K. (1989). Bootstrapped ordination: a method for estimating sampling effects in indirect gradient analysis. *Vegetatio* **80**, 153–165.

Kocherlakota, S., Kocherlakota, K., and Kirmani, S. N. U. A. (1992). Process capability indices under non-normality. *Int. J. Math. Statist.* **1**.   *(365)

Kohavi, R. (1995). A study of cross-validation and bootstrap for accuracy: assessment and model selection. Stanford University Technical Report, Department of Computer Sciences.

Kolen, M. J., and Brennan, R. L. (1987). Analytic smoothing for equipercentile equating under the common item nonequivalent population design. *Psychometrika* **52**, 43–59.

Koltchinskii, V. I. (1997). *M*-estimation convexity and quantiles. *Ann. Statist.* **25**, 435–477.

Kong, F., and Zhang, M. (1994). The Edgeworth expansion and the bootstrap approximation for a studentized *U*-statistic. In *5th Japan–China Symposium on Statistics* (M. Ichimura, S. Mao, and G. Fan, editors), **5**, 124–126.

Konishi, S. (1991). Normalizing transformations and bootstrap confidence intervals. *Ann. Statist.* **19**, 2209–2225.

Konishi, S., and Honda, M. (1990). Comparison of procedures for estimation of error rates in discriminant analysis under nonnormal populations. *J. Statist. Comput. Simult.* **36**, 105–116.

Konold, C. (1994). Understanding probability and statistical inference through resampling. In *Proceedings of the 1st Scientific Meeting — International Association for Statistical Education* (L. Brunelli and G. Cicchitelli, editors), **1**, 199–212. Universita di Perugia, Perugia.

Kononenko, I. V., and Derevyanchenko, B. I. (1995). Analysis of an algorithm for prediction of nonstationary random processes using the bootstrap. *Automatika* **2**, 88–92.

Kotz, S., and Johnson, N. L. (1992). *Breakthroughs in Statistics: Methodology and Distribution*, **2**, Springer-Verlag, New York.   *(366)

Kotz, S., and Johnson, N. L. (1993). *Process Capability Indices*. Chapman & Hall, London.   *(367)

Kotz, S., and Johnson, N. L. (editors) (1997). *Breakthroughs in Statistics*, **3**, Springer-Verlag, New York.   *(368)

Kotz, S., and Johnson, N. L., and Read, C. B. (editors) (1982). Bootstrapping. *Encyclopedia of Statistical Sciences*, **1**, 301. Wiley, New York.   *(369)

Kotz, S., Johnson, N. L., and Read, C. B. (editors) (1983). *Encyclopedia of Statistical Sciences*, **3**, 314–316. Wiley, New York.   *(370)

Kovar, J. G. (1985). Variance estimation of nonlinear statistics in stratified samples. Methodology Branch Working Paper #85–052E, Statistics Canada.   *(371)

Kovar, J. G. (1987). Variance estimation of medians in stratified samples. Methodology Branch Working Paper #87–004E, Statistics Canada.   *(372)

Kovar, J. G., Rao, J. N. K., and Wu, C. F. J. (1988). Bootstrap and other methods to measure errors in survey estimates. *Can. J. Statist.* **16**, Suppl. 25–45.   *(373)

Kreiss, J. P. (1992). Bootstrap procedures for AR(∞)-processes. In *Bootstrapping and Related Techniques. Proceedings, Trier, FRG.* (K.-H. Jockel, G. Rothe, and W. Sendler, editors). *Lecture Notes in Economics and Mathematical Systems*, **376**, 107–113. Springer-Verlag, Berlin.

Kreiss, J. P., and Franke, J. (1992). Bootstrapping stationary autoregressive moving average models. *J. Time Ser. Anal.* **13**, 297–317.

Krewski, D., and Rao, J. N. K. (1981). Inference from stratified samples: properties of the linearization, jackknife and balanced repeated replication methods. *Ann. Statist.* **9**, 1010–1019.

Krewski, D., Smythe, R. T., Dewanji, A., and Szyszkowicz, M. (1991a). Bootstrapping an empirical Bayes estimator of the distribution of historical controls in carcinogen bioassay. Unpublished manuscript.

Krewski, D., Smythe, R. T., Fung, K. Y., and Burnett, R. (1991b). Conditional and unconditional tests with historical controls. *Can. J. Statist.* **19**, 407–423.

Kuk, A. Y. C. (1987). Bootstrap estimators of variance under sampling with probability proportional to aggregate size. *J. Statist. Comput. Simul.* **28**, 303–311.   *(374)

Kuk, A. Y. C. (1989). Double bootstrap estimation of variance under systematic sampling with probability proportional to size. *J. Statist. Comput. Simul.* **31**, 73–82.   *(375)

Kulperger, P. J., and Prakasa Rao, B. L. S. (1989). Bootstrapping a finite state Markov chain. *Sankhya A* **51**, 178–191.   *(376)

Kunsch, H. (1989). The jackknife and the bootstrap for general stationary observations. *Ann. Statist.* **17**, 1217–1241.   *(377)

Lachenbruch, P. A. (1967). An almost unbiased method of obtaining confidence intervals for the probability of misclassification in discriminant analysis. *Biometrics* **23**, 639–645.   *(378)

Lachenbruch, P. A. (1975). *Discriminant Analysis.* Hafner, New York.   *(379)

Lachenbruch, P. A., and Mickey, M. R. (1968). Estimation of error rates in discriminant analysis. *Technometrics* **10**, 1–11.   *(380)

Laeuter, H. (1985). An efficient estimator of error rate in discriminant analysis. *Statistics* **16**, 107–119.

Lahiri, S. N. (1991). Second order optimality of stationary bootstrap. *Statist. Probab. Lett.* **14**, 335–341.   *(381)

Lahiri, S. N. (1992a). On bootstrapping *M*-estimators. *Sankhya A* **54**, 157–170.   *(382)

Lahiri, S. N. (1992b). Edgeworth correction by moving block bootstrap for stationary and nonstationary data. In *Exploring the Limits of Bootstrap* (R. LePage and L. Billard, editors), 183–214. Wiley, New York.   *(383)

Lahiri, S. N. (1992c). Bootstrapping *M*-estimators of a multiple linear regression parameter. *Ann. Statist.* **20**, 1548–1570.   *(384)

Lahiri, S. N. (1993a). Bootstrapping the studentized sample mean of lattice variables. *J. Multivar. Anal.* **45**, 247–256.

Lahiri, S. N. (1993b). On the moving block bootstrap under long range dependence. *Statist. Probab. Lett.* **18**, 405–413.

Lahiri, S. N. (1994). On second order correctness of Efron's bootstrap without Cramer-type conditions in linear regression models. *Math. Methods Statist.* **3**, 130–148.

Lahiri, S. N. (1995). On asymptotic behavior of the moving block bootstrap for normalized sums of heavy-tailed random variables. *Ann. Statist.* **23**, 1331–1349.   *(385)

Lahiri, S. N. (1997). Bootstrapping weighted empirical processes that do not converge weakly. *Statist. Probab. Lett.* **37**, 295–302.

Lahiri, S. N., and Koul, H. (1994). On bootstrapping *M*-estimated residual processes in multiple linear regression models. *J. Multivar. Anal.* **49**, 255–265.

Lai, T. L., and Wang, J. Q. Z. (1993). Edgeworth expansions for symmetric statistics with applications to bootstrap methods. *Statist. Sin.* **3**, 517–542.

Laird, N. M., and Louis, T. A. (1987). Empirical Bayes confidence intervals based on bootstrap samples (with discussion). *J. Am. Statist. Assoc.* **82**, 739–757.

Lake, J. A. (1995). Calculating the probability of multitaxon evolutionary trees: bootstrappers gambit. *Proc. Natl. Acad. Sci.* **92**, 9662–9666.

Lamb, R. H., Boos, D. D., and Brownie, C. (1996). Testing for effects on variance in experiments with factorial treatment structure and nested errors. *Technometrics* **38**, 170–177.

Lambert, D., and Tierney, L. (1997). Nonparametric maximum likelihood estimation from samples with irrelevant data and verification bias. *J. Am. Statist. Assoc.* **92**, 937–944.

LaMotte, L. R. (1978). Bayes linear estimators. *Technometrics* **20**, 281–290.   *(386)

Lancaster, T. (1997). Exact structural inference in optimal job-search models. *J. Bus. Econ. Statist.* **15**, 165–179.

Lange, K. L., Little, R. J. A., and Taylor, J. M. G. (1989). Robust statistical modeling using the *t* distribution. *J. Am. Statist. Assoc.* **84**, 881–905.

Lanyon, S. M. (1987). Jackknifing and bootstrapping: important "new" statistical techniques for ornithologists. *Auk* **104**, 144–146.   *(387)

Lawless, J. F. (1988). Reliability (nonparametric methods in). *Encyclopedia of Statistical Sciences*, **8**, 20–24, Wiley, New York.

Leadbetter, M. R., Lindgren, G., and Rootzen, H. (1983). *Extremes and Related Properties of Random Sequences and Processes.* Springer-Verlag, New York. *(388)

Leal, S. M., and Ott, J. (1993). A bootstrap approach to estimating power for linkage heterogeneity. *Genet. Epidemiol.* **10**, 465–470.   *(389)

LeBlanc, M., and Crowley, J. (1993). Survival trees by goodness of split. *J. Am. Statist. Assoc.* **88**, 477–485.   *(390)

LeBlanc, M., and Tibshirani, R. (1996). Combining estimates in regression and classification. *J. Am. Statist. Assoc.* **91**, 1641–1650.

Lebreton, C. M., and Visscher, P. M. (1998). Empirical nonparametric bootstrap strategies in quantitative trait loci mapping: conditioning on the genetic model. *Genetics* **148**, 525–535.

Lee, A. J. (1985). On estimating the variance of a *U*-statistic. *Commun. Statist. Theory Methods* **14**, 289–301.

Lee, K. W. (1990). Bootstrapping logistic regression models with random regressors. *Commun. Statist. Theory Methods* **19**, 2527–2539.

Lee, S. M. S. (1994). Optimal choice between parametric and nonparametric bootstrap estimates. *Math. Proc. Cambridge Philos. Soc.* **115**, 335–363.

Lee, S. M. S., and Young, G. A. (1994a). Asymptotic iterated bootstrap confidence intervals. In *26th Conference on the Interface Between Computer Science and Statistics* (J. Sall and A. Lehman, editors), **26**, 464–471. Springer-Verlag, New York.

Lee, S. M. S., and Young, G. A. (1994b). Practical higher-order smoothing of the bootstrap. *Statist. Sin.* **4**, 445–460.

Lee, S. M. S., and Young, G. A. (1995). Asymptotic iterated bootstrap confidence intervals. *Ann. Statist.* **23**, 1301–1330.

Lee, S. M. S., and Young, G. A. (1997). Asymptotics and resampling methods. In *28th Symposium on the Interface Between Computer Science and Statistics* (L. Billard and N. I. Fisher, editors), **28**, 221–227, Springer-Verlag, New York.

Leger, C., and Cleroux, R. (1992). Nonparametric age replacement: Bootstrap confidence intervals for the optimal cost. *Oper. Res.* **40**, 1062–1073.

Leger, C., and Larocque, D. (1994). Bootstrap estimates of the power of a rank test in a randomized block design. *Statist. Sin.* **4**, 423–443.

Leger, C., Politis, D. N., and Romano, J. P. (1992). Bootstrap technology and applications. *Technometrics* **34**, 378–398.   *(391)

Leger, C., and Romano, J. P. (1990a). Bootstrap adaptive estimation. *Can. J. Statist.* **18**, 297–314.

Leger, C., and Romano, J. P. (1990b). Bootstrap choice of tuning parameters. *Ann. Inst. Statist. Math.* **42**, 709–735.

Lehmann, E. L. (1986). *Testing Statistical Hypotheses*, 2nd ed. Wiley, New York. *(392)

Lehmann, E. L. (1999). *Elements of Large-Sample Theory.* Springer-Verlag, New York. *(393)

Lehmann, E. L., and Casella, E. (1998). *Theory of Point Estimation.* 2nd ed. Springer-Verlag, New York.   *(394)

Lehtonen, R and Pakkinen, E. J. (1995). *Practical Methods for the Design and Analysis of Complex Surveys.* Wiley, Chichester.

Lele, S. (1991a). Jackknifing linear estimating equations: asymptotic theory and applications in stochastic processes. *J. R. Statist. Soc. B* **53**, 253–267.   *(395)

Lele, S. (1991b). Resampling using estimating equations. In *Estimating Functions* (V. Godambe, editor), 295–304. Clarendon Press, Oxford.   *(396)

Lembregts, F., Top, J., and Neyrinck, F. (1996). Off-line synchronous resampling of vibration measurements. In *Quality Measurements: The Indispensible Bridge Between Theory and Reality. Joint Conference. IEEE Instrumentation and Measurement Technology Conference,* **1**, 748–755.

LePage, R. (1992). Bootstrapping signs. In *Exploring the Limits of Bootstrap* (R. LePage and L. Billard, editors), 215–224. Wiley, New York.   *(397)

LePage, R., and Billard, L. (editors) (1992). *Exploring the Limits of Bootstrap.* Wiley, New York.   *(398)

LePage, R., Podgorski, K. Ryznar, M., and White, A. (1998). *Bootstrapping Signs and Permutations for Regression with Heavy-Tailed Errors: A Robust Resampling.* In *A Practical Guide to Heavy Tails: Statistical Techniques and Applications* (R. J. Adler, R. E. Feldman, and M. S. Taqqu, editors), 283–310. Birkhauser, Boston.   *(399)

Li, H., and Maddala, G. S. (1996). Bootstrapping time series models (with discussion). *Econ. Rev.* **15**, 115–195.   *(400)

Li, G., Tiwari, R. C., and Wells, M. T. (1996). Quantile comparison functions in two-sample problems, with application to comparisons of diagnostic markers. *J. Am. Statist. Assoc.* **91**, 689–698.

Linder, E., and Babu, G. J. (1994). Bootstrapping the linear functional relationship with known error variance ratio. *Scand. J. Statist.* **21**, 21–39.

Lindsay, B. G., and Li, B. (1997). On second-order optimality of the observed Fisher information. *Ann. Statist.* **25**, 2172–2199.

Linhart, H., and Zucchini, W. (1986). *Model Selection.* Wiley, New York.   *(401)

Linnet, K. (1989). Assessing diagnostic tests by a strictly proper scoring rule. *Statist. Med.* **8**, 609–618.

Linssen, H. N., and Banens, P. J. A. (1983). Estimation of the radius of a circle when the coordinates of a number of points on its circumference are observed: an example of bootstrapping. *Statist. Probab. Lett.* **1**, 307–311.

Little, R. J. A. (1988). Robust estimation of the mean and covariance matrix from data with missing values. *Appl. Statist.* **37**, 23–38.

Little, R. J. A., and Rubin, D. B. (1987). *Statistical Analysis with Missing Data.* Wiley, New York.   *(402)

Liu, J. (1992). Inference from stratified samples: application of Edgeworth expansions. Ph. D. thesis, Carleton University.

Liu, J. S., and Chen, R. (1998). Sequential Monte Carlo methods for dynamic systems. *J. Am. Statist. Assoc.* **93**, 1032–1044.

Liu, R. Y. (1988). Bootstrap procedures under some non-i.i.d. models. *Ann. Statist.* **16**, 1696–1708.   *(403)

Liu, R. Y., and Singh, K. (1987). On a partial correction by the bootstrap. *Ann. Statist.* **15**, 1713–1718.   *(404)

Liu, R. Y., and Singh, K. (1995). Using i. i. d. bootstrap for general non-i. i. d. models. *J. Statist. Plann. Inf.* **43**, 67–76.   *(405)

Liu, R. Y., and Singh, K. (1992a). Efficiency and robustness in sampling. *Ann. Statist.* **20**, 370–384.

Liu, R. Y., and Singh, K. (1992b). Moving blocks jackknife and bootstrap capture weak dependence. In *Exploring the Limits of Bootstrap* (R. LePage and L. Billard, editors), 225–248. Wiley, New York.   *(406)

Liu, R. Y., and Singh, K. (1997). Notions of limiting *p* values based on data depth and the bootstrap. *J. Am. Statist. Assoc.* **92**, 266–277.

Liu, R. Y., Singh, K., and Lo, S. H. (1989). On a representation related to bootstrap. *Sankhya A* **51**, 168–177.

Liu, R. Y., and Tang, J. (1996). Control charts for dependent and independent measurements based on bootstrap methods. *J. Am. Statist. Assoc.* **91**, 1694–1700. *(407)

Lloyd, C. L. (1998). Using smoothed receiver operating characteristic curves to summarize and compare diagnostic systems. *J. Am. Statist. Assoc.* **93**, 1356–1364.

Lo, A. Y. (1984). On a class of Bayesian nonparametric estimates: I. Density estimates. *Ann. Statist.* **12**, 351–357.

Lo, A. Y. (1987). A large sample study of the Bayesian bootstrap. *Ann. Statist.* **15**, 360–375.   *(408)

Lo, A. Y. (1988). A Bayesian bootstrap for a finite population. *Ann. Statist.* **16**, 1684–1695.   *(409)

Lo, A. Y. (1991). Bayesian bootstrap clones and a biometry function. *Sankhya A* **53**, 320–333.

Lo, A. Y. (1993a). A Bayesian bootstrap for censored data. *Ann. Statist.* **21**, 100–123. *(410)

Lo, A. Y. (1993b). A Bayesian bootstrap for weighted sampling. *Ann. Statist.* **21**, 2138–2148.

Lo, S.-H. (1998). General coverage problems with applications and bootstrap method in survival analysis and reliability theory. Columbia University Technical Report.

Lo, S.-H., and Singh, K. (1986). The product limit estimator and the bootstrap: some asymptotic representations. *Probab. Theory* **71**, 455–465.

Lodder, R. A., Selby, M., and Hieftje, G. M. (1987). Detection of capsule tampering by near infra-red reflectance analysis. *Anal. Chem.* **59**, 1921–1930.

Loh, W.-Y. (1984). Estimating an endpoint of a distribution with resampling methods. *Ann. Statist.* **12**, 1543–1550.

Loh, W.-Y. (1985). A new method for testing separate families of hypotheses. *J. Am. Statist. Assoc.* **80**, 362–368.

Loh, W.-Y. (1987). Calibrating confidence coefficients. *J. Am. Statist. Assoc.* **82**, 155–162.   *(411)

Loh, W.-Y. (1991). Bootstrap calibration for confidence interval construction and selection. *Statist. Sin.* **1**, 479–495.

Lohse, K. (1987). Consistency of the bootstrap. *Statist. Dec.* **5**, 353–366.

Lokki, H., and Saurola, P. (1987). Bootstrap methods for the two-sample location and scatter problems. *Acta Ornithol.* **23**, 133–147.

Lombard, F. (1986). The change-point problem for angular data. *Technometrics* **28**, 391–397.

Loughin, T. M., and Koehler, K. (1993). Bootstrapping in proportional hazards models with fixed explanatory variables. Unpublished manuscript.

Loughin, T. M., and Noble, W. (1997). A permutation test for effects in an unreplicated factorial design. *Technometrics* **39**, 180–190.

Lovie, A. D., and Lovie, P. (1986). The flat maximum effect and linear scoring models for prediction. *J. For.* **5**, 159–168.

Low, L. Y. (1988). Resampling procedures. *Encyclopedia of Statistical Sciences* **8**, 90–93, Wiley, New York.

Lu, H. H. S., Wells, M. T., and Tiwari, R. C. (1994). Inference for shift functions in the two-sample problem with right-censored data: with applications. *J. Am. Statist. Assoc.* **89**, 1017–1026.

Lu, J.-C., Park, J., and Yang, Q. (1997). Statistical inference of a time-to-failure distribution derived from linear degradation data. *Technometrics* **39**, 391–400.

Lu, M.-C., and Chang, D. S. (1997). Bootstrap prediction intervals for the Birnbaum–Saunders distribution. *Microelectron. Reliab.* **37**, 1213–1216.

Lu, R., and Yang, C.-H. (1994). Resampling schemes for estimating the similarity measures on location model. *Commun. Statist. Simul. Comput.* **23**, 973–996.

Ludbrook, J. (1995). Issues in biomedical statistics: comparing means by computer-intensive methods. *Aust. N. Z. J. Surg.* **65**, 812–819.

Lunneborg, C. E. (1985). Estimating the correlation coefficient: The bootstrap approach. *Psych. Bull.* **98**, 209–215.   *(412)

Lunneborg, C. E. (1987). Bootstrap applications for the behavioral sciences. *Psychometrika* **52**, 477–478.   *(413)

Lunneborg, C. E. (1994). Book Review of *Resampling-Based Multiple Testing: Examples and Methods for p-Value Adjustment* (by P. Westfall and S. S. Young, editors). *Statist. Comp.* **4**, 219–220.

Lunneborg, C. E., and Tousignant, J. P. (1985). Efron's bootstrap with application to the repeated measures design. *Multivar. Behav. Res.* **20**, 161–178.

Lutz, M. W., Kenakin, T. P., Corsi, M., Menius, J. A., Krishnamoorthy, C., Rimele, T., and Morgan, P. H. (1995). Use of resampling techniques to estimate the variance of parameters in pharmacological assays when experimental protocols preclude independent replication: an example using Schild regressions. *J. Pharmacol. Toxicol. Methods* **34**, 37–46.

Magnussen, S., and Burgess, D. (1997). Stochastic resampling techniques for quantifying error propagations in forest field experiments. *Can. J. For. Res.* **27**, 630–637.

Maindonald, J. H. (1984). *Statistical Computation*. Wiley, New York.

Maitra, R. (1997). Estimating precision in functional images. *J. Comput. Graphical. Statist.* **6**, 285–299.

Maiwald, D., and Bohme, J. F. (1994). Multiple testing for seismic data using bootstrap. In *Proceedings of the 1994 IEEE International Conference on Acoustics, Speech and Signal Processing*, **6**, 89–92.

Mak, T. K., Li, W. K., and Kuk, A. Y. C. (1986). The use of surrogate variables in binary regression models. *J. Statist. Comput. Simul.* **24**, 245–254.

Makinodan, T., Albright, J. W., Peter, C. P., Good, P. I., and Heidrick, M. L. (1976). Reduced humoral immune activity in long-lived mice. *Immunology* **31**, 400–408. *(414)

Mallows, C. L. (1983). Robust methods. In *Statistical Data Analysis* (R. Gnanadesikan, editor), Chap. 3. American Mathematical Society, Providence.

Mallows, C. L., and Tukey, J. W. (1982). An overview of techniques of data analysis, emphasizing its exploratory aspects. In *Some Recent Advances in Statistics* (J. T. de Oliveira and B. Epstein, editors), 111–172. Academic Press, London.

Mammen, E. (1989a). Asymptotics with increasing dimension for robust regression with applications to the bootstrap. *Ann. Statist.* **17**, 382–400.

Mammen, E. (1989b). Bootstrap and wild bootstrap for high-dimensional linear models. Preprint. Sonderforschungsbereich 123, Universitat Heidelberg.

Mammen, E. (1990). Higher order accuracy of bootstrap for smooth functionals. Preprint. Sonderforschungsbereich 123, Universitat Heidelberg.

Mammen, E. (1992a). Bootstrap, wild bootstrap, and asymptotic normality. Preprint. Sonderforschungsbereich 123, Universitat Heidelberg.

Mammen, E. (1992b). *When Does the Bootstrap Work? Asymptotic Results and Simulations. Lecture Notes in Statistics*, **77**, Springer-Verlag, Heidelberg.   *(415)

Mammen, E. (1993). Bootstrap and wild bootstrap for high dimensional linear models. *Ann. Statist.* **21**, 255–285.   *(416)

Mammen, E., Marron, J. S., and Fisher, N. I. (1992). Some asymptotics for multimodality tests based on kernel density estimates. *Probab. Theory Relat. Fields* **91**, 115–132.

Manly. B. F. J. (1991). *Randomization and Monte Carlo Methods in Biology*. Chapman & Hall, London.   *(417)

Manly. B. F. J. (1992). Bootstrapping for determining sample sizes in biological studies. *J. Exp. Mar. Biol. Ecol.* **158**, 189–196.

Manly. B. F. J. (1993). A review of computer-intensive multivariate methods in ecology. In *Multivariate Environmental Statistics* (G. P. Patil and C. R. Rao, editors), 307–346. Elsevier Science Publishers, Amsterdam.   *(418)

Manly. B. F. J. (1997). *Randomization, Bootstrap and Monte Carlo Methods in Biology*, 2nd ed. Chapman & Hall, London.   *(419)

Mao, X. (1996). Splatting of non-rectilinear volumes through stochastic resampling. *IEEE Trans. Vis. Comput. Graphics* **2**, 156–170.

Mapleson, W. W. (1986). The use of GLIM and the bootstrap in assessing a clinical trial of two drugs. *Statist. Med.* **5**, 363–374.   *(420)

Mardia, K. V., Kent, J. T., and Bibby, J. M. (1979). *Multivariate Analysis*. Academic Press, London.   *(421)

Maritz, J. S. (1979). A note on exact robust confidence intervals for location. *Biometrika* **66**, 163–166.   *(422)

Maritz, J. S., and Jarrett, R. G. (1978). A note on estimating the variance of the sample median. *J. Am. Statist. Assoc.* **73**, 194–196.   *(423)

Maritz, J. S., and Lwin, T. (1989). *Empirical Bayes Methods*, 2nd ed. Chapman & Hall, London.   *(424)

Markus, M. T. (1994). Bootstrap confidence regions for homogeneity analysis. The influence of rotation on coverage percentages. *Proc. Comput. Statist. Symp.* **11**, 337–342.

Markus, M. T., and Visser, R. A. (1990). Bootstrap methods for generating confidence regions in HOMALS; balancing sample size and number of trials. FSW/RUL, **RR-90-02**, Department of Behavioral and Computer Science, University of Leiden, Leiden.

Markus, M. T., and Visser, R. A. (1992). Applying the bootstrap to generate confidence regions in multiple correspondence analysis. In *Bootstrapping and Related Techniques. Proceedings, Trier, FRG.* (K.-H. Jockel, G. Rothe, and W. Sendler, editors). *Lecture Notes in Economics and Mathematical Systems,* **376**, 71–75. Springer-Verlag, Berlin.

Marron, J. S. (1992). Bootstrap bandwidth selection. In *Exploring the Limits of Bootstrap* (R. LePage and L. Billard, editors), 249–262. Wiley, New York.   *(425)

Martin, M. A. (1989). On the bootstrap and confidence intervals. Unpublished Ph. D. thesis, Australian National University.

Martin, M. A. (1990a). On the bootstrap iteration for coverage correction in confidence intervals. *J. Am. Statist. Assoc.* **85**, 1105–1118.   *(426)

Martin, M. A. (1990b). On using the jackknife to estimate quantile variance. *Can. J. Statist.* **18**, 149–153.

Martin, M. A. (1994a). Book Review of *Resampling-Based Multiple Testing: Examples and Methods for p-Value Adjustment* (by P. Westfall and S. Young). *Biometrics* **50**, 1226–1227.

Martin, M. A. (1994b). Book Review of *An Introduction to Bootstrap* (by B. Efron and R. Tibshirani). *Chance* **7**, 58–60.

Martin. R. D. (1980). Robust estimation of autoregressive models. In *Reports on Directions in Time Series* (D. R. Brillinger and G. C. Tiao, editors), 228–262. Institute of Mathematical Statistics, Hayward.   *(427)

Martz, H. F., and Duran, B. S. (1985). A comparison of three methods for calculating lower confidence limits on system reliability using binomial component data. *IEEE Trans. Reliab.* **R-34**, 113–120.

Mason, D. M., and Newton, M. A. (1992). A rank statistics approach to the consistency of a general bootstrap. *Ann. Statist.* **20**, 1611–1624.

Matheron, G. (1975). *Random Sets and Integral Geometry*. Wiley, New York.   *(428)

Mattei, G., Mignani, S., and Rosa, R. (1997). Statistical resampling for accuracy estimate in Monte Carlo renormalization group. *Phys. Lett. A* **237**, 33–36.

Mazurkiewicz, M. (1995). The little bootstrap method for autoregressive model selection. *Badania Operacyjne I Decyzje* **2**, 39–53.

McCarthy, P. J. (1969). Pseudo-replication: half-samples. *Int. Statist. Rev.* **37**, 239–263. *(429)

McCarthy, P. J., and Snowden, C. B. (1985). The bootstrap and finite population sampling. In *Vital and Health Statistics (Ser. 2 No. 95) Public Health Service Publication 85–1369*. U.S. Government Printing Office, Washington, DC. *(430)

McCullough, B. D. (1994). Bootstrapping forecast intervals: an application to AR(p) models. *J. For.* **13**, 51–66. *(431)

McCullough, B. D., and Vinod, H. D. (1998). Implementing the double bootstrap. *Comput. Econ.* **12**, 79–95. *(432)

McDonald, J. A. (1982). Interactive graphics for data analysis. ORION Technical Report 011, Department of Statistics, Stanford University. *(433)

McKay, M. D., Beckman, R. J., and Conover, W. J. (1979). A comparison of three methods for selecting values of input variables in the analysis of output from a computer code. *Technometrics* **21**, 234–245. *(434)

McKean, J. W., and Schader, R. M. (1984). A comparison of methods for studentizing the sample quantile. *Commun. Statist. Simul. Comput.* **13**, 751–773.

McLachlan, G. J. (1976). The bias of the apparent error rate in discriminant analysis. *Biometrika* **63**, 239–244. *(435)

McLachlan, G. J. (1980). The efficiency of Efron's bootstrap approach applied to error rate estimation in discriminant analysis. *J. Statist. Comput. Simul.* **11**, 273–279. *(436)

McLachlan, G. J. (1986). Assessing the performance of an allocation rule. *Comput. Math. Applic.* **12A**, 261–272. *(437)

McLachlan, G. J. (1987). On bootstrapping the likelihood ratio test statistic for the number of components in a mixture. *Appl. Statist.* **36**, 318–324. *(438)

McLachlan, G. J. (1992). *Discriminant Analysis and Statistical Pattern Recognition*. Wiley, New York. *(439)

McLachlan, G. J., and Basford, K. E. (1988). *Mixture Models: Inference and Applications to Clustering*. Marcel Dekker, New York. *(440)

McLachlan, G. J., and Krishnan, T. (1997). *The EM Algorithm and Extensions*. Wiley, New York. *(441)

McLachlan, G. J., and Peel, D. (1997). On a Resampling Approach to Choosing the Number of Components in Normal Mixture Models. In *28th Symposium on the Interface Between Computer Science and Statistics* (L. Billard and N. I. Fisher, editors), **28**, 260–266. Springer-Verlag, New York.

McLeod, A. I. (1988). Simple random sampling. *Encyclopedia of Statistical Sciences*, **8**, 478–480.

McPeek, M. A., and Kalisz, S. (1993). Population sampling and bootstrapping in complex designs: demographic analysis. *Ecol. Exper.* 232–252.

McQuarrie, A. D. R., and Tsai, C.-L. (1998). *Regression and Time Series Model Selection*. World Scientific Publishing, Singapore. *(442)

Meeker, W. Q., and Escobar, L. A. (1998). *Statistical Methods for Reliability Data.* Wiley, New York. *(443)

Mehlman, D. W., Shepard, U. L., and Kelt, D. A. (1995). Bootstrapping principal components—a comment (with discussion). *Ecology* **76**, 640–645.

Meneghini, F. (1985). An application of bootstrap for spatial point patterns (in Italian). *R. St. A. Italy* **18**, 73–82.

Meyer, J. S., Ingersoll, C. G., McDonald, L. L., and Boyce, M. S. (1986). Estimating uncertainty in population growth rates: jackknife vs. bootstrap techniques. *Ecology* **67**, 1156–1166.

Mick, R., and Ratain, M. J. (1994). Bootstrap validation of pharmacodynamic models defined via stepwise linear regression. *Clin. Pharmacol. Ther.* **56**, 217–222.

Mignani, S., and Rosa, R. (1995). The moving block bootstrap to assess the accuracy of statistical estimates in Ising model simulations. *Comput. Phys. Commun.* **92**, 203–213.

Mikheenko, S., Erofeeva, S., and Kosako, T. (1994). Reconstruction of dose distribution by the bootstrap method using limited measured data. *Radioisotopes* **43**, 595–604.

Milan, L., and Whittaker, J. (1995). Application of the parametric bootstrap to models that incorporate a singular value decomposition. *Appl. Statist.* **44**, 31–49.

Miller, A. J. (1990). *Subset Selection in Regression.* Chapman & Hall, London. *(444)

Miller, R. G., Jr. (1964). A trustworthy jackknife. *Ann. Math. Statist.* **39**, 1594–1605.

Miller, R. G., Jr. (1974) The jackknife—a review. *Biometrika* **61**, 1–17. *(445)

Miller, R. G., Jr. (1981a). *Survival Analysis.* Wiley, New York. *(446)

Miller, R. G., Jr. (1981b). *Simultaneous Statistical Inference*, 2nd ed. Springer-Verlag, New York. *(447)

Miller, R. G., Jr. (1986). *Beyond ANOVA, Basics of Applied Statistics.* Wiley, New York. *(448)

Miller, R. G., Jr. (1997). *Beyond ANOVA, Basics of Applied Statistics.* 2nd ed. Chapman & Hall, New York. *(449)

Milliken, G. A., and Johnson, D. E. (1984). *Analysis of Messy Data, Volume 1: Designed Experiments.* Wadsworth, Belmont, CA. *(450)

Milliken, G. A., and Johnson, D. E. (1989). *Analysis of Messy Data, Volume 2: Nonreplicated Experiments.* Van Nostrand Reinhold, New York. *(451)

Mitani, Y., Hamamoto, Y., and Tomita, S. (1995). Use of bootstrap samples in designing artificial neural network classifiers. In *IEEE International Conference on Neural Networks Proceedings*, **4**, 2103–2106.

Miyakawa, M. (1991). Resampling plan using orthogonal array and its application to influence analysis. *Rep. Statist. Appl. Res. Union Jpn. Sci. Eng.* **38/2**, 1–10.

Moeher, M. (1987). On the estimation of the expected actual error rate in sample-based multinomial classification. *Statistics* **18**, 599–612.

Monti, A. C. (1997). Empirical likelihood confidence regions in time series models. *Biometrika* **84**, 395–405.

Montvay, I. (1996). Statistics and internal quantum numbers in the statistical bootstrap approach. Hungarian Academy of Sciences, Central Research Institute for Physics, Budapest.

Mooijart, A. (1985). Factor analysis for non-normal variables. *Psychometrika* **50**, 323–342.

Mooney, C. Z. (1996). Bootstrap statistical inference: examples and evaluations for political science. *Am. J. Political Sci.* **40**, 570–602.   *(452)

Mooney, C. Z. (1997). *Monte Carlo Simulation. Quantitative Applications in the Social Sciences*, **116**. Sage Publications, Newbury Park, CA.   *(453)

Mooney, C. Z., and Duval, R. D. (1993). *Bootstrapping: A Nonparametric Approach to Statistical Inference. Quantitative Applications in the Social Sciences* **95**. Sage Publications, Newbury Park, CA.   *(454)

Mooney, C. Z., and Krause, G. (1998). Of silicon and political science: computationally intensive techniques of statistical estimation and inference. *Br. J. Political Sci.* **27**, 83–110.

Moreau, J. V., and Jain, A. K. (1987). The bootstrap approach to clustering. In *Pattern Recognition: Theory and Applications* (P. A. Devijver and J. Kittler, editors), 63–71. Springer-Verlag, Berlin.

Morey, M. J., and Schenck, L. M. (1984). Small sample behavior of bootstrapped and jackknifed regression estimators. *Proc. ASA Bus. Econ. Statist. Sec.*, 437–442.

Morgenthaler, S., and Tukey, J. W. (editors). (1991). *Configural Polysampling: A Route to Practical Robustness.* Wiley, New York.

Morton, S. C. (1990). Bootstrap confidence intervals in a complex situation: a sequential paired clinical trial. *Commun. Statist. Simult. Comput.* **19**, 181–195.

Mosbach, O. (1988). Bootstrap-verfahren in allgemeinen linearen modellen. Universitat Bremen, Fachbereich 03: Diplomarbeit.

Mosbach, O. (1992). One-step bootstrapping in generalized linear models. In *Bootstrapping and Related Techniques. Proceedings, Trier, FRG* (K.-H. Jockel, G. Rothe, and W. Sendler, editors). *Lecture Notes in Economics and Mathematical Systems,* **376**, 143–147. Springer-Verlag, Berlin.

Mossman, D. (1995). Resampling techniques in the analysis of non-binomial ROC data. *Med. Dec. Making* **15**, 358–366.

Moulton, L. H., and Zeger, S. L. (1991). Bootstrapping generalized linear models. *Comput. Statist. Data Anal.* **11**, 53–63.

Moulton, L. H., and Zeger, S. L. (1989). Analyzing repeated measures on generalized linear models via the bootstrap. *Biometrics* **45**, 381–394.

Mueller, L. D., and Altenberg, L. (1985). Statistical inference on measures of niche overlap. *Ecology* **66**, 1204–1210.

Mueller, L. D., and Wang, J. L. (1990). Bootstrap confidence intervals for effective doses in the probit model for dose–response data. *Biom. J.* **32**, 529–544.

Mueller, P. (1986). On selecting the set of regressors. In *Classification as a Tool of Research* (W. Gaul and M. Schader, editors), 331–338. North-Holland, Amsterdam.

Murthy, V. K. (1974). *The General Point Process: Applications to Structural Fatigue, Bioscience and Medical Research.* Addison-Wesley, Boston.   *(455)

Mykland, P. (1992). Asymptotic expansions and bootstrapping distribution for dependent variables: a martingale approach. *Ann. Statist.* **20**, 623–654.

Myoungshic, J. (1986). Bootstrap method for K-spatial medians. *J. Korean Statist.* **15**, 1–8.

Nagao, H. (1985). On the limiting distribution of the jackknife statistics for eigenvalues of a sample covariance matrix. *Commun. Statist. Theory Methods* **14**, 1547–1567.

Nagao, H. (1988). On the jackknife statistics for eigenvalues and eigenvectors of a correlation matrix. *Ann. Inst. Statist. Math.* **40**, 477–489.

Nagao, H., and Srivastava, M. S. (1992). On the distributions of some test criteria for a covariance matrix under local alternatives and bootstrap approximations. *J. Multivar. Anal.* **43**, 331–350.

Navidi, W. (1989). Edgeworth expansions for bootstrapping in regression models. *Ann. Statist.* **17**, 1472–1478.

Navidi, W. (1995). Bootstrapping a method of phylogenetic inference. *J. Statist. Plann. Inf.* **43**, 169–184.

Ndlovu, P. (1993). Classical and bootstrap estimates of heritability of milk yield in Zimbabwean Holstein cows. *J. Dairy Sci.* **76**, 2013–3024.

Nelder, J. A. (1996). Statistical computing. In *Advances in Biometry* (P. Armitage and H. A. David, editors), 201–212, Wiley, New York.

Nelson, L. S. (1994). Book Review of *Permutation Tests* (by P. I. Good), *J. Qual. Technol.* **26**, 325.

Nelson, R. D. (1992). Applications of stochastic dominance using truncation, bootstrapping and kernels. *Proc. ASA Bus. Econ. Statist. Sect.*, 88–93.

Nelson, W. (1990). *Accelerated Testing: Statistical Models, and Data Analyses.* Wiley, New York.   *(456)

Nemec, A. F. L., and Brinkhurst, R. O. (1988). Using the bootstrap to assess statistical significance in the cluster analysis of species abundance data. *Can. J. Fish. Aquat. Sci.* **45**, 971–975.

Newton, M. A., and Geyer, G. J. (1994). Bootstrap recycling: a Monte Carlo alternative to the nested bootstrap. *J. Am. Statist. Assoc.* **89**, 905–912.   *(457)

Newton, M. A., and Raftery, A. E. (1994). Approximate Bayesian inference with the weighted likelihood bootstrap (with discussion). *J. R. Statist. Soc. B* **56**, 3–48.

Niederreiter, H. (1992). *Random Number Generation and Quasi-Monte Carlo Methods.* CBMS-NSF Regional Conference Series in Applied Mathematics. SIAM, Philadelphia.   *(458)

Nigam, A. K., and Rao, J. N. K. (1996). On balanced bootstrap for stratified multistage samples. *Statist. Sin.* **6**, 199–214.   *(459)

Nirel, R. (1994). Bootstrap confidence intervals for the estimation of seeding effect in an operational period. *Statist. Environ.* **2**, 109 124.

Nishizawa, O., and Noro, H. (1995). Bootstrap statistics for velocity tomography: application of a new information criterion. *Geophys. Prospect.* **43**, 157–176.

Nivelle, F., Rouy, V., and Vergnaud, P. (1993). Optimal design of neural network using resampling methods. In *Proceedings of 6th International Conference, Neural Networks and their Industrial and Cognitive Applications*, 95–106.

Nokkert, J. H. (1985). Comparison of confidence regions for a linear functional relationship based on moment, jackknife and bootstrap estimators of an asymptotic covariance matrix. *Statist. Dec. Sci.* **I.2**, 207–212.

Nordgaard, A. (1990). On the resampling of stochastic processes using the bootstrap approach. *Liu-Tek-Lic-***1990:23**, Linkoping, Sweden.

Nordgaard, A. (1992). Resampling stochastic processes using a bootstrap approach. In *Bootstrapping and Related Techniques. Proceedings, Trier, FRG* (K.-H. Jockel, G.

Rothe, and W. Sendler, editors). *Lecture Notes in Economics and Mathematical Systems,* **376**, 181–183. Springer-Verlag, Berlin.

Noreen, E. (1989). *Computer-Intensive Methods for Testing Hypotheses.* Wiley, New York.   *(460)

Nychka, D. (1991). Choosing a range for the amount of smoothing in nonparametric regression. *J. Am. Statist. Assoc.* **86**, 653–664.

Oakley, E. H. N. (1996). Genetic programming, the reflection of chaos and the bootstrap: toward a useful test for chaos. In *Genetic Programming, Proceedings of the First Annual Conference 1996* (J. R. Koza, D. E. Goldberg, D. B. Dogel, and R. Riolo, editors), 175–181.

Obgonmwan, S.-M. (1985). Accelerated resampling codes with applications to likelihood. Ph. D. Thesis, Department of Mathematics, Imperial College, London. *(461)

Obgonmwan, S.-M., and Wynn, H. P. (1986). Accelerated resampling codes with low discrepancy. Department of Statistics and Actuarial Science, City University, London.

Obgonmwan, S.-M., and Wynn, H. P. (1988). Resampling generated likelihoods. In *Statistical Decision Theory and Related Topics IV* (S. S. Gupta and J. O. Berger, editors), 133–147. Springer-Verlag, New York.

Oden, N. L. (1991). Allocation of effort in Monte Carlo simulation for power of permutation tests. *J. Am. Statist. Assoc.* **86**, 1007–1012.

Oldford, R. W. (1985). Bootstrapping by Monte Carlo versus approximating the estimator and bootstrapping exactly. *Commun. Statist. Simul. Comput.* **14**, 395–424.

Olshen, R. A., Biden, E. N., Wyatt, M. P., and Sutherland, D. H. (1989). Gait analysis and the bootstrap. *Ann. Statist.* **17**, 1419–1440.   *(462)

Ong, H.-T., Iskander, D., and Zoubir, A. (1997). Detecting a common non-Gaussian signal in two sensors using the bootstrap. *IEEE Signal Processing Workshop on Higher Order Statistics. IEEE Signal Processing Society*, 463–467.

Ooms, M., and Franses, P. H. (1997). On periodic correlations between estimated seasonal and nonseasonal components in German and U. S. unemployment. *J. Bus. Econ. Statist.* **15**, 470–481.

O'Quigley, J., and Pessione, F. (1991). The problem of a covariate-time qualitative interaction in survival study. *Biometrics* **47**, 101–115.

O'Sullivan, F. (1988). Parameter estimation in parabolic and hyperbolic equations. Technical Report 127. Department of Statistics, University of Washington, Seattle.

Overton, W. S., and Stehman, S. V. (1994). Improvement of performance of variable probability sampling strategies through application of the population space and the facsimile population bootstrap. Oregon State University, Department of Statistics Technical Report.

Owen, A. B. (1988). Empirical likelihood ratio confidence intervals for a single functional. *Biometrika* **75**, 237–249.

Owen, A. B. (1990). Empirical likelihood ratio confidence regions. *Ann. Statist.* **18**, 90–120.

Owen, A. B. (1992). A central limit theorem for Latin hypercube sampling. *J. R. Statist. Soc. B* **54**, 541–551.   *(463)

Paass, G. (1994). Assessing predictive accuracy by the bootstrap algorithm. In *Proceed-*

ings of the International Conference on Artificial Neural Networks, ICANN '94 (M. Marinaro and P. G. Morasso, editors), **2**, 823–826.

Padgett, W. J., and Thombs, L. A. (1986). Smooth nonparametric quantile estimation under censoring: simulations and bootstrap methods. *Commun. Statist. Simul. Comput.* **15**, 1003–1025.

Padmanabhan, A. R., Chinchilli, V. M., and Babu, G. J. (1997). Robust analysis of within-unit variances in repeated measurement experiments. *Biometrics* **53**, 1520–1526.

Paez, T. L., and Hunter, N. F. (1998). Fundamental concepts of the bootstrap for statistical analysis of mechanical systems. *Exp. Techniques* **22**, 35–38.

Page, J. T. (1985). Error-rate estimation in discriminant analysis. *Technometrics* **27**, 189–198.   *(464)

Pallini, A., Carletti, M., and Pesarin, F. (1994). Quasi-Monte Carlo integration methods for the bootstrap. *Proc. Ital. Statist. Soc.* **2**, 441–448.

Pallini, A., and Pesarin, F. (1994). Calibration resampling for the conditional bootstrap. *Proc. Ital. Statist. Soc.* **2**, 473–480.

Papadopoulos, A. S., and Tiwari, R. C. (1989). Bayesian bootstrap lower confidence interval estimation of the reliability and failure rate. *J. Statist. Comput. Simul.* **32**, 185–192.

Paparoditis, E. (1992). Bootstrapping some statistics useful in identifying ARMA models. In *Bootstrapping and Related Techniques. Proceedings, Trier, FRG* (K.-H. Jockel, G. Rothe, and W. Sendler, editors). *Lecture Notes in Economics and Mathematical Systems,* **376**, 115–119. Springer-Verlag, Berlin.

Paparoditis, E. (1995). A frequency domain bootstrap-based method for checking the fit of a transfer function model. *J. Am. Statist. Assoc.* **91**, 1535–1550.

Pari, R., and Chatterjee, S. (1986). Using L_2 estimation for L_1 estimators: an application of the single-index model. *Dec. Sci.* **17**, 414–423.

Parmanto, B., Munro, P. W., and Doyle, H. R. (1996a). Improving committee diagnosis with resampling techniques. In *Proceedings of the 1995 Conference: Advances in Neural Information Processing Systems* (D. S. Touretzky, M. C. Mozer, and M. E. Hasselmo, editors), **8**, 882–888.

Parmanto, B., Munro, P. W., and Doyle, H. R. (1996b). Reducing variance of committee prediction with resampling techniques. *Connect. Sci.* **8**, 405–425.

Parr, W. C. (1983). A note on the jackknife, the bootstrap and the delta method estimators of bias and variance. *Biometrika* **70**, 719–722.   *(465)

Parr, W. C. (1985a). Jackknifing differentiable statistical functionals. *J. R. Statist. Soc. B* **47**, 56–66.

Parr, W. C. (1985b). The bootstrap: some large sample theory and connections with robustness. *Statist. Probab. Lett.* **3**, 97–100.

Parzen, E. (1982). Data modeling using quantile and density-quantile functions. In *Some Recent Advances in Statistics* (J. T. de Oliveira and B. Epstein, editors), 23–52. Academic Press, London.

Parzen, M. I., Wei, L. J., and Ying, Z. (1994). A resampling method based on pivotal estimating functions. *Biometrika* **81**, 341–350.

Pavia, E. G., and O'Brien, J. J. (1986). Weibull statistics of wind speed over the ocean. *J. Climatol. Meteorol.* **25**, 1324–1332.

Peck, R., Fisher, L., and Van Ness, J. (1989). Approximate confidence intervals for the number of clusters. *J. Am. Statist. Assoc.* **84**, 184–191.

Pederson, S. P., and Johnson, M. E. (1990). Estimating model discrepancy. *Technometrics* **32**, 305–314.

Peladeau, N., and Lacouture, Y. (1993). SIMSTAT: bootstrap computer simulation and statistical program for IBM personal computers. *Behav. Res. Methods Instrum. Comput.* **25**, 410–413.

Peters, S. C., and Freedman, D. A. (1984). Some notes on the bootstrap in regression problems. *J. Bus. Econ. Statist.* **2**, 401–409.   *(466)

Peters, S. C., and Freedman, D. A. (1985). Using the bootstrap to evaluate forecast equations. *J. Forecast.* **4**, 251–262.   *(467)

Peterson, A. V. (1983). Kaplan–Meier estimator. *Encyclopedia of Statistical Sciences,* **4**, 346–352, Wiley, New York.

Pettit, A. N. (1987). Estimates for a regression parameter using ranks. *J. R. Statist. Soc. B* **49**, 58–67.

Pewsey, A. (1994). Book Review of *Exploring the Limits of Bootstrap* (R. LePage and L. Billard, editors). *Statistician* **43**, 215–216.

Pham, T. D., and Nguyen, H. T. (1993). Bootstrapping the change-point of a hazard rate. *Ann. Inst. Statist. Math.* **45**, 331–340.

Picard, R. R., and Berk, K. N. (1990). Data splitting. *Am. Statist.* **44**, 140–147.

Pictet, O. V., Dacorogna, M. M., and Muller, U. A. (1998). Hill, bootstrap and jackknife estimators for heavy tails. In *A Practical Guide to Heavy Tails: Statistical Techniques and Applications* (R. J. Adler, R. E. Feldman, and M. S. Taqqu, editors), 283–310. Birkhauser, Boston.   *(468)

Pigeot, I. (1992). Jackknifing eastimators of a common odds ratio from several 2×2 tables. In *Bootstrapping and Related Techniques. Proceedings, Trier, FRG* (K.-H. Jockel, G. Rothe, and W. Sendler, editors). *Lecture Notes in Economics and Mathematical Systems,* **376**, 204–212. Springer-Verlag, Berlin.

Pigeot, I. (1994). Special resampling techniques in categorical Sara analysis. In *25th Conference on Statistical Computing: Computational Statistics* (P. Dirschedl and R. Ostermann, editors), 159–176. Physica-Verlag, Heidelberg.

Pinheiro, J. C., and DeMets, D. L. (1997). Estimating and reducing bias in group sequential designs with Gaussian independent increment structure. *Biometrika* **84**, 831–845.

Pitt, D. G., and Kreutzweiser, D. P. (1998). Applications of computer-intensive statistical methods to environmental research. *Ecotoxicol. Environ. Saf.* **39**, 78–97.

Platt, C. A. (1982). Bootstrap stepwise regression. *Proc. ASA Bus. Econ. Statist. Sec.* 586–589.

Plotnick, R. E. (1989). Application of bootstrap methods to reduced major axis line fitting. *Syst. Zool.* **38**, 144–153.

Politis, D. N. (1998). Computer-intensive methods in statistical analysis. *IEEE Signal Process. Mag.* **15**, 39–55.   *(469)

Politis, D. N., and Romano, J. P. (1990). A nonparametric resampling procedure for multivariate confidence regions in time series analysis. *Proceedings of INTERFACE' 90, 22nd Symposium on the Interface of Computer Science and Statistics* (C. Page and R. LePage, editors). Springer-Verlag, New York.

Politis, D. N., and Romano, J. P. (1992a). A circular block-resampling procedure for stationary data. In *Exploring the Limits of Bootstrap* (R. LePage and L. Billard, editors), 263–270, Wiley, New York.   *(470)

Politis, D. N., and Romano, J. P. (1992b). A general resampling scheme for triangular arrays of α-mixing random variables with application to the problem of spectral density estimation. *Ann. Statist.* **20**, 1985–2007.

Politis, D. N., and Romano, J. P. (1993a). Estimating the distribution of a studentized statistic by subsampling. *Bull. Int. Statist. Inst. 49th Session* **2**, 315–316.   *(471)

Politis, D. N., and Romano, J. P. (1993b). Nonparametric resampling for homogeneous strong mixing random fields. *J. Multivar. Anal.* **47**, 301–328.

Politis, D. N., and Romano, J. P. (1994a). The stationary bootstrap. *J. Am. Statist. Assoc.* **89**, 1303–1313.   *(472)

Politis, D. N., and Romano, J. P. (1994b). Large sample confidence regions based on subsamples under minimal assumptions. *Ann. Statist.* **22**, 2031–2050.   *(473)

Politis, D. N., and Romano, J. P. (1994c). Limit theorems for weakly dependent Hilbert space valued random variables with applications to the stationary bootstrap. *Statist. Sin.* **4**, 461–476.

Politis, D. N., Romano, J. P., and Lai, T. L. (1992). Bootstrap confidence bands for spectra and cross-spectra. *IEEE Trans. Signal Process.* **40**, 1206–1215.   *(474)

Pollack, S., Bruce, P., Borenstein, M., and Lieberman, J. (1994). The resampling method of statistical analysis. *Psychopharmacol. Bull.* **30**, 227–234.

Pollack, S., Simon, J., Bruce, P., Borenstein, M., and Lieberman, J. (1994). Using, teaching, and evaluating the resampling method of statistical analysis. *Psychopharmacol. Bull.* **30**, 120.   *(475)

Pons, O., and de Turckheim, E. (1991a). von Mises methods, bootstrap and Hadamard differentiability for nonparametric general models. *Statistics* **22**, 205–214.

Pons, O., and de Turckheim, E. (1991b). Tests of independence for bivariate censored data based on the empirical joint hazard function. *Scand. J. Statist.* **18**, 21–37.

Portnoy, S. (1984). Tightness of the sequence of empiric c.d.f. processes defined from regression fractiles. In *Robust and Nonlinear Time Series Analysis* (J. Franke, W. Hardle, and R. D. Martin, editors), 231–246. Springer-Verlag, New York.

Prada-Sanchez, J. M., and Cotos-Yanez, T. (1997). A simulation study of iterated and non-iterated bootstrap methods for bias reduction and confidence interval estimation. *Commun. Statist. Simul. Comput.* **26**, 927–946.

Praestgaard, J., and Wellner, J. A. (1993). Exchangeably weighted bootstraps of the general empirical process. *Ann. Probab.* **21**, 2053–2086.

Praskova, Z. (1992). Empirical Edgeworth expansion and bootstrap in AR(1) models. In *11th Prague Conference on Information Theory, Statistical Decision Functions and Random Processes,* **B**, 281–294. Kluwer Academic Publishers, Boston.

Presnell, B., and Booth, J. G. (1994). Resampling methods for sample surveys. Technical Report 470, Department of Statistics, University of Florida, Gainsville.   *(476)

Presnell, B., Morrison, S. P., and Littell, R. C. (1998). Projected multivariate linear models for directional data. *J. Am. Statist. Assoc.* **93**, 1068–1077.

Press, S. J. (1989). *Bayesian Statistics: Principles, Models, and Applications.* Wiley, New York.   *(477)

Price, B., and Price, K. (1992). Sampling variability of capability indices. Technical Report, Wayne State University, Detroit.   *(478)

Priestley, M. B. (1981). *Spectral Analysis and Time Series.* Academic Press, London. *(479)

Proenca, I. (1990). Metodo bootstrap—Uma aplicacao na estimacao e previsao em modelos dinamicos. Unpublished M.Sc. dissertation, ISEG, Lisboa.

Pugh, G. A. (1995). Resampled confidence bounds on effects diagrams for signal-to-noise. *Comput. Ind. Eng.* **29**, 11–13.

Qin, J., and Zhang, B. (1997). A goodness-of-fit test for logistic regression models based on case–control data. *Biometrika* **84**, 609–618.

Quan, H., and Tsai, W.-Y. (1992). Jackknife for the proportional hazards model. *J. Statist. Comput. Simul.* **43**, 163–176.

Quenneville, B. (1986). Bootstrap procedures for testing linear hypotheses without normality. *Statistics* **17**, 533–538.

Quenouille, M. H. (1949). Approximate tests of correlation in time series. *J. R. Statist. Soc. B* **11**, 18–84.   *(480)

Quenouille, M. H. (1956). Notes on bias in estimation. *Biometrika* **43**, 353–360.

Racine, J. (1997). Consistent significance testing for nonparametric regression. *J. Bus. Econ. Statist.* **15**, 369–378.

Raftery, A. E. (1995). Hypothesis testing and model selection. In *Markov Chain Monte Carlo in Practice* (W. R. Gilks, S. Richardson, and D. J. Spiegelhalter, editors), 163–187. Chapman & Hall, London.

Rajarshi, M. B. (1990). Bootstrap in Markov sequences based on estimate of transition density. *Ann. Inst. Math. Statist.* **42**, 253–268.

Ramos, E. (1988). Resampling methods for time series. Ph.D. thesis, Department of Statistics, Harvard University.

Rao, C. R., Pathak, P. K., and Koltchinskii, V. I. (1997). Bootstrap by sequential resampling. *J. Statist.Plann. Inf.* **64**, 257–281.

Rao, C. R., and Zhao, L. (1992). Approximation to the distribution of M estimates in linear models by randomly weighted bootstrap. *Sankhya A* **54**, 323–331.

Rao, J. N. K., Kovar, J. G., and Mantel, H. J. (1990). On estimating distribution functions and quantiles from survey data using auxiliary information. *Biometrika* **77**, 365–375.

Rao, J. N. K., and Wu, C. F. J. (1985). Inference from stratified samples: second-order analysis of three methods for nonlinear statistics. *J. Am. Statist. Assoc.* **80**, 620–630.

Rao, J. N. K., and Wu, C. F. J. (1987). Methods for standard errors and confidence intervals from sample survey data: some recent work. *Bull. Int. Statist. Inst.* **52**, 5–21.

Rao, J. N. K., and Wu, C. F. J. (1988). Resampling inference with complex survey data. *J. Am. Statist. Assoc.* **83**, 231–241.   *(481)

Rao, J. N. K., Wu, C. F. J., and Yuen, K. (1992). Some recent work on resampling methods for complex surveys. *Surv. Methodology* **18**, 209–217.

Rasmussen, J. L. (1987a). Estimating correlation coefficients: bootstrap and parametric approaches. *Psychol. Bull.* **101**, 136–139.

Rasmussen, J. L. (1987b). Parametric and bootstrap approaches to repeated measures designs. *Behav. Res. Methods Instrum.* **19**, 357–360.

Raudys, S. (1988). On the accuracy of a bootstrap estimate of classification error. In *Proceedings on the 9th International Conference on Pattern Recognition*, 1230–1232.

Rayner, R. K. (1990a). Bootstrapping *p* values and power in the first-order autoregression: a Monte Carlo investigation. *J. Bus. Econ. Statist.* **8**, 251–263.

Rayner, R. K. (1990b). Bootstrap tests for generalized least squares regression models. *Econ. Lett.* **34**, 261–265.

Rayner, R. K., and Dielman, T. E. (1990). Use of the bootstrap in tests for serial correlation when regressors include lagged dependent variables. Unpublished Technical Report.   *(482)

Red-Horse, J. R., and Paez, T. L. (1998). Assessment of probability models using the bootstrap sampling method. *39th AIAA/ASME/ASCE/AHS/ASC Structures, Structural Dynamics and Materials Conference and Exhibit, Collection of Technical Papers Part 2, AIAA*, 1086–1091.

Reiczigel, J. (1996). Bootstrap tests in correspondence analysis. *Appl. Stoch. Models Data Anal.* **12**, 107–117.

Reid, N. (1981). Estimating the median survival time. *Biometrika* **68**, 601–608.   *(483)

Reid, N. (1988). Saddlepoint methods and statistical inference (with discussion). *Statist. Sci.* **3**, 213–238.   *(484)

Reiss, R.-D. (1989). *Approximate Distributions of Order Statistics with Applications to Nonparametric Statistics.* Springer-Verlag, New York.   *(485)

Reiss, R.-D., and Thomas, M. (1997). *Statistical Analysis on Extreme Values with Applications to Insurance, Finance, Hydrology and Other Fields.* Birkhauser Verlag, Basel.   *(486)

Reneau, D. M., and Samaniego, F. J. (1990). Estimating the survival curve when new is better than used of a specified age. *J. Am. Statist. Assoc.* **85**, 123–131.

Resampling Stats, Inc. (1997). *Resampling Stats User's Guide* (Peter Bruce, Julian Simon, and Terry Oswald, authors).   *(487)

Resnick, S. I. (1987). *Extreme Values, Regular Variation and Point Processes.* Springer-Verlag, New York.   *(488)

Resnick, S. I. (1997). Heavy tail modeling and teletraffic data (with discussion). *Ann. Statist.* **25**, 1805–1869.

Rey, W. J. J. (1983). *Introduction to Robust and Quasi-Robust Statistical Methods.* Springer-Verlag, Berlin.   *(489)

Rice, R. E., and Moore, A. H. (1983). A Monte-Carlo technique for estimating lower confidence limits on system reliability using pass-fail data. *IEEE Trans. Reliab.* **R-32**, 366–369.

Rieder, H. (editor) (1996). *Robust Statistics, Data Analysis, and Computer Intensive Methods. Lecture Notes in Statistics*, **109**. Springer-Verlag, Heidelberg.

Riemer, S., and Bunke, O. (1983). A note on bootstrap and other empirical procedures for testing linear hypotheses without normality. *Statistics* **14**, 517–526.

Ringrose, T. J. (1994). Bootstrap confidence regions for canonical variate analysis. In *Proceedings in Computational Statistics, 11th Symposium*, 343–348.

Ripley, B. D. (1981). *Spatial Statistics.* Wiley, New York.   *(490)

Ripley, B. D. (1987). *Stochastic Simulation.* Wiley, New York.   *(491)

Ripley, B. D. (1988). *Statistical Inference for Spatial Processes.* Cambridge University Press, Cambridge.   *(492)

Ripley, B. D. (1992). Applications of Monte Carlo methods in spatial and image analysis. In *Bootstrapping and Related Techniques. Proceedings, Trier, FRG* (K.-H. Jockel, G. Rothe, and W. Sendler, editors). *Lecture Notes in Economics and Mathematical Systems,* **376**, 47–53. Springer-Verlag, Berlin.

Rissanen, J. (1986). Stochastic complexity and modeling. *Ann. Statist.* **14**, 1080–1100.

Rissanen, J. (1989). *Stochastic Complexity in Statistical Inquiry.* World Scientific Publishing, Singapore.   *(493)

Roberts, F. S. (1984). *Applied Combinatorics.* Prentice-Hall, Englewood Cliffs, NJ.   *(494)

Robeson, S. M. (1995). Resampling of network-induced variability in estimates of terrestrial air temperature change. *Climatic Change* **29**, 213–229.   *(495)

Robinson, J. A. (1983). Bootstrap confidence intervals in location-scale model with progressive censoring. *Technometrics* **25**, 179–187.

Robinson, J. A. (1986). Bootstrap and randomization confidence intervals. In *Pacific Statistical Congress: Proceedings of the Congress* (I. S. Francis, B. F. J. Manly, and F. C. Lam, editors), 49–50. North-Holland, Amsterdam.

Robinson, J. A. (1987). Nonparametric confidence intervals in regression, the bootstrap and randomization methods. In *New Perspectives in Theoretical and Applied Statistics* (M. Puri, J. P. Vilaplana, and W. Wertz, editors), 243–256, Wiley, New York.

Robinson, J. A. (1994). Book Review of *An Introduction to the Bootstrap* (by B. Efron and R. J. Tibshirani). *Austral. J. Statist.* **36**, 380–382.

Robinson, J., Feuerverger, A., and Jing, B.-Y. (1994). On the bootstrap saddlepoint approximations. *Biometrika* **81**, 211–215.

Rocke, D. M. (1989). Bootstrapping Bartlett's adjustment in seemingly unrelated regression. *J. Am. Statist. Assoc.* **84**, 598–601.

Rocke, D. M. (1993). Almost-exact parametric bootstrap calibration via the saddlepoint approximation. *Comput. Statist. Data Anal.* **15**, 179–198.

Rocke, D. M., and Downs, G. W. (1981). Estimating the variance of robust estimators of location: influence curve, jackknife and bootstrap. *Commun. Statist. Simul. Comput.* **10**, 221–248.

Rodriguez-Campos, M. C., and Cao-Abad, R. (1993). Nonparametric bootstrap confidence intervals for discrete regression functions. *J. Econometrics* **58**, 207–222.

Romano, J. P. (1988a). A bootstrap revival of some nonparametric distance tests. *J. Am. Statist. Assoc.* **83**, 698–708.   *(496)

Romano, J. P. (1988b). On weak convergence and optimality of kernel density estimates of the mode. *Ann. Statist.* **16**, 629–647.

Romano, J. P. (1988c). Bootstrapping the mode. *Ann. Inst. Statist. Math.* **40**, 565–586.   *(497)

Romano, J. P. (1989a). Do bootstrap confidence procedures behave well uniformly in *P*? *Can. J. Statist.* **17**, 75–80.

Romano, J. P. (1989b). Bootstrap and randomization tests of some nonparametric hypotheses. *Ann. Statist.* **17**, 141–159.   *(498)

Romano, J. P. (1990). On the behavior of randomization tests of some nonparametric hypotheses. *J. Am. Statist. Assoc.* **85**, 686–692.

Romano, J. P., and Siegel, A. F. (1986). *Counterexamples in Probability and Statistics*. Wadsworth, Monterey, CA.

Romano, J. P., and Thombs, L. A. (1996). Inference for autocorrelations under weak assumptions. *J. Am. Statist. Assoc.* **91**, 590–600.

Rosa, R., and Mignani, S. (1994). Moving block bootstrap for dependent data: an application to Ising models (Italian). *Proc. Ital. Statist. Soc.* **2**, 465–472.

Rosen, O., and Cohen, A. (1995). Constructing a bootstrap confidence interval for the unknown concentration in radioimmunoassay. *Statist. Med.* **14**, 935–952.   *(499)

Ross, S. M. (1990). *A Course in Simulation*. Macmillan, New York.

Rothe, G. (1986a). Some remarks on bootstrap techniques for constructing confidence intervals. *Statist. Hefte* **27**, 165–172.

Rothe, G. (1986b). Bootstrap in generalisierten linearen modellen. *ZUMA-Arbeitsbericht* **86-11**, Mannheim.

Rothe, G., and Armingeer, G. (1992). Bootstrap for mean and covariance structure models. In *Bootstrapping and Related Techniques. Proceedings, Trier, FRG* (K.-H. Jockel, G. Rothe, and W. Sendler, editors). *Lecture Notes in Economics and Mathematical Systems, 376*, 149–155. Springer-Verlag, Berlin.

Rothery, P. (1985). Estimation of age-specific survival in hen harriers (circus c. cyaneus) in Orkney. In *Statistics in Ornithology* (B. J. T. Morgan and P. M. North, editors), 341–354. Springer-Verlag, Berlin.   *(500)

Rousseeuw, P. J. (1984). Least median of squares regression. *J. Am. Statist. Assoc.* **79**, 872–880.   *(501)

Rousseeuw, P. J., and Leroy, A. M. (1987). *Robust Regression and Outlier Detection*. Wiley, New York.   *(502)

Rousseeuw, P. J., and Yohai, V. (1984). Robust regression by means of s-estimators. In *Robust and Nonlinear Time Series Analysis* (J. Franke, W. Hardle, and R. D. Martin, editors), 256–272. Springer-Verlag, New York.

Roy, T. (1994). Bootstrap accuracy for non-linear regression models. *J. Chemometrics* **8**, 37–44.   *(503)

Rubin, D. B. (1981). The Bayesian bootstrap. *Ann. Statist.* **9**, 130–134.   *(504)

Rubin, D. B. (1983). A case study of the robustness of Bayesian methods of inference: estimating the total in a finite population using transformations to normality. In *Scientific Inference, Data Analysis, and Robustness* (G. E. P. Box, T. Leonard, and C.-F. Wu, editors), 213–244. Academic Press, London.

Rubin, D. B. (1987). *Multiple Imputation for Nonresponse in Surveys*. Wiley, New York.   *(505)

Rubin, D. B. (1996). Multiple imputation after 18+ years. *J. Am. Statist. Assoc.* **91**, 473–489.   *(506)

Rubin, D. B., and Schenker, N. (1986). Multiple imputation for interval estimation from simple random samples with ignorable nonresponse. *J. Am. Statist. Assoc.* **81**, 366–374.   *(507)

Rubin, D. B., and Schenker, N. (1991). Multiple imputation in health-care data bases: an overview and some applications. *Statist. Med.* **10**, 585–598.   *(508)

Rubin, D. B., and Schenker, N. (1998). Imputation. In *Encyclopedia of Statistical Sciences, Update Volume 2* (S. Kotz, C. B. Read, and D. L. Banks, editors), 336–342, Wiley, New York.   *(509)

Runkle, D. E. (1987). Vector autoregressions and reality (with discussion). *J. Bus. Econ. Statist.* **5**, 437–453.

Rust, K. (1985). Variance estimation for complex estimators in sample surveys. *J. Off. Statist.* **1**, 381–397.

Ryan, T. P. (1989). *Statistical Methods for Quality Improvement.* Wiley, New York. *(510)

Sager, T. W. (1986). Dimensionality reduction in density estimation. In *Statistical Image Processing and Graphics* (E. J. Wegman and D. J. DePriest, editors), 307–319. Marcel Dekker, New York.

Sain, S. R., Baggeily, K. A., and Scott, D. W. (1994). Cross-validation of multivariate densities. *J. Am. Statist. Assoc.* **89**, 807–817.

Samawi, H. M. (1994). Power estimation for two-sample tests using importance and antithetic resampling. Ph. D. thesis. Department of Actuarial Science, University of Iowa.

Sanderson, M. J. (1989). Confidence limits on phylogenies: the bootstrap revisited. *Cladistics* **5**, 113–129.   *(511)

Sanderson, M. J. (1995). Objections to bootstrapping phylogenies: a critique. *Syst. Biol.* **44**, 299–320.   *(512)

Sauerbrei, W., and Schumacher, M. (1992). A bootstrap resampling procedure for model building: application to the Cox regression model. *Statist. Med.* **11**, 2093–2109. *(513)

Sauerbrei, W. (1998). Bootstrapping in survival analysis. In *Encyclopedia of Biostatistics,* **1**, 433–436 (Peter Armitage and Theodore Colton, editors). Wiley, New York.   *(514)

Sauermann, W. (1986). Bootstrap — Verfahren in log-linearen Modellen. Dissertation. Universitat Heidelberg.

Sauermann, W. (1989). Bootstrapping the maximum likelihood estimator in high-dimensional log-linear models. *Ann. Statist.* **17**, 1198–1216.

Schader, R. M., and McKean, J. W. (1987). Small sample properties of least absolute errors. Analysis of variance. In *Statistical Data Analysis Based on the L_1 Norm and Related Methods* (Y. Dodge, editor), 307–321. North-Holland, Amsterdam.

Schafer, H. (1992). An application of the bootstrap in clinical chemistry. In *Bootstrapping and Related Techniques. Proceedings, Trier, FRG.* (K.-H. Jockel, G. Rothe, and W. Sendler, editors). *Lecture Notes in Economics and Mathematical Systems,* **376**, 213–217. Springer-Verlag, Berlin.   *(515)

Schemper, M. (1987a). Nonparametric estimation of variance, skewness and kurtosis of the distribution of a statistic by jackknife and bootstrap techniques. *Statist. Neerlandica* **41**, 59–64.

Schemper, M. (1987b). On bootstrap confidence limits for possibly skew distributed statistics. *Commun. Statist. Theory Methods* **16**, 1585–1590.

Schemper, M. (1987c). One-and two-sample tests of Kendall's τ. *Biometric J.* **29**, 1003–1009.

Schenker, N. (1985). Qualms about bootstrap confidence intervals. *J. Am. Statist. Assoc.* **80**, 360–361.   *(516)

Schervish, M. J. (1995). *Theory of Statistics.* Springer-Verlag, New York.   *(517)

Schork, N. (1992). Bootstrapping likelihood ratios in quantitative genetics. In *Exploring* *the Limits of Bootstrap* (R. LePage and L. Billard, editors), 389–396, Wiley, New York.   *(518)

Schluchter, M. D., and Forsythe, A. B. (1986). A caveat on the use of a revised bootstrap algorithm. *Psychometrika* **51**, 603–605.

Schucany, W. R. (1988). Sample reuse. *Encyclopedia of Statistical Sciences*, **8**, 235–238, Wiley, New York.

Schucany, W. R., and Bankson, D. M. (1989). Small sample variance estimators for *U*-statistics. *Aust. J. Statist.* **31**, 417–426.

Schucany, W. R., and Sheather, S. J. (1989). Jackknifing *R*-estimators. *Biometrika* **76**, 393–398.

Schucany, W. R., and Wang, S. (1991). One-step bootstrapping for smooth iterative procedures. *J. R. Statist. Soc. B* **53**, 587–596.

Schumacher, M., Hollander, N., and Sauerbrei, W. (1997). Resampling and cross-validation techniques. A tool to reduce bias caused by model building? *Statist. Med.* **16**, 2813–2828.

Schuster, E. F. (1987). Identifying the closest symmetric distribution or density function. *Ann. Statist.* **15**, 865–874.

Schuster, E. F., and Barker, R. C. (1989). Using the bootstrap in testing symmetry versus asymmetry. *Commun. Statist. Simul. Comput.* **16**, 69–84.

Scott, D. W. (1992). *Mutivariate Density Estimation: Theory, Practice, and Visualization.* Wiley, New York.   *(519)

Seber, G. A. F. (1984). *Multivariate Observations.* Wiley, New York.   *(520)

Seki, T., and Yokoyama, S. (1996). Robust parameter-estimation using the bootstrap method for the 2—parameter Weibull distribution. *IEEE Trans. Reliab.* **45**, 34–41.

Sen, A., and Srivastava, M. S. (1990). *Regression Analysis: Theory, Methods, and Applications.* Springer-Verlag, New York.   *(521)

Sen, P. K. (1988a). Functional approaches in resampling plans: a review of some recent developments. *Sankhya A* **50**, 394–435.

Sen, P. K. (1988b). Functional jackknifing: rationality and general asymptotics. *Ann. Statist.* **16**, 450–469.   *(522)

Sen, P. K. (1993). Multivariate L_1-norm estimation and the vulnerable bootstrap. In *3rd Pacific Area Statistical Conference, Makuhari, Japan,* 441–450.

Seppala, T., Moskowitz, H., Plante, R., and Tang, J. (1995). Statistical process control via the subgroup bootstrap. *J. Qual. Technol.* **27**, 139–153.   *(523)

Serfling, R. J. (1980). *Approximation Theorems of Mathematical Statistics.* Wiley, New York.   *(524)

Sezgin, N and Bar-Ness, Y. (1996). Adaptive soft limiter bootstrap separator for one-shot asynchronous CDMA channel with singular partial cross-correlation matrix. In *IEEE International Conference on Communications, Converging Technologies for Tomorrow's Applications. ICC '96*, **1**, 73–77.

Shao, J. (1987a). Bootstrap variance estimations. Ph. D. dissertation, Department of Statistics, University of Wisconsin, Madison.

Shao, J. (1987b). Sampling and resampling: an efficient approximation to jackknife variance estimators in linear models. *Chin. J. Appl. Probab. Statist.* **3**, 368–379.

Shao, J. (1988a). On resampling methods for variance and bias estimation in linear models. *Ann. Statist.* **16**, 986–1008.   *(525)

Shao, J. (1988b). Bootstrap variance and bias estimation in linear models. *Can. J. Statist.* **16**, 371–382.   *(526)

Shao, J. (1989). Bootstrapping for generalized *L*-statistics. *Commun. Statist. Theory Methods* **18**, 2005–2016.

Shao, J. (1990a). Influence function and variance estimation. *Chin. J. Appl. Probab. Statist.* **6**, 309–315.

Shao, J. (1990b). Bootstrap estimation of the asymptotic variance of statistical functionals. *Ann. Inst. Statist. Math.* **42**, 737–752.

Shao, J. (1992). Bootstrap variance estimators with truncation. *Statist. Probab. Lett.* **15**, 95–101.

Shao, J. (1993). Linear model selection by cross-validation. *J. Am. Statist. Assoc.* **88**, 486–494.

Shao, J. (1994). Bootstrap sample size in nonregular cases. *Proc. Am. Math. Soc.* **122**, 1251–1262.

Shao, J. (1996). Bootstrap model selection. *J. Am. Statist. Assoc.* **91**, 655–665.   *(527)

Shao, J., and Sitter, R. R. (1996). Bootstrap for imputed survey data. *J. Am. Statist. Assoc.* **91**, 1278–1288.   *(528)

Shao, J., and Tu, D. (1995). *The Jackknife and Bootstrap.* Springer-Verlag, New York.   *(529)

Shao, J., and Wu, C. F. J. (1987). Heteroscedasticity-robustness of jackknife variance estimators in linear models. *Ann. Statist.* **15**, 1563–1579.

Shao, J., and Wu, C. F. J. (1989). A general theory for jackknife variance estimation. *Ann. Statist.* **17**, 1176–1197.

Shao, J., and Wu, C. F. J. (1992). Asymptotic properties of the balanced repeated replication method for sample quantiles. *Ann. Statist.* **20**, 1571–1593.

Shao, Q., and Yu. H. (1993). Bootstrapping the sample means for stationary mixing sequences. *Stoch. Proc. Their Appl.* **48**, 175–190.   *(530)

Shaw, F. H., and Geyer, C. J. (1997). Estimation and testing in constrained covariance component models. *Biometrika* **84**, 95–102.

Sheather, S. J. (1987). Assessing the accuracy of the sample median: estimated standard errors versus interpolated confidence intervals. In *Statistical Data Analysis Based on the L_1 Norm and Related Methods* (Y. Dodge, editor), 203–215. North-Holland, Amsterdam.

Sheather, S. J., and Marron, J. S. (1990). Kernel quantile estimators. *J. Am. Statist. Assoc.* **85**, 410–416.

Shen, C. F., and Iglewicz, B. (1994). Robust and bootstrap testing procedures for bioequivalence. *J. Biopharmacol. Stat.* **4**, 65–90.   *(531)

Sherman, M. (1996). Variance estimation for statistics computed from spatial lattice data. *J. R. Statist. Soc. B* **58**, 509–523.

Sherman, M. (1997). Subseries methods in regression. *J. Am. Statist. Assoc.* **92**, 1041–1048.

Sherman, M., and Carlstein, E. (1997). Omnibus confidence intervals. Preprint. Department of Statistics, Texas A&M University.   *(532)

Sherman, M., and Carlstein, E. (1996). Replicate histograms. *J. Am. Statist. Assoc.* **91**, 566–576.

Sherman, M., and le Cessie, S. (1997). A comparison between bootstrap methods and generalized estimating equations for correlated outcomes in generalized linear models. *Commun. Statist. Simul. Comput.* **26**, 901–925.

Shi, X. (1984). The approximate independence of jackknife pseudo-values and the bootstrap methods. *J. Wuhan Inst. Hydra. Elect. Eng.* **2**, 83–90.

Shi, X. (1986a). A note on bootstrapping *U*-statistics. *Chin. J. Appl. Probab. Statist.* **2**, 144–148.

Shi, X. (1986b). Bootstrap estimate for *m*-dependent sample means. *Kexue Tongbao (Chin. Bull. Sci.)* **31**, 404–407.

Shi, X. (1987). Some asymptotic properties of bootstrapping *U*-statistics. *J. Sys. Sci. Math. Sci.* **7**, 23–26.

Shi, X. (1991). Some asymptotic results for jackknifing the sample quantile. *Ann. Statist.* **19**, 496–503.

Shi, X., and Liu, K. (1992). Resampling method under dependent models. *Chin. Ann. Math. B* **13**, 25–34.

Shi, X., Tang, D., and Zhang, Z. (1992). Bootstrap interval estimation for reliability indices from type II censored data (in Chinese). *J. Sys. Sci. Math. Sci.* **12**, 215–220.

Shi, X., and Shao, J. (1988). Resampling estimation when the observations are *m*-dependent. *Commun. Statist. Theory Methods* **17**, 3923–3934.

Shi, X., Wu, C. F. J., and Chen, J. (1990). Weak and strong representations for quantile processes from finite populations with application to simulation size in resampling inference. *Can. J. Statist.* **18**, 141–148.

Shimabukuro, F. I., Lazar, S., Dyson, H. B., and Chernick, M. R. (1984). A quasi-optical method for measuring the complex permittivity of materials. *IEEE Trans. Microwave Theory Tech.* **32**, 659–665.   *(533)

Shipley, B. (1996). Exploratory path analysis using the bootstrap with applications in ecology and evolution. *Bull. Ecol. Soc. Am. Suppl. Part 2* **77**, 406.   *(534).

Shiue, W.-K., Xu, C.-W., and Rea, C. B. (1993). Bootstrap confidence intervals for simulation outputs. *J. Statist. Comput. Simul.* **45**, 249–255.

Shorack, G. R. (1982). Bootstrapping robust regression. *Commun. Statist. Theory Methods* **11**, 961–972.   *(535)

Shorack, G. R., and Wellner, J. A. (1986). *Empirical Processes with Applications to Statistics*. Wiley, New York.   *(536)

Siegel, A. F. (1986). Rarefaction curves. *Encyclopedia of Statistical Sciences*, **7**, 623–626.

Silverman, B. W. (1981). Using kernel density estimates to investigate multimodality. *J. R. Statist. Soc. B* **43**, 97–99.

Silverman, B. W. (1986). *Density Estimation for Statistics and Data Analysis*. Chapman & Hall, London.   *(537)

Silverman, B. W., and Young, G. A. (1987). The bootstrap: to smooth or not to smooth? *Biometrika* **74**, 469–479.   *(538)

Simar, L., and Wilson, P. W. (1998). Sensitivity analysis of efficiency scores: how to bootstrap in nonparametric frontier models. *Manage. Sci.* **44**, 49–61.

Simon. J. L. (1969). *Basic Research Methods in Social Science*. Random House, New York.   *(539)

Simon. J. L., and Bruce, P. (1991). Resampling: a tool for everyday statistical work. *Chance* **4**, 22–32.   *(540)

Simon. J. L., and Bruce, P. (1995). The new biostatistics of resampling. *MD Comput.* **12**, 115–121.   *(541)

Simonoff, J. S. (1986). Jackknifing and bootstrapping goodness-of-fit statistics in sparse multinomials. *J. Am. Statist. Assoc.* **81**, 1005–1011.   *(542)

Simonoff, J. S. (1994a). Book Review of *Computer Intensive Statistical Methods: Validation, Model Selection and Bootstrap* (by J. S. U. Hjorth). *J. Am. Statist. Assoc.* **89**, 1559–1560.

Simonoff, J. S. (1994b). Book Review of *An Introduction to the Bootstrap* (by B. Efron and R. Tibshirani), *J. Am. Statist. Assoc.* **89**, 1559–1560.

Simonoff, J. S. (1996). *Smoothing Methods in Statistics.* Springer-Verlag, New York.   *(543)

Simonoff, J. S., and Reiser, Y. B. (1986). Alternative estimation procedures for $\Pr(X < Y)$ in categorized data. *Biometrics* **42**, 895–907.

Simonoff, J. S., and Tsai, C.-L. (1988). Jackknife and bootstrapping quasi-likelihood. *J. Statist. Comput. Simul.* **30**, 213–232.

Singh, K. (1981). On the asymptotic accuracy of Efron's bootstrap. *Ann. Statist.* **9**, 1187–1195.   *(544)

Singh, K. (1996). Breakdown theory for bootstrap quantiles. TR # 96-015. Department of Statistics, Rutgers University.

Singh, K. (1997). Book Review of *The Jackknife and Bootstrap* (by J. Shao and D. Tu), *J. Am. Statist. Assoc.* **92**, 1214.

Singh, K., and Babu, G. J. (1990). On the asymptotic optimality of the bootstrap. *Scand. J. Statist.* **17**, 1–9.

Singh, K., and Liu, R. Y. (1990). On the validity of the jackknife procedure. *Scand. J. Statist.* **17**, 11–21.

Sinha, A. L. (1983). Bootstrap algorithms for parameter and smoothing state estimation. *IEEE Trans. Aerosp. Electron. Syst.* **19**, 85–88.

Sitnikova, T. (1996). Bootstrap method of interior-branch test for phylogenetic trees. *Mol. Biol. Evol.* **13**, 605–611.   *(545)

Sitnikova, T., Rzhetsky, A., and Nei, M. (1995). Interior-branch and bootstrap tests of phylogenetic trees. *Mol. Biol. Evol.* **12**, 319–333.   *(546)

Sitter, R. R. (1992a). A resampling procedure for complex survey data. *J. Am. Statist. Assoc.* **87**, 755–765.   *(547)

Sitter, R. R. (1992b). Comparing three bootstrap methods for survey data. *Can. J. Statist.* **20**, 135–154.   *(548)

Sitter, R. R. (1997). Variance estimation for the regression estimator in two-phase sampling. *J. Am. Statist. Assoc.* **92**, 779–787.

Sitter, R. R. (1998). Balanced resampling using orthogonal multiarrays. In *Encyclopedia of Statistical Sciences, Update Volume 2* (S. Kotz, C. B. Campbell, and D. L. Banks, editors), 46–50, Wiley, New York.   *(549)

Sivaganesan, S. (1994). Book Review of *An Introduction to the Bootstrap* (by B. Efron and R. J. Tibshirani). *SIAM Rev.* **36**, 677–678.

Skinner, C. J., Holt, D., and Smith, T. M. F. (editors) (1989). *Analysis of Complex Surveys.* Wiley, New York.

Smith, A. F. M., and Gelfand, A. E. (1992). Bayesian statistics without tears: a sampling–resampling perspective. *J. Am. Statist. Assoc.* **46**, 84–88.

Smith, L. A., and Sielken, R. L. (1988). Bootstrap bounds for "safe" doses in the multistage cancer dose–response model. *Commun. Statist. Simul. Comput.* **17**, 153–175.   *(550)

Smith, L. R., Harrell, F. E., and Muhlbaier, L. H. (1992). Problems and potentials in modeling survival. In *Medical Effectiveness Research Data Methods.* (M. L. Grady and H. A. Schwarts, editors), 151–159. U.S. Department of Health and Human Services, Agency for Health Care Policy Research.

Smith, P. W. F., Forster, J. J., and McDonald, J. W. (1996). Monte Carlo exact tests for square contingency tables. *J. R. Statist. Soc. A* **159**, 309–321.

Smith, S. J. (1997). Bootstrap confidence limits for groundfish trawl survey estimates of mean abundance. *Can. J. Fish. Aquat. Sci.* **54**, 616–630.

Snapinn, S. M., and Knoke, J. D. (1984). Classification error rate estimators evaluated by unconditional mean squared error. *Technometrics* **26**, 371–378.   *(551)

Snapinn, S. M., and Knoke, J. D. (1985a). An evaluation of smoothed classification error rate estimators. *Technometrics* **27**, 199–206.   *(552)

Snapinn, S. M., and Knoke, J. D. (1985b). Improved classification error rate estimation: bootstrap or smooth? Unpublished report.   *(553)

Snapinn, S. M., and Knoke, J. D. (1988). Bootstrap and smoothed classification error rate estimates. *Commun. Statist. Simul. Comput.* **17**, 1135–1153.   *(554)

Solka, J. L., Wegman, E. J., Priebe, C. E., Poston, W. L., and Rogers, G. W. (1995). A method to determine the structure of an unknown mixture using the Akaike Information Criterion and the bootstrap. George Mason University Technical Report.

Solomon, H. (1986). Confidence intervals in legal settings. In *Statistics and the Law* (M. H. DeGroot, S. E. Fienberg, and J. B. Kadane, editors), 455–473, Wiley, New York.

Solow, A. R. (1985). Bootstrapping correlated data. *J.I.A. Math. Geo.* **17**, 769–775.   *(555)

Solow, A. R., and Caswell, H. (1996). Sequential methods in randomization and bootstrap tests. *Bull. Ecol. Soc. Am. Suppl. Part 2* **77**, 417.

Soofi, E. S., and Gensch, D. H. (1994). Bootstrap method for measuring accuracy of probabilistic-choice models. *Eur. J. Operational Res.* **76**, 321–330.

Sorum, M. (1972). Three probabilities of misclassification. *Technometrics* **14**, 309–316.   *(556)

Sparks, T. H., and Rothery, P. (1996). Resampling methods for ecotoxicological data. *Ecotoxicology* **5**, 197–207.

Sprent, P. (1998). *Data Driven Statistical Methods.* Chapman & Hall, London.   *(557)

Srivastava, M. S. (1987a). Bootstrapping Durbin–Watson statistics. *Indian J. Math.* **29**, 193–210.

Srivastava, M. S. (1987b). Bootstrap methods in ranking and slippage problems. *Commun. Statist. Theory Methods* **16**, 3285–3299.

Srivastava, M. S., and Carter, E. M. (1983). *An Introduction to Applied Multivariate Statistics.* Elsevier Science Publishing, New York.   *(558)

Srivastava, M. S., and Chan, Y. M. (1989). A comparison of bootstrap methods and Edgeworth expansion in approximation of the distribution of sample variance: One sample and two sample cases. *Commun. Statist. Simul. Comput.* **18**, 339–361.

Srivastava, M. S., and Lee, G. C. (1984). On the distribution of the correlation coefficient when sampling from a mixture of two bivariate normal densities: robustness and outliers. *Can. J. Statist.* **2**, 119–133.   *(559)

Srivastava, M. S., and Singh, B. (1989). Bootstrapping in multiplicative models. *J. Econ.* **42**, 287–297.   *(560)

Stangenhaus, G. (1987). Bootstrap and inference procedures for L_1 regression. In *Statistical Data Analysis Based on the L_1 Norm and related Methods* (Y. Dodge, editor), 323–332. North-Holland, Amsterdam.

Staudte, R. G., and Sheather, S. J. (1990). *Robust Estimation and Testing.* Wiley, New York.   *(561)

Stauffer, D. F., Garton, E. O., and Steinhorst, R. K. (1985). A comparison of principal components from real and random data. *Ecology* **66**, 1693–1698.

Stefanski, L. A., and Cook, J. R. (1995). Simultaneous extrapolation: the measurement error jackknife. *J. Am. Statist. Assoc.* **90**, 1247–1256.

Steinberg, S. M. (1983). Confidence intervals for functions of quantiles using linear combinations of order statistics. Ph. D. dissertation. Department of Statistics, University of North Carolina, Chapel Hill.

Stein, M. (1987). Large sample properties of simulations using Latin hypercube sampling. *Technometrics* **29**, 143–151.   *(562)

Stewart, T. J. (1986). Experience with a Bayesian bootstrap method incorporating proper prior information. *Commun. Statist. Theory Methods* **15**, 3205–3225.

Stine, R. A. (1982). Prediction intervals for time series. Ph. D. dissertation, Princeton University, Princeton.

Stine, R. A. (1985). Bootstrap prediction intervals for regression. *J. Am. Statist. Assoc.* **80**, 1026–1031.   *(563)

Stine, R. A. (1987). Estimating properties of autoregressive forecasts. *J. Am. Statist. Assoc.* **82**, 1072–1078.   *(564)

Stine, R. A. (1992). An introduction to bootstrap methods. *Sociol. Methods Res.* **18**, 243–291.   *(565)

Stine, R. A., and Bollen, K. A. (1993). Bootstrapping goodness-of-fit measures in structural equation models. In *Testing Structural Equation Models.* Sage Publications, Beverly Hills.

Stoffer, D. S., and Wall, K. D. (1991). Bootstrapping state-space models: Gaussian maximum likelihood estimation and Kalman filter. *J. Am. Statist. Assoc.* **86**, 1024–1033.   *(566)

Strawderman, R. L., Parzen, M. I., and Wells, M. T. (1997). Accurate confidence limits for quantiles under random censoring. *Biometrics* **53**, 1399–1415.

Strawderman, R. L., and Wells, M. T. (1997). Accurate bootstrap confidence limits for the cumulative hazard and survivor functions under random censoring. *J. Am. Statist. Assoc.* **92**, 1356–1374.

Stromberg, A. J. (1997). Robust covariance estimates based on resampling. *J. Statist. Plann. Inf.* **57**, 321–334.

Stuart, A., and Ord, K. (1993). *Kendall's Advanced Theory of Statistics, Vol. 1,* 6th. ed. Edward Arnold, London.   *(567)

Stute, W. (1990). Bootstrap of the linear correlation model. *Statistics* **21**, 433–436.

Stute, W. (1992). Modified cross-validation in density estimation. *J. Statist. Plann. Inf.* **30**, 293–305.

Stute, W., and Grunder, B. (1992). Bootstrap approximations to prediction intervals for explosive AR(1)-processes. In *Bootstrapping and Related Techniques. Proceedings, Trier, FRG.* (K.-H. Jockel, G. Rothe, and W. Sendler, editors). *Lecture Notes in Economics and Mathematical Systems,* **376**, 121–130. Springer-Verlag, Berlin.

Stute, W., and Wang, J. (1994). The jackknife estimate of a Kaplan–Meier integral. *Biometrika* **81**, 602–606.

Stute, W., Manteiga, W. C., and Quindimil, M. P. (1993). Bootstrap based goodness-of-fit tests. *Metrika* **40**, 243–256.

Sun, L., and Muller-Schwarze, D. (1996). Statistical resampling methods in biology: a case study of beaver dispersal patterns. *Am. J. Math. Manage. Sci.* **16**, 463–502. *(568)

Sutton, C. D. (1993). Computer-intensive methods for tests about the mean of an asymmetric distribution. *J. Am. Statist. Assoc.* **88**, 802–810.

Swanepoel, J. W. H. (1983). Bootstrap selection procedures based on robust estimators. *Commun. Statist. Theory Methods* **12**, 2059–2083.

Swanepoel, J. W. H. (1985). Bootstrap selection procedures based on robust estimators. In *The Frontiers of Modern Statistical Inference Procedures* (E. J. Dudewicz, editor), 45–64. American Science Press, Columbus.

Swanepoel, J. W. H. (1986a). A note on proving that the (modified) bootstrap works. *Commun. Statist. Theory Methods* **15**, 1399–1415.

Swanepoel, J. W. H., and van Wyk, J. W. J. (1986). The bootstrap applied to power spectral density function estimation. *Biometrika* **73**, 135–141.   *(569)

Swanepoel, J. W. H., van Wyk, J. W. J., and Venter, J. H. (1983). Fixed width confidence intervals based on bootstrap procedures. *Seq. Anal.* **2**, 289–310.

Swift, M. B. (1995). Simple confidence intervals for standardized rates based on the approximate bootstrap method. *Statist. Med.* **14**, 1875–1888.

Takeuchi, L. R., Sharp, P. A., and Ming-Tung, L. (1994). Identifying data generating families of probability densities: a bootstrap resampling approach. In *1994 Proceedings Decision Sciences,* **2**, 1391–1393.

Tambour, M., and Zethraeus, N. (1998). Bootstrap confidence intervals for cost-effectiveness ratios: some simulation results. *Health Econ.* **7**, 143–147.   *(570)

Tamura, H., and Frost, P. A. (1986). Tightening CAV (DUS) bounds by using a parametric model. *J. Acct. Res.* **24**, 364–371.

Taylor, C. C. (1989). Bootstrap choice of smoothing parameter in kernel density estimation. *Biometrika* **76**, 705–712.

Taylor, M. S., and Thompson, J. R. (1992). A nonparametric density estimation based resampling algorithm. In *Exploring the Limits of Bootstrap* (R. LePage and L. Billard, editors), 399–403, Wiley, New York.   *(571)

ter Braak, C. J. F. (1992). Permutation versus bootstrap significance tests in multiple regression and ANOVA. In *Bootstrapping and Related Techniques. Proceedings, Trier, FRG* (K.-H. Jockel, G. Rothe, and W. Sendler, editors). *Lecture Notes in Economics and Mathematical Systems,* **376**, 70–85. Springer-Verlag, Berlin.

Theodossiou, P. T. (1993). Predicting shifts in the mean of a multivariate time series process: an application in predicting business failures. *J. Am. Statist. Assoc.* **88**, 441–449.

Therneau, T. (1983). Variance reduction techniques for the bootstrap. Ph. D. dissertation. Department of Statistics, Stanford University, Stanford.   *(572)

Thisted, R. A. (1988). *Elements of Statistical Computing: Numerical Computation.* Chapman & Hall, New York.   *(573)

Thomas, G. E. (1994). Book Review of *Resampling-Based Multiple Testing: Examples and Methods for p-Value Adjustment* (by P. H. Westfall and S. S. Young). *Statistician* **43**, 467–486.

Thombs, L. A., and Schucany, W. R. (1990). Bootstrap prediction intervals for autoregression. *J. Am. Statist. Assoc.* **85**, 486–492.   *(574)

Thompson, J. R. (1989). *Empirical Model Building.* Wiley, New York.   *(575)

Tibshirani, R. (1985). Bootstrap computations. *SAS SUGI* **10**, 1059–1063.

Tibshirani, R. (1986). Bootstrap confidence intervals. *Proc. Comput. Sci. Statist.* **18**, 267–273.

Tibshirani, R. (1988). Variance stabilization and the bootstrap. *Biometrika* **75**, 433–444.

Tibshirani, R. (1992). Some applications of the bootstrap in complex problems. In *Exploring the Limits of Bootstrap* (R. LePage and L. Billard, editors), 271–277, Wiley, New York.   *(576)

Tibshirani, R. (1996). Regression shrinkage and selection via the lasso. *J. R. Statist. Soc.* B **58**, 267–288.

Tibshirani, R. (1997a). Who is the fastest man in the world? *Am. Statist.* **51**, 106–111.

Tibshirani, R. (1997b). The lasso method for variable selection in the Cox model. *Statist. Med.* **16**, 385–395.

Tingley, M., and Field, C. (1990). Small-sample confidence intervals. *J. Am. Statist. Assoc.* **85**, 427–434.   *(577)

Titterington, D. M., Smith, A. F. M., and Makov, U. E. (1985). *Statistical Analysis of Finite Mixture Distributions.* Wiley, Chichester.   *(578)

Tivang, J. G., Nienhuis, J., and Smith, O. S. (1994). Estimation of sampling variance of molecular marker data using the bootstrap procedure. *Theor. Appl. Genet.* **89**, 259–264.   *(579)

Tomasson, H. (1995). Risk scores from logistic regression: unbiased estimates of relative and attributable risk. *Statist. Med.* **14**, 1331–1339.

Tong, H. (1983). *Threshold Models in Non-linear Time Series Analysis, Lecture Notes in Statistics,* **21**. Springer-Verlag, New York.   *(580)

Tong, H. (1990). *Non-linear Time Series: A Dynamical System Approach.* Clarendon Press, Oxford.   *(581)

Tourassi, G. D., Floyd, C. E., Sostman, H. D., and Coleman, R. E. (1995). Performance evaluation of an artificial neural network for the diagnosis of acute pulmonary embolism using the cross-validation, jackknife and bootstrap methods. *WCNN '95* **2**, 897–900.

Toussaint, G. T. (1974). Bibliography on estimation on misclassification. *IEEE Trans. Inform. Theory* **20**, 472–479.   *(582)

Tran, Z. V. (1996). The bootstrap: verifying mean changes in cholesterol from resistive exercise. *Med. Sci. Sports Exerc. Suppl.* **28**, 187.   *(583)

Troendle, J. F. (1995). A stepwise resampling method of multiple hypothesis testing. *J. Am. Statist. Assoc.* **90**, 370–378.

Tsay, R. S. (1992). Model checking via parametric bootstraps in time series. *Appl. Statist.* **41**, 1–15.   *(584)

Tsodikov, A., Hasenclever, D., and Loeffler, M. (1998). Regression with bounded outcome score: estimation of power by bootstrap and simulation in a chronic myelogenous leukemia clinical trial. *Statist. Med.* **17**, 1909–1922.   *(585)

Tsumoto, S., and Tanaka, H. (1995). Automated selection of rule induction methods based on recursive iteration of resampling methods and multiple statistical testing. In *1st International Conference on Knowledge Discovery and Data Mining* (U. M. Fayyad and R. Uthurusamy, editors), **1**, 312–317. AAAI Press.

Tsumoto, S., and Tanaka, H. (1996a). Automated acquisition of medical expert system rules based on rough sets and resampling methods. In *Proceedings of the Third World Congress on Expert Systems* (J. K. Lee, J. Liebowitz, and Y. M. Chae, editors), **2**, 877–884.

Tsumoto, S., and Tanaka, H. (1996b). Induction of expert system rules from databases based on rough set theory and resampling methods. In *Foundations of Intelligent Systems, 9th International Symposium ISMIS '96 Proceedings* (Z. W. Ras and M. Michalewicz, editors), 128–138. Springer-Verlag, Berlin.

Tu. D. (1986). Bootstrapping of *L*-statistics. *Kexue Tongbao (Chin. Bull. Sci.)* **31**, 965–969.

Tu, D. (1988a). The kernel estimator of conditional *L*-functional and its bootstrapping statistics. *Acta Math. Appl. Sin.* **11**, 53–68.

Tu, D. (1988b). The nearest neighbor estimate of the conditional *L*-functional and its bootstrapping statistics. *Chin. Ann. Math. A* **8**, 345–357.

Tu, D. (1989). *L*-functional and nonparametric *L*-regression estimates: asymptotic distributions and bootstrapping approximations. Technical Report 89-51. Center for Multivariate Analysis, Pennsylvania State University, State College.

Tu, D. (1992). Approximating the distribution of a general standardized functional statistic with that of jackknife pseudovalues. In *Exploring the Limits of Bootstrap* (R. LePage and L. Billard, editors), 279–306, Wiley, New York.   *(586)

Tu, D., and Cheng, P. (1989). Bootstrapping untrimmed *L*-statistics. *J. Sys. Sci. Math. Sci.* **9**, 14–23.

Tu, D., and Gross, A. J. (1994). Bias reduction for jackknife skewness estimators. *Commun. Statist. Theory Methods* **23**, 2323–2341.

Tu, D., and Gross, A. J. (1995). Accurate confidence intervals for the ratio of specific occurrence/exposure rates in risk and survival analysis. *Biometrics J.* **37**, 611–626.

Tu, D., and Shi, X. (1988). Bootstrapping and randomly weighting the *U*-statistics with jackknifed pseudo values. *Math. Statist. Appl. Probab.* **3**, 205–212.

Tu, D., and Zhang, L. (1992a). On the estimation of skewness of a statistic using the jackknife and the bootstrap. *Statist. Hefte* **33**, 39–56.

Tu, D., and Zhang, L. (1992b). Jackknife approximations for some nonparametric confidence intervals of functional parameters based on normalizing transformations. *Comput. Statist.* **7**, 3–15.

Tu, D., and Zheng, Z. (1987). On the Edgeworth's expansion of random weighting method. *Chin. J. Appl. Probab. Statist.* **3**, 340–347.

Tu, D., and Zheng, Z. (1991). Random weighting: another approach to approximate the unknown distributions of pivotal quantities. *J. Comb. Info. Sys. Sci.* **16**, 249–270

Tu, X. M., Burdick, D. S., and Mitchell, B. C. (1992). Nonparametric rank estimation using bootstrap resampling and canonical correlation analysis. In *Exploring the Limits of Bootstrap* (R. LePage and L. Billard, editors), 405–418. Wiley, New York. *(587)

Tucker, H. G. (1959). A generalization of the Glivenko–Cantelli theorem. *Ann. Math. Statist.* **30**, 828–830. *(588)

Tukey, J. W. (1958). Bias and confidence in not quite large samples (abstract). *Ann. Math. Statist.* **29**, 614. *(589)

Turnbull, B. W., and Mitchell, T. J. (1978). Exploratory analysis of disease prevalence data from survival/sacrifice experiments. *Biometrics* **34**, 555–570. *(590)

Turner, S., Myklebust, R. L., Thorne, B. B., Leigh, S. D., and Steel, E. B. (1996). Airborne asbestos method: bootstrap method for determining the uncertainty of asbestos concentration. National Institute of Standards Technical Report.

Ueda, N., and Nakano, R. (1995). Estimating expected error rates of neural network classifiers in small sample size situations: a comparison of cross-validation and bootstrap. In *1995 IEEE International Conference on Neural Networks Proceedings.* **1**, 101–104. *(591)

Upton, G. J. G., and Fingleton, B. (1985). *Spatial Data Analysis by Example, Volume 1: Point Pattern and Quantitative Data.* Wiley, Chichester. *(592)

Upton, G. J. G., and Fingleton, B. (1989). *Spatial Data Analysis by Example, Volume 2: Categorical and Directional Data.* Wiley, Chichester. *(593)

van der Burg, E., and de Leeuw, J. (1988). Use of the multinomial, jackknife and bootstrap in generalized nonlinear canonical correlation analysis. *Appl. Stoch. Models Data Anal.* **4**, 159–172.

van der Kloot, W. (1996). Statistics for studying quanta at synapses: resampling and confidence limits on histograms. *J. Neurosci. Methods* **65**, 151–155.

van der Vaart, A. W., and Wellner, J. A. (1996). *Weak Convergence and Empirical Processes with Applications to Statistics.* Springer-Verlag, New York. *(594)

van Dongen, S. (1995). How should we bootstrap allozyme data? *Heredity* **74**, 445–447.

van Dongen S., and Backeljau, T. (1995). One- and two-sample tests for single-locus inbreeding coefficients using the bootstrap. *Heredity* **74**, 129–135.

van Dongen, S., and Backeljau, T. (1997). Bootstrap tests for specific hypotheses at single locus inbreeding coefficients. *Genetica* **99**, 47–58.

van Zwet, W. (1989). Hoeffding's decomposition and the bootstrap. Talk given at Conference on Asymptotic Methods for Computer-Intensive Procedures in Statistics, Oberwolfach, Germany.

Veall, M. R. (1989). Applications of computationally-intensive methods to econometrics. *Proceedings of the 47th Session of the International Statistical Institute,* Paris 75–88.

Ventura, V. (1997). Likelihood inference by Monte Carlo methods and efficient nested bootstrapping. D. Philos. thesis. Department of Statistics, Oxford University. *(595)

Ventura, V., Davison, A. C., and Boniface, S. J. (1997). Statistical inference for the effect of magnetic brain stimulation on a motoneurone. *Appl. Statist.* **46**, 77–94. *(596)

Vinod, H. D., and McCullough, B. D. (1995). Estimating cointegration parameters: an application of the double bootstrap. *J. Statist. Plann. Inf.* **43**, 147–156.

Vinod, H. D., and Raj, B. (1988). Econometric issues in Bell System diverstiture: a bootstrap application. *Appl. Statist.* **37**, 251–261.

Visscher, P. M., Thompson, R., and Haley, C. S. (1996). Confidence intervals in QTL mapping by bootstrapping. *Genetics* **143**, 1013–1020.

Wacholder, S., Gail, M. H., Pee, D., and Brookmeyer, R. (1989). Alternative variance and efficiency calculations for the case–cohort design. *Biometrika* **76**, 117–123.

Waclawiw, M. A., and Liang, K. (1994). Empirical Bayes estimation and inference for random effects model with binary response. *Statist. Med.* **13**, 541–551.   *(597)

Wagner, R. F., Chan, H.-P., Sahiner, B., Petrick, N., and Mossoba, J. T. (1997). Finite-sample effects and resampling plans: applications to linear classifiers in computer-aided diagnosis. *Proc. SPIE* **3034**, 467–477.

Wahrendorf, J., Becher, H., and Brown, C. C. (1987). Bootstrap comparison of non-nested generalized linear models: applications in survival analysis and epidemiology. *Appl. Statist.* **36**, 72–81.

Wahrendorf, J., and Brown, C. C. (1980). Bootstrapping a basic inequality in the analysis of joint action of two drugs. *Biometrics* **36**, 653–657.   *(598)

Wallach, D., and Goffinet, B. (1987). Mean squared error of prediction in models for studying ecological and agronomic systems. *Biometrics* **43**, 561–573.

Walther, G. (1997). Granulometric smoothing. *Ann. Statist.* **25**, 2273–2299.

Wang, M.-C. (1986). Re-sampling procedures for reducing bias of error rate estimation in multinomial classification. *Comput. Statist. Data Anal.* **4**, 15–39.

Wang, J.-L., and Hettmansperger, T. P. (1990). Two-sample inference for median survival times based on one-sample procedures for censored survival data. *J. Am. Statist. Assoc.* **85**, 529–536.

Wang, S. J. (1989). On the bootstrap and smoothed bootstrap. *Commun. Statist. Theory Methods* **18**, 3949–3962.

Wang, S. J. (1990). Saddlepoint approximations in resampling analysis. *Ann. Inst. Statist. Math.* **42**, 115–131.

Wang, S. J. (1992). General saddlepoint approximations in the bootstrap. *Statist. Probab. Lett.* **13**, 61–66.

Wang, S. J. (1993a). Saddlepoint expansions in finite population problems. *Biometrika* **80**, 583–590.

Wang, S. J. (1993b). Saddlepoint methods for bootstrap confidence bands in nonparametric regression. *Aust. J. Statist.* **35**, 93–101.

Wang, S. J. (1995). Optimizing the smoothed bootstrap. *Ann. Inst. Statist. Math.* **47**, 65–80.

Wang, S. J., Woodward, W. A., Gray, H. L., Wiechecki, S., and Sain, S. R. (1997). A new test for outlier detection from a multivariate mixture distribution. *J. Comput. Graphical Statist.* **6**, 285–299.

Wang, X., and Mong, J. (1994). Resampling-based estimator in nonlinear regression. *Statist. Sin.* **4**, 187–198.

Wang, Y. (1996). A likelihood ratio test against stochastic ordering in several populations. *J. Am. Statist.* **91**, 1676–1683.

Wang, Y., Prade, R. A., Griffith, J., Timberlake, W. E., and Arnold, J. (1994). Assessing the statistical reliability of physical maps by bootstrap resampling. *Comput. Appl. Biosci.* **10**, 625–634.

Wang, Y., and Wahba, G. (1995). Bootstrap confidence intervals for smoothing splines and their comparison to Bayesian confidence intervals. *J. Statist. Comput. Simul.* **51**, 263–279.   *(599)

Ware, J. H., and De Gruttola, V. (1985). Multivariate linear models for longitudinal data: a bootstrap study of the GLS estimate. In *Biostatistics in Biomedical, Public Health and Environmental Sciences* (P. K. Sen, editor), 424–434. North-Holland, Amsterdam.

Wasserman, G. S., Mohsen, H. A., and Franklin, L. A. (1991). A program to calculate bootstrap confidence intervals for process capability index, C_{pk}. *Commun. Statist. Simul. Comput.* **20**, 497–510.   *(600)

Watson, G. S. (1983). The computer simulation treatment of directional data. In *Proceedings of the Geological Conference, Kharagpur India, India. Indian J. Earth Sci.* 19–23.

Weber, N. C. (1984). On resampling techniques for regression models. *Statist. Probab. Lett.* **2**, 275–278.   *(601)

Weber, N. C. (1986). On the jackknife and bootstrap techniques for regression models. In *Pacific Statistical Congress: Proceedings of the Congress* (I. S. Francis, B. F. J. Manly, and F. C. Lam, editors), 51–55. North-Holland, Amsterdam.

Weinberg, S. L., Carroll, J. D., and Cohen, H. S. (1984). Confidence regions for IDSCAL using the jackknife and bootstrap techniques. *Psychometrika* **49**, 475–491.

Weiss, G. (1975). Time-reversibility of linear stochastic processes. *J. Appl. Probab.* **12**, 831–836.   *(602)

Weiss, I. M. (1970). A survey of discrete Kalman–Bucy filtering with unknown noise covariances. *AIAA Guidance, Control and Flight Mechanics Conference, AIAA Paper No. 70-955.* American Institute of Aeronautics and Astronautics, New York. *(603)

Weissfeld, L. A., and Schneider, H. (1987). Inference based on the Buckley–James procedure. *Commun. Statist. Theory Methods* **16**, 1773–1787.

Welch, B. L., and Peers, H. W. (1963). On formulae for confidence points based on integrals of weighted likelihoods. *J. R. Statist. Soc. B* **25**, 318–329.   *(604)

Welch, W. J. (1990). Construction of permutation tests. *J. Am. Statist. Assoc.* **85**, 693–698.

Wellner, J. A., and Zhan, Y. (1996). Bootstrapping *z*-estimators. Technical Report, University of Washington, Department of Statistics.

Wellner, J. A., and Zhan, Y. (1997). A hybrid algorithm for computation of the nonparametric maximum likelihood estimator from censored data. *J. Am. Statist. Assoc.* **92**, 945–959.

Wells, M. T., and Tiwari, R. C. (1994). Bootstrapping a Bayes estimator of a survival function with censored data. *Ann. Inst. Statist. Math.* **46**, 487–495.

Wendel, M. (1989). Eine anwendung der bootstrap—Methode auf nicht-lineare autoregressive prozesse erster ordnung. Diploma thesis, Technical University of Berlin, Berlin.

Weng, C.-S. (1989). On a second order asymptotic property of the Bayesian bootstrap mean. *Ann. Statist.* **17**, 705–710.   *(605)

Wernecke, K.-D., and Kalb, G. (1987). Estimation of error rates by means of simulated bootstrap distributions. *Biom. J.* **29**, 287–292.

Westfall, P. (1985). Simultaneous small-sample multivariate Bernoulli confidence intervals. *Biometrics* **41**, 1001–1013.   *(606)

Westfall, P., and Young, S. S. (1993). *Resampling-Based Multiple Testing: Examples and Methods for p Value Adjustment.* Wiley, New York.   *(607)

Willemain, T. R. (1994). Bootstrapping on a shoestring: resampling using spreadsheets. *Am. Statist.* **48**, 40–42.

Withers, C. S. (1983). Expansion for the distribution and quantiles of a regular functional of the empirical distribution with applications to nonparametric confidence intervals. *Ann. Statist.* **11**, 577–587.

Wolfe, J. H. (1967). NORMIX: computational methods for estimating the parameters of multivariate normal mixtures of distributions. Research Memo. SRM 68-2. U. S. Naval Personnel Research Activity, San Diego.   *(608)

Wolfe, J. H. (1970). Pattern clustering by multivariate mixture analysis. *Mult. Behav. Res.* **5**, 329–350.   *(609)

Wong, M. A. (1985). A bootstrap testing procedure for investigating the number of subpopulations. *J. Statist. Comput. Simul.* **22**, 99–112.

Woodroof, J. B. (1997). Statistically comparing user satisfaction instruments: an application of the bootstrap using a spreadsheet. *Proceedings of the 13th Hawaii International Conference on System Sciences* **3**, 49–56.

Woodroofe, M., and Jhun, M. (1989). Singh's theorem in the lattice case. *Statist. Probab. Lett.* **7**, 201–205.

Worton, B. J. (1994). Book Review of *Computer Intensive Statistical Methods: Validation, Model Selection and Bootstrap* (by J. S. U. Hjorth). *J. R. Statist. Soc. A* **157**, 504–505.

Worton, B. J. (1995). Modeling radio-tracking data. *Environ. Ecol. Statist.* **2**, 15–23.

Wu, C. F. J. (1986). Jackknife, bootstrap and other resampling plans in regression analysis (with discussion). *Ann. Statist.* **14**, 1261–1350.   *(610)

Wu, C. F. J. (1990). On the asymptotic properties of the jackknife histogram. *Ann. Statist.* **18**, 1438–1452.

Wu, C. F. J. (1991). Balanced repeated replications based on mixed orthogonal arrays. *Biometrika* **78**, 181–188.

Xie, F., and Paik, M. C. (1997). Multiple imputation methods for the missing covariates in generalized estimating equations. *Biometrics* **53**, 1538–1546.

Yandell, B. S., and Horvath, L. (1988). Bootstrapped multi-dimensional product limit process. *Aust. J. Statist.* **30**, 342–358.

Yang, S. S. (1985a). On bootstrapping a class of differentiable statistical functionals with application to *L*- and *M*-estimates. *Statist. Neerlandica* **39**, 375–385.

Yang, S. S. (1985b). A smooth nonparametric estimator of a quantile function. *J. Am. Statist. Assoc.* **80**, 1004–1011.

Yang, S. S. (1988). A central limit theorem for the bootstrap mean. *Am. Statist.* **42**, 202–203.   *(611)

Yang, Z. R., Zwolinski, M., and Chalk, C. D. (1998). Bootstrap, an alternative to Monte Carlo simulation. *Electron. Lett.* **34**, 1174–1175.

Yeh, A. B., and Singh, K. (1997). Balanced confidence regions based on Tukey's depth and the bootstrap. *J. R. Statist. Soc. B* **59**, 639–652.

Yokoyama, S., Seki, T., and Takashina, T. (1993). Bootstrap for the normal parameters. *Commun. Statist. Simul. Comput.* **22**, 191–203.

Young, G. A. (1986). Conditioned data-based simulations: some examples from geometrical statistics. *Int. Statist. Rev.* **54**, 1–13.

Young, G. A. (1988a). A note on bootstrapping the correlation coefficient. *Biometrika* **75**, 370–373.   *(612)

Young, G. A. (1988b). Resampling tests of statistical hypotheses. In *Proceedings of the Eighth Biannual Symposium on Computational Statistics* (D. Edwards and N. E. Raum, editors), 233–238. Physica-Verlag, Heidelberg.

Young, G. A. (1990). Alternative smoothed bootstraps. *J. R. Statist. Soc. B* **52**, 477–484.

Young, G. A. (1993). *Book Review of The Bootstrap and Edgeworth Expansion* (by P. Hall). *J. R. Statist. Soc. A* **156**, 504–505.

Young, G. A. (1994). Bootstrap: More than a stab in the dark? (with discussion). *Statist. Sci.* **9**, 382–415.   *(613)

Young, G. A., and Daniels, H. E. (1990). Bootstrap bias. *Biometrika* **77**, 179–185.   *(614)

Youyi, C., and Tu, D. (1987). Estimating the error rate in discriminant analysis: by the delta, jackknife and bootstrap method (in Chinese). *Chin. J. Appl. Probab. Statist.* **3**, 203–210.

Yu, Z., and Tu, D. (1987). On the convergence rate of bootstrapped and randomly weighted m-dependent means. Research Report. Institute of Systems Science, Academia Sinica, Beijing.

Yuen, K. C., and Burke, H. D. (1997). A test of fit for a semiparametric additive risk model. *Biometrika* **84**, 631–639.

Yule, G. U. (1927). On a method of investigating periodicities in disturbed series, with special reference to Wolfer's sunspot numbers. *Philos. Trans. A* **226**, 267–298.   *(615)

Zecchi, S., and Camillo, F. (1994). An application of bootstrap to the Rv coefficient in the analysis of stock markets (Italian). *Quand. Statist. Mat. Appl.* **14**, 99–109.

Zelterman, D. (1993). A semiparametric bootstrap technique for simulating extreme order statistics. *J. Am. Statist. Assoc.* **88**, 477–485.

Zeng, Q., and Davidian, M. (1997). Bootstrap-adjusted calibration confidence intervals for immunoassay. *J. Am. Statist. Assoc.* **92**, 278–290.

Zeng, Q., and Davidian, M. (1997). Testing homogeneity of intra-run variance parameters in immunoassay. *Statist. Med.* **16**, 1765–1776.

Zhan, Y. (1996). Bootstrapping functional M-estimators. Unpublished Ph. D. dissertation. University of Washington, Department of Statistics.

Zhang, J., and Boos, D. D. (1992). Bootstrap critical values for testing homogeneity of covariance matrices. *J. Am. Statist. Assoc.* **87**, 425–429.

Zhang, J., and Boos, D. D. (1993). Testing hypotheses about covariance matrices using bootstrap methods. *Commun. Statist. Theory Methods* **22**, 723–739.

Zhang, L., and Tu, D. (1990). A comparison of some jackknife and bootstrap procedures in estimating sampling distributions of studentized statistics and constructing confidence intervals. Technical Report 90-28. Center for Multivariate Analysis, Pennsylvania State University, State College.

Zhang, Y., Hatzinakos, D., and Venetsanopoulos, A. N. (1993). Bootstrapping techniques in the estimation of higher-order cumulants from short data records. *Proceedings of the 1993 IEEE International Conference on Acoustics, Speech and Signal Processing*, **4**, 200–203.

Zhao, L. (1986). Bootstrapping the error variance estimate in linear model (in Chinese). *A. Ma. Peking* **29**, 36–45.

Zharkikh, A., and Li, W.-H. (1992). Statistical properties of bootstrap estimation of phylogenetic variability from nucleotide sequences: II. Four taxa without a molecular clock. *J. Mol. Evol.* **35**, 356–366.   *(616)

Zharkikh, A., and Li, W.-H. (1995). Estimation of confidence in phylogeny: the complete-and-partial bootstrap technique. *Mol. Phylogenet. Evol.* **4**, 44–63. *(617)

Zheng, X. (1994). Third-order correct bootstrap calibrated confidence bounds for nonparametric mean. *Math. Methods Statist.* **3**, 62–75.

Zheng, Z. (1985). The asymptotic behavior of the nearest neighbour estimate and its bootstrap statistics. *Sci. Sin. A.* **28**, 479–494.

Zhou, M. (1993). Bootstrapping the survival curve estimator when data are doubly censored. Technical Report 335. University of Kentucky, Department of Statistics.

Zhu, L. X., and Fang, K. T. (1994). The accurate distribution of the Kolmogorov statistic with applications to bootstrap approximation. *Adv. Appl. Math.* **15**, 476–489.

Zhu, W. (1997). Making bootstrap statistical inferences: a tutorial. *Res. Q. Exerc. Sport.* **68**, 44–55.

Ziari, H. A., Leatham, D. J., and Ellinger, P. N. (1997). Developments of statistical discriminant mathematical programming model via resampling estimation techniques. *Am. J. Agric. Econ.* **79**, 1352–1362.

Ziegel, E. (1994). Book Review of *Bootstrapping: A Nonparametric Approach to Statistical Inference* (by C. Z. Mooney and R. Duvall). *Technometrics* **36**, 435–436.

Ziliak, J. P. (1997). Efficient estimation with panel data when instruments are predetermined: an empirical comparison of moment-condition estimators. *J. Bus. Econ. Statist.* **15**, 419–431.

Zoubir, A. M. (1994a). Bootstrap multiple tests: an application to optimal sensor location for knock detection. *Appl. Signal Process.* **1**, 120–130.

Zoubir, A. M. (1994b). Multiple bootstrap tests and their application. *1994 IEEE International Conference on Acoustics, Speech and Signal Processing* **6**, 69–72.

Zoubir, A. M., and Ameer, I. (1997). Bootstrap analysis of polynomial amplitude and phase signals. In *Proceedings of IEEE TENCON '97, IEEE Region 10 Annual Conference, Speech and Image Technologies for Computing and Telecommunications* (M. Deriche, M. Moody, and M. Bennamoun, editors), **2**, 843–846.

Zoubir, A. M., and Boashash, B. (1998). The bootstrap and its application in signal processing. *IEEE Signal Process. Mag.* **15**, 56–76.   *(618)

Zoubir, A. M., and Bohme, J. F. (1995). Bootstrap multiple tests applied to sensor location. *IEEE Trans. Signal Process.* **43**, 1386–1396.   *(619)

Zoubir, A. M., and Iskander, D. R. (1996). Bispectrum based Gaussianity test using the bootstrap. *Proceedings of the IEEE International Conference on Acoustics, Speech and Signal Processing*, **5**, 3029–3032.

Zoubir, A. M., Iskander, D. R., Ristac, B., and Boashash, B. (1994). Instantaneous frequency estimation: confidence bounds using the bootstrap. In *Conference Record of the Twenty-Eighth Asilomar Conference on Signals, Systems and Computers* (A. Singh, editor), **2**, 55–59.

Author Index

Subject Index

WILEY SERIES IN PROBABILITY AND STATISTICS
ESTABLISHED BY WALTER A. SHEWHART AND SAMUEL S. WILKS

Editors
Vic Barnett, Noel A. C. Cressie, Nicholas I. Fisher,
Iain M. Johnstone, J. B. Kadane, David G. Kendall, David W. Scott,
Bernard W. Silverman, Adrian F. M. Smith, Jozef L. Teugels;
Ralph A. Bradley, Emeritus, J. Stuart Hunter, Emeritus

Probability and Statistics Section

*ANDERSON · The Statistical Analysis of Time Series
ARNOLD, BALAKRISHNAN, and NAGARAJA · A First Course in Order Statistics
ARNOLD, BALAKRISHNAN, and NAGARAJA · Records
BACCELLI, COHEN, OLSDER, and QUADRAT · Synchronization and Linearity:
   An Algebra for Discrete Event Systems
BASILEVSKY · Statistical Factor Analysis and Related Methods: Theory and
   Applications
BERNARDO and SMITH · Bayesian Statistical Concepts and Theory
BILLINGSLEY · Convergence of Probability Measures, *Second Edition*
BOROVKOV · Asymptotic Methods in Queuing Theory
BOROVKOV · Ergodicity and Stability of Stochastic Processes
BRANDT, FRANKEN, and LISEK · Stationary Stochastic Models
CAINES · Linear Stochastic Systems
CAIROLI and DALANG · Sequential Stochastic Optimization
CONSTANTINE · Combinatorial Theory and Statistical Design
COOK · Regression Graphics
COVER and THOMAS · Elements of Information Theory
CSÖRGŐ and HORVÁTH · Weighted Approximations in Probability Statistics
CSÖRGŐ and HORVÁTH · Limit Theorems in Change Point Analysis
DETTE and STUDDEN · The Theory of Canonical Moments with Applications in
   Statistics, Probability, and Analysis
DEY and MUKERJEE · Fractional Factorial Plans
*DOOB · Stochastic Processes
DRYDEN and MARDIA · Statistical Analysis of Shape
DUPUIS and ELLIS · A Weak Convergence Approach to the Theory of Large Deviations
ETHIER and KURTZ · Markov Processes: Characterization and Convergence
FELLER · An Introduction to Probability Theory and Its Applications, Volume 1,
   Third Edition, Revised; Volume II, *Second Edition*
FULLER · Introduction to Statistical Time Series, *Second Edition*
FULLER · Measurement Error Models
GHOSH, MUKHOPADHYAY, and SEN · Sequential Estimation
GIFI · Nonlinear Multivariate Analysis
GUTTORP · Statistical Inference for Branching Processes
HALL · Introduction to the Theory of Coverage Processes
HAMPEL · Robust Statistics: The Approach Based on Influence Functions
HANNAN and DEISTLER · The Statistical Theory of Linear Systems
HUBER · Robust Statistics
IMAN and CONOVER · A Modern Approach to Statistics
JUREK and MASON · Operator-Limit Distributions in Probability Theory
KASS and VOS · Geometrical Foundations of Asymptotic Inference

*Now available in a lower priced paperback edition in the Wiley Classics Library.

*Now available in a lower priced paperback edition in the Wiley Classics Library.

*Now available in a lower priced paperback edition in the Wiley Classics Library.

*Now available in a lower priced paperback edition in the Wiley Classics Library.

*Now available in a lower priced paperback edition in the Wiley Classics Library.

Texts and References (Continued)

KOTZ and JOHNSON (editors) · Encyclopedia of Statistical Sciences: Supplement
  Volume
KOTZ, REED, and BANKS (editors) · Encyclopedia of Statistical Sciences: Update
  Volume 1
KOTZ, REED, and BANKS (editors) · Encyclopedia of Statistical Sciences: Update
  Volume 2
LAMPERTI · Probability: A Survey of the Mathematical Theory, *Second Edition*
LARSON · Introduction to Probability Theory and Statistical Inference, *Third Edition*
LE · Applied Categorical Data Analysis
LE · Applied Survival Analysis
MALLOWS · Design, Data, and Analysis by Some Friends of Cuthbert Daniel
MARDIA · The Art of Statistical Science: A Tribute to G. S. Watson
MASON, GUNST, and HESS · Statistical Design and Analysis of Experiments with
  Applications to Engineering and Science
MURRAY · X-STAT 2.0 Statistical Experimentation, Design Data Analysis, and
  Nonlinear Optimization
PURI, VILAPLANA, and WERTZ · New Perspectives in Theoretical and Applied
  Statistics
RENCHER · Methods of Multivariate Analysis
RENCHER · Multivariate Statistical Inference with Applications
ROSS · Introduction to Probability and Statistics for Engineers and Scientists
ROHATGI · An Introduction to Probability Theory and Mathematical Statistics
RYAN · Modern Regression Methods
SCHOTT · Matrix Analysis for Statistics
SEARLE · Matrix Algebra Useful for Statistics
STYAN · The Collected Papers of T. W. Anderson: 1943–1985
TIERNEY · LISP-STAT: An Object-Oriented Environment for Statistical Computing
  and Dynamic Graphics
WONNACOTT and WONNACOTT · Econometrics, *Second Edition*

WILEY SERIES IN PROBABILITY AND STATISTICS
ESTABLISHED BY WALTER A. SHEWHART AND SAMUEL S. WILKS

Editors
*Robert M. Groves, Graham Kalton, J. N. K. Rao, Norbert Schwarz,
Christopher Skinner*

Survey Methodology Section

BIEMER, GROVES, LYBERG, MATHIOWETZ, and SUDMAN · Measurement
  Errors in Surveys
COCHRAN · Sampling Techniques, *Third Edition*
COUPER, BAKER, BETHLEHEM, CLARK, MARTIN, NICHOLLS, and O'REILLY
  (editors) · Computer Assisted Survey Information Collection
COX, BINDER, CHINNAPPA, CHRISTIANSON, COLLEDGE, and KOTT (editors) ·
  Business Survey Methods
*DEMING · Sample Design in Business Research
DILLMAN · Mail and Telephone Surveys: The Total Design Method

*Now available in a lower priced paperback edition in the Wiley Classics Library.

*Now available in a lower priced paperback edition in the Wiley Classics Library.